AF477177

Academic Writing

(Including UGC Syllabi for Research and Publication Ethics)

Academic Writing

(Including UGC Syllabi for Research and Publication Ethics)

Ajay Semalty

M. Pharm., MBA, Ph. D., PDF (Japan)
Course Coordinator- India's Most popular SWAYM MOOC: Academic Writing
H.N.B. Garhwal University (A Central University)
Srinagar (Garhwal), Uttarakhand
Visiting Scientist
Faculty of Pharmacy, Meijo University Nagoya, Japan

BSP BS Publications
A unit of **BSP Books Pvt. Ltd.**
4-4-309/316, Giriraj Lane, Sultan Bazar,
Hyderabad - 500 095 - T.S.

Published by

BS Publications
A unit of **BSP Books Pvt. Ltd.**
4-4-309/316, Giriraj Lane, Sultan Bazar,
Hyderabad - 500 095
Phone : 040 - 23445688
e-mail : info@bspbooks.net
www.bspbooks.net

ISBN: 978-93-91910-02-0

Dedicated to my Mother........
Late Mrs. Laxmi Semalty
(04 Nov. 1949- 21 Aug. 2019)

Preface

Academic writing is the utmost requirement of an academic and research career. Academic writing is an art as well as a science. You may learn science in classes, but the art is to be inculcated within ourselves. To present your hard work in the form of publication you must need to be familiar with the fundamentals of academic writing. You have done your work and now you want to publish it. In this regard, you have to present it most simply and effectively for the reader/journal or editor.

Despite of being a vital requirement in an academic & research career, there is no comprehensive set up of learning academic writing in the knowledge domain. Only in some PG courses and all pre-PhD courses a bit of content is provided and that too without the practical aspects. This book aims to fill this gap by providing the fundamental knowledge required for effective and result oriented academic writing.

After the huge success of my MOOC "Academic Writing" on the Ministry of Education, Government of India's SWAYAM platform, I hereby present this book to all the researchers and students. In the aegis of UGC, the Academic Writing MOOC was developed and run for the very first time in July-Dec 2019 and got more than 12500 learners in the very first run. The course ranked #1 in exam registration with the global top 30 rankings of MOOCs. In the online course, the entire syllabi have been covered in 15 weeks, in 150 lectures (10-15 minutes each). The course is available for credit transfer through SWAYAM MOOC for students. We got a huge response across the globe for our online course and the fourth cycle in Jan-Apr 2021 had more than 6600 learners. More than 8000 learners from more than 95 countries are enrolled in the course to date including all the last four cycles. On behalf of the National Coordinator-Consortium of Educational Communication (CEC) and course team, I welcome all the learners to join the free online course at the SWAYAM platform (www.swayam.gov.in).

The book provides basic know-hows for researchers. The book covers all the basic aspects of Academic Writing. Planning, executing, reporting, documenting, and presenting the research work are well illustrated in the book in a very simple and effective way. Apart from all the technical aspects of various kinds of publications (review, research, book, chapter, thesis, project proposal, etc) the psychological aspects like team management, communication with supervisor, field/empirical studies, and presentation skills have also been given in sufficient detail. The annexures provide some very vital information on various guidelines and important links. The book is very useful for all researchers, young scientists, students, and teachers. The book aims to trigger an effective academic writer in you.

The content is power-packed with easy to learn presentation for readers. The effective use of infographics, to the point presentation, quality content, reader friendliness, and important links for further readings are USPs of the book. Moreover, the book also covers the entire syllabi of UGC's recently introduced new subject of "Research and Publication Ethics".

I hope the book will be useful for all the stakeholders of academic and industrial research. I welcome suggestions for improvement and feedback about the book.

Dr. Ajay Semalty

Acknowledgments

I would like to thank all the persons who helped in designing, planning, and finishing this book. Firstly, I would like to thank my family members for their continued encouragement, support, and assistance. Without them, it would not have been possible to complete the book.

"Research is not just a job or group of tasks. It is a passion and a commitment to give your best for the sustainable development of science and humanity." The research adds value to the quality of the teaching and learning process. In the process of researching with teaching and learning, I learned researching by maximum utilization of scarce resources available.

Learning is a never-ending process that will positively enhance our awareness and knowledge and will positively contribute to our abilities as teachers. I started this mission with the determination to share (with serious learners), whatever I know. The sharing develops the capacity exponentially. So, in advance, I am thankful to all the respected teachers who read, refer, share and recommend this book to their students.

I warmly acknowledge the contribution of Dr. Mona Semalty, Assistant Professor, Department of Pharmaceutical Sciences, H.N.B. Garhwal University (A Central University) Srinagar (Garhwal) and Prof Rajat Agrawal, Department of Management Studies, Indian Institute of Technology Roorkee, Roorkee in the "Academic Writing" SWAYAM MOOC. The role of Prof Rajat is vital in the sense that he is the gentleman who triggered this urge (in me) to excel in the field of MOOCs. I will always be indebted for his valuable guidance. Thanks are due to Dr. Lokesh Adhikari, Senior Project Associate, UGC-DAE CSR project, Department of Pharmaceutical Sciences, H.N.B. Garhwal University (A Central University) Srinagar (Garhwal) for his vital role in designing the Academic Writing MOOC as Instructional designer as well as a Teaching Assistant. I extend many thanks to all the reviewers Prof Diwan Singh Rawat, University of Delhi, Prof Joseph Koyippally, Central University of Kerala, India, and Prof. MSM Rawat, Adviser (Higher Education), Govt of Uttarakhand.

I cannot express my gratitude in words to Dr. Pankaj Mittal Ma'am, General Secretary, Association of Indian Universities. She has been the powerhouse of inspiration and motivation for developing the Academic Writing course and for its success. With special thanks to Dr. N Gopukumar (Joint Secretary, UGC), Dr. Diksha Rajput (Deputy Secretary, UGC), and Mr. Abhishek (Project Officer, UGC), I humbly extend my deep gratitude to the entire UGC SWAYAM team. I also thank to all my fellow Course Coordinators specially

Prof. Harel Thomas, Dr. HS Gaur University Sagar MP; Prof. VK Garg, Central University of Punjab, Bhatinda; Prof K Kusuma, Jamia Milia Islamia, New Delhi; Prof Rekha Sharma, RTM Nagpur University, Nagpur, and Prof. Sajna Jaleel MG University Kottayam, Kerala for the motivation and moral support.

We started vocal for local in 2018, with giving the production responsibility to Dobhal brothers of Pandavaas Creations Pvt Ltd. It was a memorable journey. Thanks a lot to the production team of Dobhal Brothers for the wonderful and dedicated work in producing a world-class MOOC from the lap of the Himalayas.

I humbly acknowledge the Ministry of Education, Govt. of India, UGC, SWAYAM National Coordinator CEC, New Delhi, and the Team EMRC Roorkee for providing grants and vital support in the development and running of the Academic Writing online course.

I do remember the teachings of our beloved sir Late Prof. Y. S. Tanwar, B. N. College of Pharmacy, B.N. University Udaipur (Raj.). I shall always be thankful to him for incubating me as a researcher and blessing the research temperament in me.

I also thank Prof. P.K. Sahoo, Director-Research (DPSR University New Delhi), Prof Munish Ahuja (Department of Pharmaceutical Sciences, Guru Jambheshwar University of Science and Technology Hisar), Prof. Vandana Panda (K.M. Kundanani College of Pharmacy), Mumbai, Dr. Usha Ganeshan, IIM Bangalore, Prof K Srinivas (NIEPA, New Delhi), Dr. Sameer S Sahasrabudhe (Director EMRC Pune), Dr Jayakrishnan M (IIT Madras), Mr. Girish VG (IIM Bangalore), Dr. Rajkumar Bhardwaj (Director EMRC Roorkee), Prof Neeraj and Prof Vandana Punia, UGC-HRDC GJU Hisar, and all the other well-wishers who always keep triggering and supporting me time to time for the work like this.

I am thankful to my Project Associates (Dr Lokesh Adhikari, Mr. Mukesh Pandey and Mr. Himanshu Mishra) and all the staff members of the Department of Pharmaceutical Sciences, H.N.B. Garhwal University Srinagar Garhwal for their moral support. Dr Lokesh Adhikari played a vital role in every step of incubation, designing, production and executing this course. I also extend my gratitude to the entire University family of Garhwal University with special thanks to Hon'ble Vice Chancellor, Prof Annpurna Nautiyal for the continuous support and blessings.

I humbly acknowledge Mr. Anil Shah, and his entire Publishing team of BSP Books Hyderabad, India, with special thanks to Mr. Naresh Davergave for their efforts, guidance, and timely support in the preparation of this work.

I am always indebted to my Guru, Prof M.S.M. Rawat, Former Vice-Chancellor and currently working as Advisor, Department of Higher Education Government of Uttarakhand for his continuous support, guidance, and blessings.

Last but not the least, I thank my mother Late Mrs. Laxmi Semalty, and God almighty for giving strength, patience, and determination to complete this work on time in the present form.

Dr. Ajay Semalty

Contents

About the Author

Dr. Ajay Semalty, is Course Coordinator of India's most popular SWAYAM MOOC- Academic Writing and working as a faculty member in the Department of Pharmaceutical Sciences, H.N.B. Garhwal University (A Central University) Srinagar (Garhwal), Uttarakhand for more than the last 17 years with Academic, research, and industrial experience of a total of more than 18 yrs. He is also working as the University SWAYAM Coordinator and SPOC NPTEL Local chapter of Garhwal University. He is M. Pharm. (Pharmaceutics), MBA (Operations Management); Ph. D., PDF Meijo University, Japan; He has been included **in the World's top 2 % Scientist of Pharmacy and Pharmacology** (by Stanford University research, 2020) recently. He has completed 11 research projects funded by various agencies; He has published 80 papers, 6 books (including the best seller "Essentials of Pharmaceutical Technology" by Pharma Med Press), 2 patents (01 granted). He is a highly cited researcher of the field (h index 26). He has presented papers in more than 100 conferences; Awarded Daiko Foundation Post-Doctoral Research Fellowship and invited by Meijo University, Nagoya, Japan as visiting scientist for research for FY 2011-12; Awarded with the prestigious national Research Award by UGC, New Delhi, and Young Scientist Award of the year 2007 by Uttarakhand Council of Science & Technology (UCOST); Expert reviewer of Iceland Research Foundation Iceland; Israel Science Foundation, Israel and Scientific Expert (REPRISE) for Ministry of Institution and University Research (MIUR), Italy, Elsevier Books; Member of The Asian Council of Science Editors, APTI and IPA; Editorial Board Member and Peer reviewer of various national and international reputed journals; supervised 25 M. Pharm research projects and supervising 02 Ph.D. candidates.

He has delivered more than 200 invited lectures (on OERs, academic writing, nanoparticle research, research methodology, research ethics, software in research, MOOC, blended/ flipped learning, research proposal writing, academic integrity, plagiarism, etc.) in various training programs in Faculty Development Centers (FDCs), Human Resource Development Centres (HRDCs) and other institutions.

He has developed more than 100 hrs. of e-content for SWAYAM, SWAYAMPRABHA, and ARPIT courses. He has developed two SWAYAM Massive Open Online Courses (MOOCs): "Academic Writing" (listed in top 30 world MOOCs; appreciated by PIB and MoE, GoI; ranked #1 in exam registration in 2019 and 2020) and Industrial Pharmacy I (First and only Pharmacy MOOC in India) with more than 43000 learners from more than 95 countries including both the courses.

His research interest includes cyclodextrin and lipid complexes, Small-angle neutron scattering (SANS), nanoparticles, natural products for drug delivery like obesity and hair loss.

CHAPTER 1

Academic and Research Writing: Introduction

- Dr Ajay Semalty
H.N.B Garhwal University (A Central University)
Srinagar Garhwal-246174

Learning Outcome

After completing this chapter, you would be able

- To understand the basic importance of writing and AW skill
- How is AW different from simple writing?
- What are the types of AW?
- What are the basic traits required for AW?
- When to learn AW?
- What actually it is?
- Required tools or components for AW
- Facts and actions required to be a successful academic writer
- Basic flow of action in AW

Introduction

We, the Humans, are the master species or the God's best creation. Why? Apart from other reasons, because, we can WRITE…The writing skill makes us the unique species. The writing can change the perceptions; it can change the history. It has the almighty power.

When we talk about the academic writing. It can change the generations. Remember the example of Einstein. You know, when Einstein submitted his thesis **"A New Determination of Molecular Dimensions"** (a piece of academic writing obviously) of just 24 pages. It was just returned in first instance with the comment that *"It's fine, but it is too short to be considered as thesis."* He was well ahead of time from that point only…. And he proved to the world later with his inventions and his writing. In many world class research institutes the thesis is not more than 40- 50 pages. The effective communication matters, not the length…

Your research is incomplete, if you just did the experiments, collected the results and understood it. You have to make a detailed report and submit it. Then only it will serve the purpose. Your report is the basic piece of AW. Moving ahead, until you present the work, in the form of a research or review article, it won't be recognized or be useful for global research fraternity. At this moment I remember the famous quote from Edison

"Anything that won't sell, I don't want to invent. It's sale is proof of utility, and utility is success."

- Edison

Yes… THE UTILITY or the need-based research is the demand or the focal point of the current era. Though basic and applied research are equally important, the more focus is there on the applied research. The question is "how can you apply the basic knowledge for the development of humanity and for improving the overall quality of life of human beings?"

Here is the answer. The AW… AW is leading and steering the progress. It is not just about the recognition. Recognition is OK. But actually, we are focusing the recognition of value and quality of life of human being. We are doing the work for sustainability of humanity in this planet. That's why, the dissemination of knowledge and exchange of ideas is required and for that the AW plays a very vital role.

Being the best species in the planet, It is our responsibility to do the sincere efforts to ensure the sustainable development of humanity through the research and academic writing. Do you remember the famous dialogue from the Hollywood Movie Spiderman?

"With great power comes the great responsibility".

So, to use the writing for academic and research purpose, it requires great sense of responsibility.

AW has the power to drive and delegate the future. We were not talking about the deforestation, global warming etc., 3-4 decades ago. We came to know about the vital importance of these issues when the academic writing highlighted these issues in global platform. Now these issues are the hot topic, only because these were presented in an effective and convincing manner with the effective presentation of evidences supporting these things. So, it's clear that without the effective presentation of these issues, in the form of academic writing the issues would have been unexplored and unattended.

As far the knowledge dissemination is concerned the Sanskrit shloka.

न चौरहार्यं न च राजहार्यं न भ्रातृभाज्यं न च भारकारी ।

व्यये कृते वर्धते एव नित्यं विद्याधनं सर्वधन प्रधानम् ॥

"The wealth that cannot be stolen, neither abducted by state, nor can be divided amongst brothers, neither it is burdensome to carry, the wealth that increases by giving. That wealth is education and is supreme of all possessions"

It describes it as the supreme wealth that increases by sharing..........

In continuation to the previous discussion on introducing Academic Writing, we will discuss here

- How is AW different from simple writing?
- What are the types of AW?
- What are the basic traits required for AW?
- When to learn AW?
- What actually it is?

Let's begin with a simple example…. See these two sentences

       "Life is nowhere."   *"Life is now here."*

These sentences contain same alphabets but the one sound spessimistic about life while the other sounds optimistic. This proves how the proper writing can change the scenario of the thought processing. Now let's take another example.

Mango: A fruit, juicy, sweet.   (Writing)

Mango: Juicy stone fruit (drupe) from numerous species of tropical trees belonging to the flowering plant genus <u>Mangifera</u>, cultivated mostly for their edible fruit.   (Academic writing)

   It's different.

   Guess, how? Try to give the words to the difference…We will answer this after going through the types of AW

Types of Academic Writing

In academics, it is a well-established trend and responsibility to share our gathered knowledge via writing essay, passages, dissertation, thesis, research/review articles, short notes, books, abstract, digital writing/ OERs etc.

For all types of writing the basic traits are same.

The clarity: In all these forms you must be very clear in communicating your idea to the reader.

Completeness: Each communication should be complete in totality… we cannot left it open ended like in literature

Logic: The logic must be there

Technicality: Technical terms are always used depending on the field of research.

Sequencing: Sequence of presenting the information is very important for smooth transition from one point to the next point.

Unambiguity: There is no place for ambiguity in AW.

The reader friendliness is the key point.

These were some of the special traits of academic writing.

Learning AW on Time

Most of us don't understand the requirements of academic writing in time and it delays our professional growth and demotivates us. Writing wrong or making mistake in academic writing is considered as ethical crime.

Take an example of ignorance about the ethical crime at the highest level...

In 2014, it was found that a Prof. who was Vice Chancellor of an Indian University was co-author in three papers published between 2007 and 2014 and these three papers were found plagiarized. As per the statement of the Professor "the plagiarized content was a mistake and it won't happen again."

The most important question is "How one can learn academic writing?"

The most possible answers you will get are:

- "I had to learn AW on my own."
- "I learned AW when I was working on my dissertation."
- "Academic writing was very much varying style according to my professors. I picked up a little something from each professor I worked with in graduate school."
- "I am still learning what academic writing is—it seems to change according to journal, colleague, and discipline!"

Are you satisfied with these answers? Or you really want to learn academic writing?

For generations we have wrapped academic writing in mystery— keeping quiet about our own writing issues and publicly shaming those who visibly struggle with theirs. This has to change. Our trouble with writing is not evidence of our unfitness for the profession. It is not some secret sign of unworthiness or ineptitude. It is nothing to be ashamed of. When our writing isn't happening, we need to become willing to admit this and ask for help.

Not even the students but the senior faculty members even run away from publishing their work just due to lack of basic knowledge and confidence of AW.

And we are here with the solution in the form of this MOOC on Academic Writing. Shall we start from tomorrow? NO... NO... NO.... Tomorrow never comes....let's start today itself.........

So, first of all we have to understand what academic writing actually is:

Community exercise: Academic writing is a community exercise. As it is what people do together, it is a community exercise.

Reader friendliness: The readership is always the prime focus of academic writing. With the purpose of explaining or persuading the knowledge; one starts writing. In this process the judgment of right and wrong, appropriate or inappropriate is purely defined by the readers. Other students, lecturers or examiners are the judges in the case of academic writing.

All are nonnatives for academic English: Academic writing is like a different culture of writing with its own language.

"It is as it is because that is the way it has developed through centuries of use by practitioners. For that reason, it has to be learned. No-one speaks (or writes) academic English as a first language (Bourdieu & Passeron, 1994, p. 8)."

So leave the fear of being non-native speaker. Now, this learning can be achieved by using observation, study and experiment as the tools.

In the next lecture we will be discussing the basic steps of AW.

Dear learners, after completing this lecture you will be able to understand

- Tools/required components for AW
- How AW is different
- Facts and actions required to be a successful academic writer
- Basic flow of action in AW

Let's move to the basic requirements and steps of AW one by one.

- **Critical reading:** The first and foremost step of academic writing process is critical reading. In this process the reader gets engaged with the text deeply and in a complex manner. This engagement promotes readers **critical questioning** with the context of the text. It also develops the **understanding of judgment** about the **mechanism** of effective communication. **In reading, we just absorb the idea** or understand it while **in critical reading we go through a process of analysis, interpretation and evaluation.** Another difference between reading and critical reading is that in reading our direction of understanding is towards the direction of the text while in **critical reading** our direction is just opposite as we try to **question every assumption and argument** available. After **reading** we conclude the text as **summary** but in case of **critical reading we interpret, describe** the text.

- **Language:** Language is the second important part of academic writing. The most interesting fact about language is that your doing with language (output) is a reflection of language you have absorbed (input). **So to write better, you have to read the best.** Reading the academic write-ups is the first step in the process of writing. Learning one language is a different thing, than getting that language into our awareness. **Academic writing requires the awareness to the language.**

- **Good knowledge of grammar**, vocabulary and mechanics is also a main part of academic writing which helps in clear demonstration of our ideas in a sophisticated and precise way.

- **Other unique required traits of AW are**

- Rational, logical, sequential flow and reader friendliness

- The purpose of writing should be argumentary.

- Academic credibility of content is very important.

- Connecting ideas and cohesive writing style is a prime requirement in academic writing.

We will discuss these traits in detail in coming section. We have summarized the facts and actions required to be a successful a writer (Fig. 1.1 & Table 1.1).

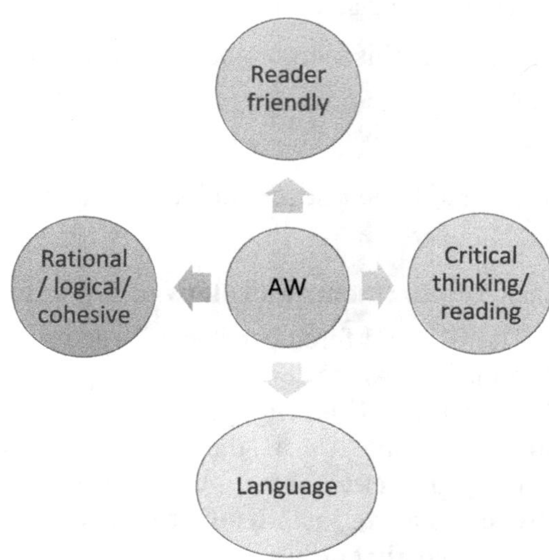

Fig. 1.1 Facts and actions required to be a successful Academic writer

Table 1.1 Facts and actions required to be a successful a writer

Facts	Action required
The journey of being an academic writer is quite personalized.	Adaptation of attitude for understanding the difference between academic writing and other forms of writing.
Academic writing requires specific language skills and writing styles.	Attentive reading of academic write-ups for better exposure to writing language and styles.
There is a significant similarity as well as difference in every type of academic writing.	Observation of macro-level organizational patterns in texts. Comparison of parts of different types of academic writing.
Being a successful writer in other context does not ensure your success in academic writing.	Acceptance of novelty of academic writing with new challenges. Strategic efforts in reading to improve the understanding of academic writing.
From learner to a successful academic writer.	Approach towards the "good" samples/assignments of academic write-ups. Practice and patience are the keys.

Flow of action in AW

Process of academic writing consists the following steps. Each individual step has its own significance and its necessary to be followed cautiously (Fig. 1.2 & Table 1.2).

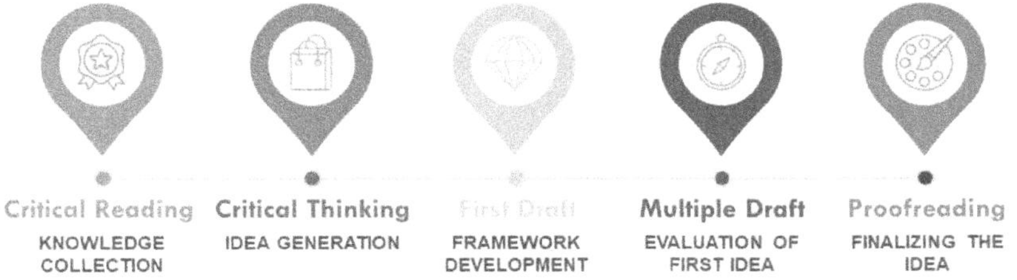

Fig. 1.2 Flow of action in AW

Table 1.2 Flow of action in AW

Critical Reading	Critical Thinking	Step Two Generating Outlines	Step Three Writing First Draft	Step Four Writing Multiple Draft	Step Five Writing Final Draft	Step Six Publishing
• Analyze • Evaluate	• Interpret • Literature review	• Build framework around the idea • Sequencing of idea and sub-idea	• Flexible ideas • Revising outline • Visual representation • Listing sources	• Revision of write-up • Major changes • Feedback	• Formatting the final write-up • Proof reading	• Follow-up publishers guideline

Before you start to write, you can **prepare yourself** by

Developing the connection between ideas: It will allow you to construct the conceptual framework.

Updating the knowledge of current research trends in your discipline.

By familiarize yourself with the guidelines of the top journals of your field for their stylistic preferences.

Dear learners! don't be hesitant to admit to learn.

In coming weeks, we will be discussing and covering various aspects of AW and will try to trigger and incubate an effective academic writer in you.

Please do spare some time for the activity. Lastly, please go through these links to update and improve your knowledge.

Further Reading

1. An Introduction to Academic Writing,
   https://www.youtube.com/watch?v=MyTLosz6aHA
2. https://study.com/academy/lesson/what-is-academic-writing-definition-examples-quiz.html
3. https://integrity.mit.edu/handbook/academic-writing/summarizing

References

1. https://doi.org/10.3929/ethz-a-000565688
2. https://books.google.co.in/books?id=ET4-QAAQBAJ&pg=PA5&lpg=PA5&dq=•%09
3. Bourdieu P, Passeron JC, Martin MS, Academic Discourse Linguistic Misunderstandings and Professorial Power Polity Press, 1994.

Activity Time

- Critically read two or three related articles of your choice
- Think critically
- Write a concise critical essay out of your understanding
- Check it yourself for consistency
- Get it checked from friends/ mentors

CHAPTER 2

Importance of Academic Writing

- Dr Ajay Semalty
H.N.B Garhwal University (A Central University),
Srinagar Garhwal-246174

Welcome dear learners! In this chapter, we will discuss why Academic writing (AW) is important.

Learning Outcome	Learning Plan

Learning Outcome

After going through this chapter, you will be able to understand the importance of AW for

- Academicians/teachers,
- Students,
- Researchers and
- Institutions.

Learning Plan

A. The importance of AW… "WHY" and "HOW" it is important for

- Academicians
- Students
- Researchers
- Institutions

B. Current Challenges

AW is as important as your basic testimonials. It is the way to communicate effectively your hard work with the global academic and researcher fraternity. Like we feel proud of our Class 10 th certificate, or 10+2 certificate being an academician you feel proud of your writing work. As we have discussed in previous lecture, there are so many types of academic writing each type of AW has its own domain of importance.

The importance of AW

If we split the importance with respect to importance for stakeholders, the different stakeholders may have one or more different point of importance.

For academicians

If you are a teacher you might be well aware of the importance of AW. Can you imagine an academician without AW? I think this simple question is itself the answer. Just to list out, for teachers, AW is

1. **Effective and well accepted medium to disseminate the knowledge:** Apart from your classroom teaching, if you go for publications you will

reach to more learners and students and will be able to disseminate your knowledge globally. So why to be local be global......

2. **A tool for assessing the eligibility for fresh academic/ research position or promotions:** All institutions give huge weightage to publications for recruiting and promoting faculty members. For this, you all know that the quality publications published in reputed journals play a vital role.

3. **A bench mark to assess the academic proficiency**: Without any access to CV of an academician the publications are always reliable indicator of his or her expertise. For example do you get the reviewer's invitation by reputed journals? If yes ... surely you might be having good publication record in that area of research.

4. **A tool of Intellectual contribution to Knowledge domain:** Teacher is ought to give significant intellectual contribution to the knowledge domain. Even if you are not an active researcher you need to write to contribute in learning process. So... why do you want to limit yourself to your few students? Come up and contribute....

How good you are as a teacher is generally assessed and approved by AW.

For students:

In this digital era students hardly go for writing lecture notes or some other basic writing. So when it comes to deliver a piece of AW it becomes a hurricane task for them. Then students start putting the efforts. So don't just wait for the last moment for delivering. Practice by preparing concise, logical, to the point and quality notes whatever you have been taught in the class. And always get it checked or reviewed by your mentor. For students the academic writing is important for getting these vital benefits.

1. **Opens up your mind:** When you write your brain works more effectively and new areas of brain start working. You can refer an article from The New York Times dated June 20 2014 (https://www.nytimes.com/2014 /06/19/science/researching-the-brain-of-writers.html).

   Gwendolyn Bounds, How Handwriting Trains the Brain, Oct 5, 2010; https://www.wsj.com/articles/SB10001424052748704631504575553193275 4922518#articleTabes%3Darticle

   Learning to read and write alters brain wiring within months, even for adults (Washington Post-28-May-2017); https://www.washingtonpost.com /national/health-science/learning-to-read-and-write-alters-brain-wiring-within-months-even-for-adults/2017/05/26/0ba3b3f2-4153-11e7-8c25-44d09ff5a4a8_story.html?utm_term=.30f97a28c1c0

2. **Better understanding of the topic for effective communication:** If you know AW you can effectively express or communicate your level of understanding.

3. **Triggering the analytical thinking:** AW trigger the analytical thinking. And being analytical is important for academic and research. You are required to analytically study two or more related studies for getting the essence out of the work as your future plan of work. You are ought to study and present two or more work analytically rather than doing simple description of previous works.

4. **Triggering critical and objective thinking:** Learning is not complete until it is thought critically and objectively. Without critical thinking information can not be framed to build knowledge. (Information $\rightarrow$ Knowledge $\rightarrow$ Wisdom)

5. **Learning Focused and framed writing:** As you have to be more formal and bound to some framework or style in AW. It enables you to deliver the best in the required style/template/framework (as per the requirement of thesis/articles etc).

6. **Fulfilling the mandatory requirement:** in the form of PG dissertations, Ph D thesis, and required papers for Ph D work AW is vital for students.

For researchers

If you are a researchers AW is very- very vital and indispensible for you.... Let see why...it is important for

1. **Getting your work evaluated for free:** Can you get your research work evaluated with peers without any cost of time and money? Just write a paper communicate to a good journal. Even if your paper is rejected you will get the vital inputs, comments and suggestions from the experts FOR FREE. And you will find that many a times even your supervisor can't give these vital suggestions.

2. **Fulfilling the mandatory requirement of publications with Ph D thesis:** Almost in all institutions published articles out of the work are to be submitted for getting the permission to submit Ph D thesis.

3. **Sharing the research output with national and international researchers:** you publish your work in reputed journals and by this you present your work (in the form of research or review articles etc.) globally.

4. **Getting recognition/ international approval for your work :** When you publish a paper it is but obvious that it has gone through the rigorous peer review process from the experts of the field. So it is itself recognition as well as the international approval of your work.

5. **Giving weight to your CV and getting weightage in the academic and research jobs:** publications are the heart of a CV of a researcher. This gives the direct impression of your proficiency in research. Author metric like h

index, total citations etc give direct message to the world how effective researcher you are…. We will discuss these in coming chapters.

6. **Planning future research**: AW lays a foundation for the future work. When you plan, design and draft a manuscript at that time only you will find that this this this point are to be kept in mind in future while doing the experiments. Or you will be able to chalk out the next level of work for further stage of your research.

7. **Getting project grants:** Only when you have the prior publications on the field of your project proposal you have the chances to get the grants from funding agencies.

For Institutions

As Performance indicator of Institutions: Many well known agencies do the survey and use number of publications and citations as measure of performance for their ranking. Times higher education world university ranking (www.thewur.com) ranks global institutions on the basis of several factors including total number of publications by the institutes.

https://www.timeshighereducation.com/; https://youtu.be/3h6gtgaUK6U

In national level India is also having similar ranking of institutions through National Institutional Ranking Framework (NIRF). (https://www.nirfindia. org/2018/Ranking2018.html)

All institutes show case there total publications on homepage of their website so as to show their excellence in research.

For getting the funds from funding agencies: After patents and technology development almost all the institutes give prime importance to AW. The one who performs well gets more funds from government agencies or has the better chances to get funds.

Attracting Prospective students, Researchers and foreign collaborations

Publications of an institute are the marker of its reliability as state of art research institute. This helps to attract prospective students, researchers and foreign collaborators. The collaborations further take the research to next level and give recognition globally.

Building goodwill and prestige: Directly or indirectly it also builds goodwill of the institute in the academic, society and market. This helps in attracting more campus placement of students.

Challenges

We have discussed a lot about the importance and positive side of the AW. But let's have a look on the challenges or other side of the coin.

Global trend of papers: It has been found that China has ranked one in total publications output in 2018 Jan (*Nature* **553**, 390 (2018); *doi: 10.1038/d41586-018-00927-4*) as compared to 2016 jan (*Nature;* doi:10.1038/ nature.2016.19198) and Jan 2014 (*Nature;d*oi:10.1038/ nature.2014.14684). In 2014 China was on rank 3, in 2016 it was on rank 2 and in 2018 China is on top of the world on total number of publications.

Fig. 2.1 New featured in Nature Nature 553, 390 (2018); doi: 10.1038/d41586-018-00927-4; https://www.nature.com/articles/d41586-018-00927-4

As per the Nature's News (Fig. 2.1)*"For the first time, China has overtaken the United States in terms of the total number of science publications, according to statistics compiled by the US National Science Foundation (NSF)."*

"China published more than 426,000 studies in 2016, or 18.6% of the total documented in Elsevier's Scopus database. That compares with nearly 409,000 by the United States. India surpassed Japan, and the rest of the developing world continued its upward trend."

But the question is "Does only the number matters?" or "Shall it really help with the quality?"

Let's see another news (Fig. 2.2)

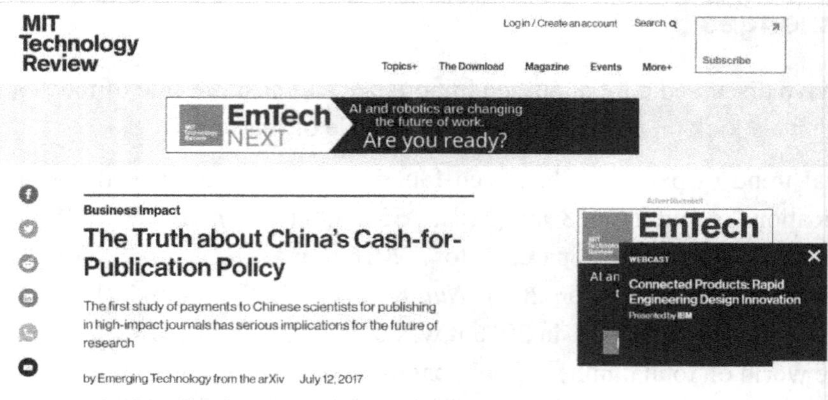

Fig. 2.2 Cash for publication policy of China in News

China launched cash-per-publication policy way back in 1990.

Starting from $25 for each published paper in 1990 the researchers are paid upto $165,000 per publications depending on the impact factor of the journal. But this policy has caused the change in behavior of some researchers. The plagiarism, academic dishonesty, ghost-written papers, and fake peer-review scandals are on the increase in China, as is the number of mistakes.

So "these financial rewards have raised more serious questions about the credibility of work published with this kind of incentive and the integrity of Chinese science in general"

Focus on Quality and not the Quantity

In Indian context the flood of TDH (Tom Dick and Harry) journals has also affected the credibility of Indian research globally. To get more and more score in Academic performance indicator many researchers are just publishing for numbers.

In the study (Fig. 2.3) authors analyzed 1009 journals after excluding 327 indexed in Scopus/Web of Science. About 34.5% - disqualified (under the basic criteria because of incorrect or non-availability of essential information such as address, website details and names of editors), 52.3% of these provided false information like incorrect ISSN, false impact factor claims, false indexing claims, poor credentials of editors.

It was concluded that over 88% of the non-indexed journals in the university source component of the UGC-approved list (included on the basis of suggestions from different universities) could be of low quality.

Please remember the trend has come........ Many state of art institute and Universities just ask for the papers published in SCI indexed journals. Then all your efforts and money to publish your papers and to have a long list of papers in those TDH journals will go all in vain.

RESEARCH COMMUNICATIONS

A critical analysis of the 'UGC-approved list of journals'

Bhushan Patwardhan[1,*], Shubhada Nagarkar[2], Shridhar R. Gadre[3], Subhash C. Lakhotia[4], Vishwa Mohan Katoch[5] and David Moher[6]

[1]Interdisciplinary School of Health Sciences,
[2]Department of Library and Information Science, and
[3]Interdisciplinary School of Scientific Computing,
Savitribai Phule Pune University, Ganeshkhind, Pune 411 007, India
[4]Department of Zoology, Banaras Hindu University, Varanasi 221 005, India
[5]Rajasthan University of Health Sciences, Jaipur 302 033, India
[6]Centre for Journalology, Clinical Epidemiology Program, Ottawa Hospital Research Institute, Ottawa, ON, K1H 8L6, Canada

Scholarly journals play an important role in maintaining the quality and integrity of research by what they publish. Unethical practices in publishing are leading to an increased number of predatory, dubious and low-quality journals worldwide. It has been reported that the percentage of research articles published in predatory journals is high in India. The University with peers and others. This has also fuelled unprecedented commercial interests in publication of research journals, so that major publishers across the globe indulge in aggressive publication efforts and policies. The competitive market of research publications has witnessed undesirable and unhealthy publication practices. The widespread 'publish or perish' policies have given rise to a breed of 'predatory journals', whose main objective is to make money by publishing 'anything' in the name of a research paper for a 'fee' commonly known as article/author processing charge (APC)[1]. Such unethical practices and the unscrupulous business of publishing have rapidly grown during the last decade. It is common to receive unsolicited, dubious e-mails inviting articles, promoting special issues, editorial board memberships and speaker invitations from predatory journals, publishers and conference organizers. The pioneering effort known as Beall's list of 'potential, possible, or probable predatory' publishers and journals[2] was closed down in January 2017, depriving researchers across the world of some cautionary advice.

The global concern of researchers and other stakehold-

Fig. 2.3 Critical analysis of UGC approved list of Journals

Dear learners don't fall in the trap of these TDH journals just to increase the numbers. Publishing in these substandard journals may be disgraceful for you and your institute.

Be honest in AW.

Suggested Readings

1. https://www.nirfindia.org/2018/Ranking2018.html

2. https://www.timeshighereducation.com/

3. The best universities in the world 2019,   https://www.youtube.com/watch?v=9GNVbF140F8

4. The Handbook of Academic Writing: A Fresh Approach By Rowena Murray; Sarah Moore Open University Press, 2006

References

1. Faber J, Writing scientific manuscripts: most common mistakes, Dental Press J Orthod. 2017; 22(5): 113–117. doi: 10.1590/2177-6709.22.5.113-117.sar

2. Ohwovoriole AE, Writing biomedical manuscripts part I: fundamentals and general rules. West Afr J Med. 2011 May-Jun;30(3):151-7.

3. Ohwovoriole AE, West Afr J Med. Writing biomedical manuscripts part II: standard elements and common errors.2011 Nov-Dec;30(6):389-99.

4. https://www.nytimes.com/2014/06/19/science/researching-the-brain-of-writers.html

5. Gwendolyn Bounds, How Handwriting Trains the Brain, Oct 5, 2010; https://www.wsj.com/articles/SB10001424052748704631504575553193275 4922518#articleTabes%3Darticle

6. Nature News, Nature, 2018: doi: 10.1038/d41586-018-00927-4

7. Nature News, Nature; 2016:doi:10.1038/nature.2016.19198

8. Nature News, Nature; 2014:doi:10.1038/nature.2014.14684

9. arxiv.org/abs/1707.01162 : Publish Or Impoverish: An investigation of the monetary reward system of science in China (1999-2016)

10.Patwardhan, B. et al. A critical analysis of the 'UGC-approved list of journals'. Curr. Sci. 2018;114: 1299-1303.

Activity

1. Search for NIRF ranking of your institute https://www.nirfindia.org /2018/Ranking2018.html

2. Go through Times higher education world university ranking (www.thewur.com and https://www.timeshighereducation.com/

3. Search the researcher with maximum number of publications and citations from google scholar from your institute or from your field of research.

Basic Rules of Academic Paper Writing

- Dr Ajay Semalty
H.N.B Garhwal University (A Central University),
Srinagar Garhwal-246174

Learning Outcome

After completing this chapter, you will be able to have a quick overview of the basics of academic writing.

Learning Plan

In previous chapter, we have discussed "why" to write and "How" it is important. Here we will discuss:

- When to write
- What to write
- How to plan
- How to write
- Where to send
- How to address review comments
- How to do proof reading

Have Rationality

- **Patent or publish:** You should first see whether your work is innovative or not. If innovation has the merit for patentability, then don't plan the publications before trying for patent. If the innovation is not worthy to be patented, then go for publications.

- **Novelty of the work:** The novelty of work is the driving force for publications. If you are first in reporting a work, then it's a win-win condition for publications.

- **Contribution to the state of knowledge:** Either in the form of research or in the form of review there must be something in your work which can contribute to the state of knowledge in that domain.

Plan Well on Time

- **Start writing early:** Plan the article well before the completion of your research. So that you can plan some studies if required during writing.

- **For review writing:** Plan review article immediately after exhaustive literature review. See if you have enough literature survey and your prior expertise or experience in the research area.

- **Plan the outline on time:** Plan outline of paper with co-authors, PI or supervisor, well ahead of the draft writing.

Target a Journal Wisely

- Assess your work wisely: Neither overestimate nor underestimate your quality of work. See the journal to be targeted and the level of articles in that and then as per the quality of your work plan the submission.

- *It would not be a wise decision to target a journal like Nature or Science for a very simple a basic research like reporting antioxidant activity of a plant.*

- Target quality journals or journals with good impact or indexing: Target the best possible journal for your article. Best in the sense of impact factor.

- Matching with the scope of the journal: Target a journal which publishes the articles similar to your research area. You can check "Scope of journal" section of the target journal for checking the same.

- See Article processing charges (APC): yes or no?, affordability

- Check how much time the journal takes for the peer review process: This is very important to check the time taken by a journal in review process. Target a journal which takes less time for the same. Surely, you cannot wait for years if you are planning article with your Ph D thesis submission. In general, the minimum time is 2to 3 months taken by a journal's review process. Sometime journal offers fast track processing of article with a fee. If you can afford, avail it.

Chose the Type of Article

As per the scope of your study you can plan your article as a full-length research article, short communication, rapid communication or letter to the editor:

Drafting a Manuscript

- A tight or well-defined outline of the article helps a lot in drafting.
- Stick to author guidelines of the journal while drafting the manuscript
- Use either British or American English
- Avoid any typo and grammatical errors.
- Use simple and reader friendly language with good clarity.
- Maintain a systematic flow in writing and presenting information.
- Don't forget to give due attribution.
- Avoid plagiarism: Do not copy
- Plan a precise aim, split it into objectives and tell what methods were adopted for getting results and then discuss them properly.

Focus Every Component of an Article

- Title: I suggest plan it at last. Keep in mind that it should be simple, catchy, reader friendly, suitable for the journal to be targeted and representative of your work.
- "Antidiabetic potential of a medicinal plant"   -- X
- "Anti-diabetic activity of *Withania somnifera* fruit extract in streptozocin induced diabetic rats "   -- √
- Authorship: Only persons who have given actual contribution in planning, designing, execution, completion, and drafting of study should be given authorship. It should **not** be used as a gift or any favor.
- Affiliation: apart from your institutional address give email which you use more frequently rather than the one which you do not use normally.
- Abstract: In drafting stage, you should plan the abstract after the completion of the paper. Make it structured or unstructured as per requirement of the journal. Give to the point aim/objective, methods, results and conclusion; and make it concise (generally 300 words).
- Keywords: select the keywords which are not in the title. Right selection of keywords makes it easy to get your work searched easily on internet or search engines. Add keywords which are related to your study and which are more likely to be used by the people to search for the similar study.
- IMRAD style for research articles: Though we will discuss it in detail in later chapters, we will just give very basic tips for the same here (Table 1).

Table 3.1 IMRaD Format

Section of Paper	Content
Abstract	The central idea
Introduction	Defining problem and objectives; rationality of work
Materials and methods	The way to solve the problem
Results	Findings of the experiments
Discussion	Analytical study of obtained results
Conclusion	Concluding remark on Outcome
Acknowledgment	Support received
References	Sources of study
Infographics	Tables and figures

- **Introduction:** Try to answer: What is the problem and what is your hypothesis:
  - Give background or origin of idea
  - Logically, sequentially, systematically introduce your problem and give your hypothesis.
  - Focus on justifying the rationality of your work.
  - Give proper citation to latest and important previous studies.
  - Don't use very old references until they are indispensible or they are must to discuss.

- **Materials and methods:** Try to answer: How the problem was studied following your hypothesis.
  - Discuss the methodology logically and in concise way.
  - Don't miss discussing about any modification made by you in standard method.

- **Results:** Try to answer: What was the outcome of problem treatment from your hypothesis.
  - This section provides the evidence that leads to the answers of the study to the question you posed at the start.
  - The reader should be guided to these findings by using the text of the results along with a judicious use of tables and illustrations.
  - State the results without any bias.
  - Don't exaggerate the results.
  - Don't be afraid of reporting negative results.
  - Report any paradox.
  - Use good statistics to show the results
  - Don't repeat the text in figure or table or vice versa

- o Cite the table or figure in text.
- o Make the tables and figures self-explanatory with suitable legends or footnotes.
- o Equations and special characters: Give attention to presentation of equations. Its better to write the equations in Microsoft equation (a function in MS Word). The special characters like micron (μ), beta (β), Rho (ρ), gamma (γ) etc. should be checked specially for their accurate presentation. Sometime they get changed automatically upon pasting from a pdf file to word file.

- **Discussion:** Try to answer: How was the outcome of problem treatment correlated and/or contradicted with the previous studies.
  - o Discussion may be separate section or joined as Results and discussion section as per the journal's requirement.
  - o Take the results one by one and discuss them.
  - o Discuss the results in light of existing knowledge.
  - o Be critical.
  - o Don't be biased.
  - o Discuss the results with clarity and reader friendliness.
  - o Use the references freely to support discussion.

- **Conclusion:** Conclude your study outcome in concise manner. The conclusion should not just be copied from abstract. Give future direction also.

- **Acknowledgements:** Try to answer: Who helped you and supported you but not eligible to be author on merit.
  - o Acknowledge only to project grants,
  - o Gift samples,
  - o Analytical or other service providers
  - o Logistic support etc.
  - o Do not use this section for greasing
  - o Donot acknowledge senior authors and direct stakeholders like department, principal/head etc.

References

- Stick to uniform formatting and style as prescribed by the journal.
- Don't unnecessarily increase the references.

- Spell author's name properly (he/she may be one of the reviewers; none wants his name printed wrong);
- Don't miss the pioneer, most important and most relevant studies for your work.
- Donot forget to cite latest references.
- Limit the references as per the need of your article
- Review article are expected to give as much as references.
- Editing & Revising
  - Major Changes: Fill the missing links, correct the flow or the logic, rewrite/ reorganize the text for putting in the logical sequence.
  - Polish the style: Refine the content followed by corrections in grammar and spelling.
  - Formatting: Give time to your text so that it is more attractive and easy to read

Do the work in the given sequence. Otherwise, you will be wasting your time in revising the things and then deleting the same thing later.

- **Self-Revision by the Author(s):** Revision of your writing is an on-going process from the time you begin until the final copy is submitted. You must write the first draft and then leave it for a day. Then you can come back to the first draft and begin revising again and check the things in following order.
- Check
  - The sequence of content and idea in each section for logical flow
  - Whether there is a strong relationship of ideas between the introduction and the discussion or not.
  - Each paragraph has a main sentence for coherence or not.
  - In paragraph, other sentences support the main or lead sentence or not?
  - The smooth, natural and logical transitions between paragraphs.
  - The accuracy of terminology used.
  - For possibility of active voice framing of an existing passive voice.
  - For complete removal of any colloquial language.
  - For any redundancy of content
  - Clarity and brevity of sentences. Try to use fewer words for same sentence.
  - Correct citation and formatting.
  - The sequence of your infographics.
  - The line spacing, font size in text and infographics.

- o The page breaks to ensure unbroken infographics.
- o The accuracy of names and affiliation of authors
- Reading aloud: Reading the paper aloud is the best and the simplest strategy to check the error of writing.
- Use spell check/thesaurus: Using spell checker helps a lot in checking the most typos and spell errors. But watch the replacements carefully do not allow automatic replacement. Because it may lead to a right but completely non-contextual/ nonsense word.
- Don't rely on thesaurus and grammar checker: It may lead to non-contextual/ nonsense words.

Get feedback on your manuscript and then revise your manuscript again

You must get the feedback to improve your article. Ensure that your co-authors have read and commented on the draft or had the chance of it. Then, when it is ready, give the manuscript to some colleagues. Tell them, when you would like to receive their comments, and what levels of information you would like (e.g., comments on the science, logic, language, and/or style). After you get their comments, revise your manuscript to address their concerns. Do not submit the manuscript until you feel it is completely OK. Once it is accepted, further changes in your manuscript will be difficult and may also be costly.

Final Revision

Allow you reviewer does, examine the paper again, if it is possible. Co-authors should check and approve the final draft to ensure that their inputs were included, and the quality is good now. If all the changes have been done to everyone's expectations and satisfaction, make the last check of overall appearance of the manuscript and check any break in natural flow of content at grammatical or idea level.

- **Submitting Manuscript & Follow-ups:** Once you have prepared final version of manuscript, you should send the manuscript to the editor of the journal as earlier as possible. The following steps are needed for the same.
  - o Identifying mode of submission: online or offline:
  - o Preparing a Cover Letter: Prepare a simple, short but effective cover letter for the manuscript submission addressing the editor of the journal with short description of your study.
  - o Providing Copyright
  - o Submitting/uploading the manuscript: Upload the suitable files in desired format.

- o   Address the comments of reviewers point to point.
- o   Understand what the flaw/fault is and try to overcome it.
- o   Plan and write a good rebuttal letter.
- o   Be alert on submission of revised manuscript deadline.
- o   Be on time to return the sent final version received for final checking.
- o   Be careful in proof reading.
- o   Once more check typo errors, consistency of text, table and figures in final pdf version.

So, to summarize:

Lastly, here are some suggestions for improving a manuscript:

- Write simple language with clarity.
- Rational of the study must be clear and be focused in the entire study.
- Aims and methodology should be clearly explained
- Provide the results systematically and in totality;
- Discuss the results critically and with logical reasoning without any general and vague statements;
- Conclude the study as per the objective of the study and provide future trend (if any)
- Provide the additional data or spectra also in support of our study.
- Always stick to manuscript guidelines prescribed by the journal with respect to formatting and all other aspects of the research article.

Further Readings

- McMillan, V.E. 2001. Writing papers in the biological sciences. 3rd Edition. Bedford/St. Martin's Press, Boston and New York. 207 p.
- Faber J, Writing scientific manuscripts: most common mistakes, Dental Press J Orthod. 2017; 22(5): 113–117. doi: 10.1590/2177-6709.22.5.113-117.sar
- Borja A, Writing the first draft of your science paper — some dos and don'ts, https://www.elsevier.com/connect/writing-a-science-paper-some-dos-and-donts
- https://www.springer.com/gp/authors-editors/journal-author/journal-author-academy
- https://www.apa.org/pubs/authors/new-author-guide.pdf
- https://projects.ncsu.edu/labwrite/index.html

- Maxim S. Pshenichnikov, Academic skills, file:///C:/Users/DELL/Desktop/Writing-paper-SEPOMO3.pdf

References

- Ohwovoriole AE, Writing biomedical manuscripts part I: fundamentals and general rules. WestAfr J Med. 2011 May-Jun;30(3):151-7.
- Ohwovoriole AE, West Afr J Med. Writing biomedical manuscripts part II: standard elements and common errors. 2011 Nov-Dec;30(6):389-99.
- https://writingcenter.unc.edu/tips-and-tools/scientific-reports/

CHAPTER 4

English in Academic Writing-I

- Dr Ajay Semalty
H.N.B Garhwal University (A Central University),
Srinagar Garhwal-246174

In this chapter, we will discuss English as a language in Academic Writing (AW). Non-compliance with the standards and conventions of AW is a major reason for the rejection of many manuscript.

Learning Outcome

After completing this chapter, you will be able

- To understand the importance of effective use of English in AW.

- To solve the problems faced by nonnative speakers in AW.

- To effectively use the most common elements for effective AW.

- Dear learners, we are not going to discuss the English grammar here. Rather, we will just focus on the most common practices for effective AW.

Learning Plan

Here we will discuss:

- Importance of English in AW

- Basic rules of effective English AW

- Problems and solutions for Nonnative speakers

- Precise use of subject, verb and adjective

- Active and passive voices

- Judicious use of articles

- Proper use of punctuation

Importance of English in AW

- **Universal/ International language:** English is an international/global language. As most of the journals are in English, it is crucial for every researcher or academician to learn effective writing.

- **For Exchange of ideas:** English is needed to communicate and exchange your ideas with the international experts.

- **It's different:** The use of English for scientific writings is somewhat different from that of any other kind of communication. The main challenge in AW is to write complex things in simple and clear manner.

- **For Effective communication:** You must have a good command on the language for scientific communication.

- **For better understanding:** Effective use of language is necessary for the better understanding and clarity of the topic being addressed.

Basic Rules of Effective English in AW

- Keep it simple....
- Keep it clear/ non-ambiguous
- Keep it short/concise
- Keep it reader friendly
- Do not try to impress by ornamental English, just communicate for better understanding.

"Readers of scientific papers do not read them to assess them, they read them to learn from them. What is needed is more simplicity, not more sophistication!" Aim "to inform, not to impress."

- JeanMargaret Perttunen, quoted by Carolyn Brimley Norris, Language Services, University of Helsinki, 2018 in Academic Writing in English.

- Do not use passive voice unnecessarily: Until and unless passive voice is more comfortable and common than that of active voice, donot use it.

- Do not use much Latin words: like *modus operandi*, *prima facia* etc., if they are not absolutely essential.

- Avoid using &, @, ASAP like abbreviations or other slang.

- Avoid punctuations errors: even a missing a comma "," can alter the meaning of a sentence.

- Use articles judiciously: Some authors habitually use "the" carelessly anywhere in their manuscript even if it is not required. And same holds true for "a" and "an".

- Avoid very long sentences: Try to break the long sentences because very long sentences always lose the attention of readers.

Problems and Solutions for Nonnative Speakers

- **Practice hearing and reading more....**

Writing English is almost always a major hurdle for non-native speakers. It is especially so for the beginners:

Why?

According to Prof. S. P. Pani from Utkal University:

"….For the simple reason we face the problem in learning a foreign language…."

In general, our natural learning follows the trend

"We hear first, then we speak, and then we write"   -   *(Mother tongue)*

BUT for a foreign language the trend is opposite from our natural learning process

"We Write first, then we speak, and then we hear"   - *(Foreign language)*

So,

the problem lies here. We hear less and try to deliver it naturally.

"The most important thing is to read as much as you can, like I did. It will give you an understanding of what makes good writing and it will enlarge your vocabulary."- J. K. Rowling

So, What to do?

Practice to hear and read more….

It will enrich your vocabulary and improve your understanding about the foreign language. Moreover, you will recognize errors yourself if you read your own writing loudly.

Be Formal: You should not write merely by listening to a native speaker. Spoken English is casual speech. For academic writing, you need formal expressions. So, in writing, we can adopt only the suitable elements even from a native speaker's speech.

- **Be the generator not the translator:** You are supposed to generate a sentence. If you prefer to translate from your language to English, instead, it is likely to invite errors, ambiguity and loss of meaning. So just try to think in English and write it in simple words. Avoid translating.

- **Be reader friendly:** Adopt the "Reader first!!" policy. Clarity, unambiguity and simplicity are the keys of reader friendliness. Nobody enjoys reading an unclear text. So academic writing must be clear from the first reading itself.

- **Either use British or American English:** Non-native speakers merely focus on the correctness of language. They do not subscribe to any one verity of English. You may use either British or American English. E.g. COLOR (American); COLOUR (British). However, you must choose one. Otherwise it would be confusing for readers and reviewers.

Precise use of Subject, Verb and Adjective

If you are in this course certainly you don't require English grammar. However, from the point of view of AW, it is certainly good to recall some of its basic points. From the point of view of grammar, language has to be accurate. From the point of view of AW, it should be precise. In order to improve the quality of your manuscript, merely focusing on the accuracy of your content is not enough. You need to make it precise in terms of language also. Let's see how....

- Agreement of subject and verb must always be there.

  Singular subject → singular verb

  Plural subject → plural verb

  (Just read your sentence aloud if it is wrong it will sound awkward to you. The Microsoft grammar checker always marks it red.)

  e.g. *"Most of the synthetic **drugs** used for anti-obesity **have** one or more side effects."*

  - Semalty et al., Obesity and herbal drug research: exploring the safer alternative

  and lead molecule, Current Traditional Medicine, 2017, 3, 74-92

  "The article focuses on the critical review and meta-analysis of these two techniques for improving aqueous solubility and dissolution." –Semalty A, Cyclodextrin and phospholipid complexation in solubility and dissolution enhancement: A critical and meta-analysis, Expert Opin. Drug Deliv. (2014) 11(8)

- **Sometimes, the subject is not required for obvious reasons:** In scientific studies, some things are obvious and known: for example, we know that the authors / researchers have done a work. In such cases it is not required to mention the subject.

  Let us see these sentences.

  "Animals were sacrificed"

  "Animals were sacrificed by the authors."

  Or

  "Authors sacrificed the animals."

  All these sentences are grammatically correct. But from the point of view of academic writing, it is but obvious that authors have sacrificed the animals. So the first sentence is the precise one to be used.

- **For emphasis, nonliving objects may become the subject of the sentence:** We often use sentences with the nonliving as subjects, for the sake of emphasis, as in: "The result showed that....'", "The article aims to......" etc.

- **Use the most appropriate verb:** It is the use of verb which gives the emphasis to the sentence. But be cautious in replacing the verb for emphasis. The meaning should not be distorted. Synonyms do not have the same meaning.

    e.g."Observe" in place of "see"

    "Determine" in place of "calculate"

    "Possess" in place of "have"

- **Avoid vague adjectives:** To emphasize the degree of results never use the adjectives in rough or vague manner.

- **E.g.: Donot use the language like "very good results or very excellent results ... were obtained"."The tablets were too strong".** You are supposed to precisely mention, compare or present the results statistically, not by plain adjectives.

- **Do not use unnecessary words**

- Avoid common or casual words which are used often in spoken English. E.g. actually, in fact, by the way....

    "The secret of good writing is to strip every sentence to its cleanest components. Every word that serves no function, every long word that could be a short word, ever adverb that carries the same meaning that's already in the verb... these are the thousands and one adulterants that weaken the strength of a sentence.

    - William Zinser "On writing Well"

Active and Passive Voice

- **It is better to be Active .....**

    Until passive is more comfortable and common, frame the sentences in active voice. "Active voice gives more clarity."

    But the classic academic writing preferred passive voice. Modern communicative writing prefers active voice. Where the demarcating line lies is still a debated question.

Active voice

Subject + verb (+ object)

e.g. He prepared it.

Passive voice

Object + auxiliary verb + III form of verb

e.g. It was prepared (by him).

- Make passive if you want
  - o to emphasize on the object or objective rather than the subject:
  - o e.g. "Nanotechnology was coined by Eric Drexler."
  - o to illustrate a universal or general truth or fact.
  - o e.g. "Animal study protocols were approved by Ethical committee of institute"
  - o to be polite in tone
  - o Passive voice is polite in tone or less harsh. E.g. "All the rats were sacrificed", "The formulations with poor results were dropped out for the in vivo study",
  - o "The effect was terminated"
  - o to avoid using first person

In the AW, the focus is on the facts, methods, or procedures involved, **so passive voice is preferred.**

- **Most of the journals discourage writing in first person to create an impersonal tone.**

  Avoid use of "I" and "we", and frame a passive voice construction instead. Skip the subject and if it is unavoidable, use "authors". For example, "The effect was studied on diabetic rats." Or "In previous studies, authors studied the effect of drug on ……."

- **Have a well balance of both the voices:** Don't over use a single type of voice consistently. Have a good balance and avoid monotony in writing. You won't find an article or even a minor section of an article with one single voice.

- See the journal's guidelines properly. For example British Medical Journal (BMJ) states that

"Write in the active and use the first person where necessary." - BMJ

Use of articles judiciously

Many authors use articles carelessly. This is one of the most common problems with nonnative writers. Let's see some examples

- Articles "a" and "an" are called indefinite articles.

("a" is used before a singular noun beginning with a consonant while "an" is used before a singular noun beginning with a vowel sound)

Generally, we do not make mistake in using indefinite articles.
Article "the" is called definite article.

"The" is used before a singular or plural noun. It is used

o   to denote a specific thing or a particular member of a group e.g. "The formulation 1 showed better activity."

o   before superlatives (the highest, the most…)

o   before the geographical areas/places (the US, the Europe, the Tajmahal etc.)

o   before the natural things (the Ganges, The sun, the moon, the Himalaya)

o   before a noun which has been mentioned earlier. E.g. "A particle was observed in the test tube. The particle was removed from the test tube."

Key message: Use the articles judiciously. Unnecessary and inappropriate use of articles may be very awkward and it may also lead to rejection of your article.

Proper use of punctuation

Punctuation errors are the most common errors in a manuscript. It is also the most common cause of rejection of an article in initial phase. While drafting a manuscript, we busy ourselves with technical issues and keep on resolving and shaping the same until the final version of manuscript is prepared. Meanwhile, we ignore the punctuation errors. But even before sending it to the reviewer, sometimes, editors themselves send it back to authors because of the spelling and punctuation errors in the article. So, special attention and meticulous checking should be done to check punctuation errors.

Among the various punctuation marks, we are generally comfortable in the use offull stop (.), question mark (?), exclamation mark (!), apostrophe (') etc. Mistakes occur in our use of comma (,), hyphen (-), colon (:), and semicolon (;). Please refer the very simple and effective text from the free e-content available on the website of Commonwealth of Learning (www.col.org).

"The Comma (,) Commas separate the following:

two or more adjectives describing one noun: He has a large, boisterous dog.

individual items in a list: I shall buy books, pens, paper and a birthday cake from the shop.

They also mark off the following words, phrases or clauses from the rest of the sentence:

A parenthesis: the party, held in the community centre, was a great success.

The name of the person addressed: Susan, please come here.

Sentence adverbs: Spain, for example, is the ideal location."

[Adapted from: "OER 4: English Second Language" fromwww.col.org/CourseMaterials. CC by SA]

Tips

- Use comma (,) after sentence adverbs like so, therefore, however, consequently, thereupon, for example, for instance, otherwise, nevertheless, perhaps, after all, in the meantime, indeed, moreover, and so on.
- Check the brackets for completeness. Never skip the closing bracket.
- Never place any punctuation in front of any parenthesis. e.g. The results were in agreement to the previous study, (10) and with the official limits."
- Quote the words or sentences you have copied. "The effect of "
- Do not rely on the spell checker for use of hyphen in scientific writing. Technically it should be convincing. E.g. "antidiabetic" is OK while the spell checker suggests it to be "anti-diabetic".
- Spell checker or autocorrecting must not be relied on completely. The spelling of word may be right but it may be totally irrelevant for your context.

  e.g.: The **patience** was given the dosage.   (here patient was required).

  The **manger** ordered the executives to complete the task.   [manager not manger (container)].

Take Away Message

Let's summarize some key points: Use formal language, write simple English, keep it concise, keep it unambiguous, do not rely on spell checker, avoid grammatical errors; see that the verb agrees with the subject, use passive or active voice and articles in the right way, and take care of punctuation errors.

In this chapter, we learned the important and basic rules of effective English in Academic Writing. Non-native speakers must take special care of language in the manuscript.

Further Reading

- "OER 4: English Second Language" from www.col.org/CourseMaterials. CC by SA
- Beach David, Graduate Academic Writing, https://www.youtube.com/watch?v=SNJAMxiOmkU&t=0s&index=12&list =LL-b7qstPT37YEWLECEltHuw
- Using Articles (a, an, the) in Academic Writing, https://www.youtube.com/watch?v=uJ0092Tqku0

References

- Carolyn Brimley Norris, Language Services, University of Helsinki, 2018 in Academic Writing in English
- "OER 4: English Second Language" fromwww.col.org/CourseMaterials.
- Faber J, Writing scientific manuscripts: most common mistakes, Dental Press J Orthod. 2017; 22(5): 113–117. doi: 10.1590/2177-6709.22.5.113-117.sar

CHAPTER 5

English in Academic Writing-II

- **Dr Ajay Semalty**
H.N.B Garhwal University (A Central University),
Srinagar Garhwal-246174

In continuation of the previous chapter, let's discuss some different domains of language part in academic writing.

Learning Outcome	Learning Plan
After completing this chapter, you will be able to	Here we will discuss:

Learning Outcome

After completing this chapter, you will be able to

- Structure information and Link your ideas logically in your manuscript
- Take practical steps to improve cohesiveness in manuscript
- Use tools of improving cohesiveness

Learning Plan

Here we will discuss:

- Concept of cohesion and coherence
- Tools of Cohesive Writing
- Conjunctions
- Technical words/ Lexis for Lexical cohesion
- References
- Paragraph development

We have discussed the grammatical, vocabulary and punctuation part in the previous chapter. Now, what next? Do you think, a person who has done the research work and is good in vocabulary, grammar and punctuation can write a good manuscript?

No! It requires a precise use of all the things put together so that the write up is reader friendly. Remember, the fact, that the reader is of course familiar with the field of study and that's why he or she is reading the same. So, factually correctness is the foremost thing.

Domains of AW

An Academic writing work, in gross, can be classified in three domains:

- Language
- Technical
- Formatting (as per the requirement)

We will be discussing the technical and formatting domain, later. Now, let's discuss the vital domain of Language and the components of language acceptability. And the first and foremost thing is Cohesion and coherence in academic writing.

What is cohesion or coherence?

Cohesion and coherence are the attributes of a good and effective piece of writing. It is derived from word "cohere" means to join together.

Let me explain you with an example:

Do you remember what is required to build a house? AW may be considered like building a house. It requires:

A foundation: Your thesis/ research data

A structure: Your presentation or discussion

And

Cohesion: to keep everything joined together

Imagine this wall as a paragraph of your piece of AW and the bricks as your words and phrases. Now the question is: What is keeping these bricks together?

YES! It is cement which is called coherence/ cohesion or cohesiveness in AW.

So, Cohesiveness is "The logical connections between sentences."

Coherence v/s Cohesion

"Coherence means the connection of ideas at the idea level, and cohesion means the connection of ideas at the sentence level."

- https://www.uwb.edu/wacc/what-we-do/eslhandbook/coherence

So, coherence is technical or "rhetorical" (developing, discussing, supporting, integrating, organizing and clarifying ideas) aspect while cohesion is grammatical aspect.

"Rhetoric may be defined as the effective organization of text."

Let's discuss why to go for cohesion and coherence?

"Writing is a reflection of our thinking"   - Tim Ferriss

If we are confused, we can not write a piece of writing with clarity. The clarity in mind is utmost essential for delivering good piece of writing with effective and logical flow of the idea.

So, let me come back to the question **"Why Cohesion and coherence needed?"**

- For presenting the objectives or the idea in easy to understand and easy to follow manner

- For connecting the ideas or the main points for making them easy to understand and follow.
- For reducing the interpretive burden of readers
- For following the outline of manuscript or roadmap of the idea

So, a cohesive writing is easy to follow, reader friendly, and effective in communicating the research idea.

"When we talk about coherence in writing, we talk about how the piece of writing moves through ideas."

Achieving Cohesion and coherence in AW

Cohesion and coherence should be achieved at three different levels in manuscript (Fig. 5.1).

- within a sentence
- between sentences (within a paragraph)
- between paragraphs

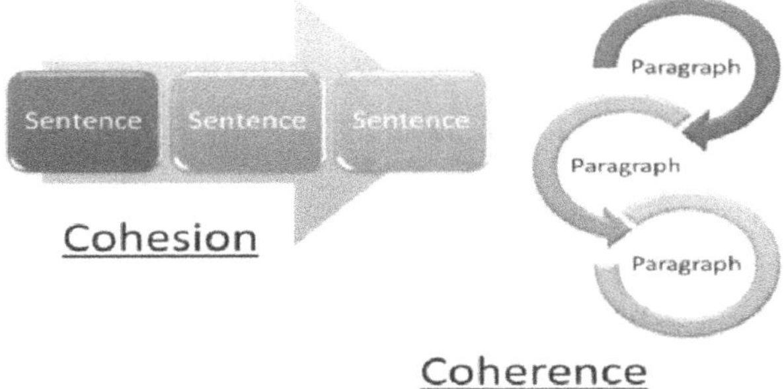

Fig. 5.1 Cohesion V/S Coherence

Within a sentence: You begin with a theme or the main point followed by the new idea. It is your choice to whom to make a theme or the focal point and then give new information. Remember that you have to, link this sentence further with the next sentence, so the new information may be the theme/main point for the next sentence for ensuring cohesiveness.

For example: "In the current era, the focus of research is on nanotechnology. The nanotechnology is about using the unique properties of particles and devices of nanometer size."

Here the focus of research is the main point

and

Nanotechnology is the new information.

In the second sentence the nanotechnology becomes main point or theme which is giving new information again.

And this way logical connection must be made within the sentence and between sentences.

Linking sentences: A single sentence should lead to the new idea which may be developed in the following sentence. The selection of **theme** and **new information** and their positioning in the sentence affect the flow and cohesiveness. So, it is very critical to chose the theme and new information and their combination for each sentence and the forthcoming sentences, for developing the whole idea.

(Cohesive Writing Chapter: Unit 2. Independent Learning Resources)

So, what is the first step in ensuring cohesiveness

Step 1: Distinguishing between theme and the new information

Step 2: Arranging the theme or the main point effectively.

Step 3: Recognizing the theme and new patterns

- AB:AC
- AB:BC
- Combined

Can you remember what pattern was there in our previous example?

"In the current era, the focus of research is on nanotechnology. The nanotechnology is about using the unique properties of particles and devices of nanometer size."

Yes!! You got it …

It is AB:BC pattern.

After understanding g the basic concept of cohesion and coherence, their importance and basic patterns of ensuring cohesiveness, let's discuss the tools of cohesive writing

Tools of Cohesive Writing

- **Conjunctions**
- **Technical words/ Lexis for Lexical cohesion**
- **References**
- **Paragraph development**

A. Conjunctions: The conjunctions are the words that join two clauses, two sentences or two paragraphs. The effective use of conjunctions establishes the logical relationship of ideas and create the cohesion. E.g. However, moreover, next, then, lastly, while etc.

On the basis of type of relationship conjunctions are used.

If the relationship is

Cause and effect type: *so, as a result, because*

Temporal (a time sequence) type: *then, first, next*

Contrastive type   *on the other hand, however*

Simple additive type   *moreover, in addition*

Conjunctions may be

- Prepositional phrases (in spite of, on the contrary)
- A preposition introducing a noun phrase (besides the various evidences)
- A preposition introducing an 'ing' verb or adverb (besides increasing the release…)

Logical relationship	Transitional expression
Similarity	also, in the same way, just as … so too, likewise, similarly
Exception/Contrast	but, however, in spite of, on the one hand … on the other hand, nevertheless, nonetheless, notwithstanding, in contrast, on the contrary, still, yet
Sequence/Order	first, second, third, … next, then, finally
Time	after, afterward, at last, before, currently, during, earlier, immediately, later, meanwhile, now, recently, simultaneously, subsequently, then
Example	for example, for instance, namely, specifically, to illustrate
Emphasis	even, indeed, in fact, of course, truly
Place/Position	above, adjacent, below, beyond, here, in front, in back, nearby, there
Cause and Effect	accordingly, consequently, hence, so, therefore, thus
Additional Support or Evidence	additionally, again, also, and, as well, besides, equally important, further, furthermore, in addition, moreover, then
Conclusion/Summary	finally, in a word, in brief, briefly, in conclusion, in the end, in the final analysis, on the whole, thus, to conclude, to summarize, in sum, to sum up, in summary

Adapted from https://writingcenter.unc.edu/tips-and-tools/transitions/

Let's see the use of conjunctions for establishing cohesion with this example.

"**In spite of** questionable effectiveness **and** one or more side effects, hair growth products, available in the form of many synthetic and alternative medicines, contribute the million dollar market globally."

The same set of information can be given in this way also.

"Hair growth products, available in the form of many synthetic and alternative medicines have the million-dollar market globally. **But** the effectiveness of these medicines is questionable. **Besides,** questionable effectiveness, the synthetic drugs have one or more side effects also."

Have you noted? The first form of the write up is long single sentence. Using conjunctions help to split the same message in smaller sentences with effective cohesion and do not allow the reader to lose his or her attention.

The use of conjunctions becomes very important in AW when you are discussing the results in a critical way. But remember not to overuse the conjunctions. Otherwise it would look like spoken English.

Now let's focus on the other tool: lexical cohesion

After discussing the cohesion and coherence and conjunction as a tool for the same, let's discuss some other tools.

B. Technical words for Lexical cohesion

Halliday & Hasan (1976)* —Cohesion in English (Pioneer in introducing the term lexical cohesion)

Halliday & Hasan (1976) introduced two classes of cohesion:

- grammatical
- lexical.

*Halliday, M. & Hasan, R. (1976). *Cohesion in English.* London: Longman.

Grammatical cohesion consists of:

- conjunction: connecting clauses, sentences and/or para
- reference: anaphoric and cataphoric reference
- substitution: replacing a word
- ellipsis: removal or omission of words for contextual meaning

"Lexis creates cohesion using synonyms, hyponyms, and super ordinates. The use of lexical chains creates variety in writing and avoids monotony."

https://www.enago.com/academy/coherence-academic-writing-tips-strategies/

"Lexical cohesion may be defined as the relations between lexical items in a text that contribute to textual continuity."

It is less systematic and not much simple as that of grammatical cohesion.

"lexical cohesion occurs when related word pairs join together to form larger groups of related words that can extend freely over sentence boundaries"
-Morris and Hirst (2004)

Morris, J. (2004). Readers' perceptions of lexical cohesion in text. Proceedings of the 32nd Annual Conference of the Canadian Association for Information Science, Winnipeg, 2-5

Now let's see simply

Lexical cohesion = words that go together

Lexical items = the nouns, verbs, adjectives and adverbs in a text come repeatedly and related in one or other way.

This relation may be in terms of

- **repetition (same word repeated),**
- **synonymy (similar meaning word repeated)**
- **collocation (expected words together)**

So, either you repeat the word to refer the thing you are talking about. Or you may give synonym of the same i.e. a word with the same meaning, or if the link is a 'kind of' something else, or 'part of' something else. In both the way it is lexical cohesion.

"Hair are important part of our personality. But nowadays, hair fall has become a very common problem. Hair fall in medical terms is called Alopecia. Alopecia may not be a major health issue but certainly it is a psychosocial problem."

Can you see we used "hair" and "hair fall" as repetition type of lexical cohesion.

While using alopecia for hair fall is synonymy

Collocation is a type of **'expectancy relation'** between lexical items. You expect another word with a certain word almost always.

Factors- affecting/ governing

Disease- treatment, medicine, patient

Depending on your field of research you can see the how you can adapt a smooth and natural lexical cohesion.

C. References:

Firstly, be clear in your mind this reference tool is different. It is a language tool, not your usual reference or citations which you put at the end of your manuscript.

It is about introducing a noun andkeep tracking of the same. This can be achieved by the use of pronouns. Using pronouns allow us not repeating it again and again in the text.

Reference creates cohesion by using different type of pronouns: possessive pronouns (e.g. your, their, etc.), pronouns (e.g. she, me, etc.), and determiners (e.g. those, these, etc.).

The references may be anaphoric or cataphoric.

- **Anaphoric reference:** Backward pointing reference. This is meant to inform the reader that current point has something which was told earlier.

  e.g. A total of 36 Wistar albino rats were used for the study. They were given sufficient time for acclimatization.

- **Cataphoric reference:** Forward pointing reference. This is meant to inform the reader that there is more to come. (Cataphoric reference occurs when a word or phrase refers to something mentioned later in the discourse.)

  e.g. In spite of **its** simple and effective construction, the **ABC instrument** is not very much used in the industry.

  Can you come up with some more examples of anaphoric and cataphoric reference in discussion forum.

- **Substitution:** It is the replacement of a word or phrase with a filler word (like: same, do, so etc.) to avoid repetition.

  e.g. "I bought a designer bag today. She did the same."

- **Ellipsis** is the removal or omission of one or more words of which meaning is drawn automatically by the reader through context.

  e.g. He goes to yoga classes in the afternoon. I hope I can too.

  Or

  The blood pressure came down with in one hour for group one. The group 4 followed the trend.

So, have you got the essence? Apart from conjunctions, it is the pronoun, in one or other way, which must be used effectively throughout the manuscript to ensure cohesiveness. It may be reference or lexis.

This was all about the cohesion now let's move towards the coherence through paragraph development

D. Paragraph development

The tools so far, we discussed were dealing with the cohesion. We learned how we can connect the sentences using conjunctions, lexis and references and ensuring cohesion. Now lets move from cohesion to coherence.

Let me remind you

"Coherence means the connection of ideas at the idea level, and cohesion means the connection of ideas at the sentence level."

So, after ensuring the grammatical cohesion within the sentence we have to develop the major ideawith a natural flow and progression through sentence to sentence and paragraph to paragraph.

So, let us first define the paragraph **"a group of sentences or a single sentence that forms a unit or support one main idea"**. Paragraphs are the **building blocks** of your manuscript or thesis. In general, a paragraph has 4-5 sentences. In AW, very long sentence which itself looks like a paragraph should be avoided. The very long sentence may lose the attention of the reader.

Properties of a strong paragraph (5 Cs)

- **Centered (idea):** All sentences in paragraph should be closely related to the main idea.
- **Clarity:** The sentences must have clarity to communicate and developing the idea
- **Connectivity to main idea:** The sentences must be connected to main idea
- **Coherent:** The idea must be developed in a natural and logical way.
- **Complete:** Each paragraph must develop the main subidea (controlling idea of the paragraph) completely.

Steps involved in paragraph development

Now let's see the steps involved in paragraph development

I. Planning a tight outline: After critical thinking and considering all the aspects to be covered, the major idea must be subdivided in to various sub ideas. Each sub idea must be discussed in a single paragraph in such a way that all sub ideas may logically lead to effective presentation of the whole idea.

II. Developing a subidea in a single paragraph: A paragraph must support sub idea of the one main idea. And the idea must be developed sentence by sentence. And finally, should lead to the next main idea to be developed in the next paragraph. Remember, what we discussed in the last section, the sentences linking patterns

- AB:AC

- AB:BC
- Combined

The topic or the theme sentence (addressing main idea) should be ideally in the beginning of a paragraph. It may be immediately after the paragraph linking phrase for ensuring coherence.

For developing the sub-idea you can

- Narrate (tell a story, sequentially to develop the idea)
- Describe (describe the idea)
- Process (Step by step process the idea)
- Classify (classify or subdivide the idea)
- Illustrate(explain with the help of one or more examples)

III. Completing sub idea and transitioning to the next sub idea

After sufficient and effective description of the sub idea, you must close the statement to give it completeness within the scope of sub idea. But in closing the idea you must remember that the next paragraph must be linked to ensure the coherence. The transition of idea from one paragraph to another paragraph must be smooth and logical. The last lines of the paragraph should be planned to anticipate the next sub idea or the next paragraph.

A paragraph should also be connected to the previous paragraph. Use transitional words and phrases like "Having discussed; Besides the methods discussed earlier, etc."

e.g.

"Having defined the solubility, this paper now addresses the methods of improving it."

Tips for building strong and coherent paragraph

- Only one main idea per paragraph
- Main idea addressed in a topic sentence
- Theme sentence near beginning of paragraph
- Five to eight sentences per paragraph
- Developing the idea sentence by sentence
- Use narration, description, processing, classification and illustration
- Smooth transition from sentence to sentence and paragraph to paragraph

Let me give you a simple tip for checking the connection between the sentences in a paragraph. Just split the separate the sentences and tell someone to unified the sentences to build a paragraph. If your paragraph is cohesive and coherent anyone can remake the paragraph to its original form. If not, then you have to make extra effort to make the paragraph strong.

The language services and software

The language improvement services are available on payment basis. Nowadays many journal publishers offer or recommend the paid language services. If you can pay this amount ranging from 100 to 300 US$, its fine.

Alternatively, apart from **MSOffice**, various software are available which help you in language editing they are again either paid or free. But please remember, no program will ever be as good as a human being to edit. Editing services are great but costly and less secure.

Grammarly: Free online Grammar Checker;www.grammarly.com/

SWAN: The scientific writing assistant can be used offline, available free at: https://cs.joensuu.fi/swan/

Ginger; Online editor and Free Grammar Checker; available at; www.gingersoftware.com

Hemingway editor 3; online grammar editor; online free desktop version paid; www.hemingwayapp.com

There are so many such software please Google and find the most suitable one for your work.

Activity resources

1. Practice developing cohesiveness by doing the exercises available in this cohesive writing chapter: https://sydney.edu.au/stuserv/ documents/ learning_centre/Cohesive2.pdf
2. To understand the examples of paragraph development refer https://writingcenter.unc.edu/tips-and-tools/paragraphs/

Take Away Message: To present your idea in reader friendly manner, you need to ensure cohesion and coherence in your writing. What is cohesion? It is about grammatical aspect and coherence is about technical aspect. "Coherence means the connection of ideas at the idea level, and cohesion means the connection of ideas at the sentence level."Use the most suitable conjunctions, lexis, references and coherent paragraphs to develop your main idea into your manuscript.

Further Reading

- Wall work, Adrian, English for Writing Research Papers, Springer, 2011.
- Paper cohesion and flow;
  http://writing.umn.edu/sws/quickhelp/style/flow.html
- Coherence: How Writing Clearly Facilitates Manuscript Acceptance;
  https://www.enago.com/academy/coherence-academic-writing-tips-strategies/

References

- Cohesive writing https://www.uow.edu.au/content/groups/public/@web/@stsv/@ld/documents/doc/uow195607.pdf
- Cohesive writing chapter https://www.uow.edu.au/content/groups/public/@web/@stsv/@ld/documents/doc/uow195610.pdf
- Cohesive writing;
  https://sydney.edu.au/stuserv/documents/learning_centre/Cohesive2.pdf
- Transitions, The Writing Center, University of North Carolina at Chapel Hil, https://writingcenter.unc.edu/tips-and-tools/transitions/
- Halliday, M. & Hasan, R. (1976). Cohesion in English. London: Longman.
- Morris, J. (2004). Readers' perceptions of lexical cohesion in text. Proceedings of the 32nd Annual Conference of the Canadian Association for Information Science, Winnipeg, 2-5
- Paragraphs, https://writingcenter.unc.edu/tips-and-tools/paragraphs/

CHAPTER 6

English in Academic Writing-III

-Dr Ajay Semalty
H.N.B Garhwal University (A Central University)
Srinagar Garhwal-246174

In the previous chapters, we discussed about effective academic writing in English by avoiding grammatical errors, punctuation errors and by improving cohesion and coherence in the writing. Now, we will focus how the language part should be taken care of in each section of a manuscript, major do's and don'ts and styles of writing.

Learning outcome

After completing this chapter, you will learn

- Improving language of each section of manuscript
- Writing styles and tips
- Avoiding most common mistakes of manuscript writing
- Features of different standard styles of research writing

Lesson Plan

Here we will discuss:

- The section wise language improvement tips
- Most common mistakes to be avoided
- Major Don'ts
- Types of styles
- APA/ MLA/ CMS style

Article Section Wise Tips (with respect to English language)

Though we will discuss the technical part of each and every section in later chapters, we will discuss the language part of the sections here.

Title

As discussed earlier, title should be catchy, representative of your work and free from any spell error. Double check the spelling, if the title is in upper case. Because spell check does not work in upper case. Follow the word limit, if prescribed in author's guidelines.

Abstract

In this section, we give the central idea of our work and main finding in brief so as to attract the prospective reader.

- **Make the abstract simple, effective and catchy:** You are aiming to attract the prospective reader so the presentation must be very effective.

- **Make short sentences:** It is itself a small section, so using long sentences will be awkward.

- **Make it self-explanatory and concise:** Do not give the reader THE INTRPRETIVE BURDEN.

- **Follow word limit as per the journal's requirement:** Many journals put a limit of words ranging from 200-300 words. And if you have planned a longer abstract, the online submission system would not accept it. In that case you have to reframe the abstract and concise it. And when you cut short it in hurry, it results in disaster.

- **Avoid abbreviations and complex sentences:** You cannot start using the abbreviations in the abstract itself. Using abbreviations without reference will just confuse the readers. And the same holds true with the complex sentences.

- **Make it structured or unstructured as per the "Instructions to authors" prescribed**: Structured abstract means writing the abstract divided in some heads like background/ Aim/objective, Method, Results, Conclusion. If required by the journal make it structured.

Introduction

Introduction section gives problem definition and acquaints the readers about the idea of research.

When we write introduction following points must be taken care of

- **Give brief background:** Giving background is just like taking some steps back like a tiger before the final attack. Or in another way taking a run up before delivering a ball in cricket match. But have you seen any bowler starting the run up from pavilion or delivering the bowl from non- striking end standing itself? So what is the point? Give the background of the study in a concise way.

- **Give a power packed story:** Tell the idea like a complete story but do not make it too long. Long introductions lose their focus.

- **Ensure cohesion and coherence:** Cohesion at grammatical level and coherence at idea level must be ensured for giving the sequential flow of information in a logical way. So, the reader should be able to predict about the very next sentence.

- **Present tense followed by the past:** Introduction may be started in present tense to share the current knowledge or status of the problem followed by background and problem/ hypothesis which should be given in past tense.

Materials and Methods

- Use past tense throughout. Because you have already done the work and you reporting it.
- Clarify the method fully. It may be decisive in review process.
- Use good balance of passive and active voice.
- Many a times, in method passive voice is more common.
- Do not start a sentence with a digit.
- Use punctuations carefully.

Results & Discussion

- Follow a pattern or the logical sequence to present the results and their discussion.
- Give due weightage to present every particular result.
- Avoid exaggeration or the vague statements in discussion.
- Avoid repeating the results already mentioned intable/figure.
- Be very clear in discussing and correlating the study with other studies.
- Make crisp sentences.
- Convince reader with your logic on the basis of evidence based discussion of results.

Conclusion

- Concisely give concluding remark.
- Conclude as per the laid down objectives.

Acknowledgements

- Be humble and straightforward in acknowledgements.

References

- Follow punctuation as per the journal's guidelines strictly.

We will discuss the technical part of each of these sections in later chapters. In the next lecture we will quickly focus the most common language mistakes committed by authors in AW.

Most Common Language Mistakes

We have discussed the grammatical part, cohesion, coherence and section wise language tips earlier. Now let us focus some of the most common language mistakes in AW.

Now, let us discuss the most common writing errors. This list of the most common errors was compiled by Bob Connors and Andrea Lunsford after surveying 21,000 essays.

1. **Apostrophe confusion:** Rule: do not use the contractions. Its shows possession; It's is used for "it is" or "it has" but meaning is different.

2. **Misplaced/dangling modifier:** Rule: Put modifier close to the word they modify.

   Example: Noting down carefully people answered the questionnaire. (wrong)

   People answered the questionnaire and we noted down carefully.

   We noted down carefully when people were answering the questionnaire.

3. **Fused/run on sentences: Rule:** Independent clauses must be separated with a comma and a coordinating conjunction (FANBOYS: For, And, Nor, But, Or, Yet and So).

   Boys liked the movie girls did not like it. (wrong)

   Boys liked the movie, BUT girls did not like it   (right)

4. **Unnecessary comma(s): Rule:** Donot place commas around elements that restrict or define the meaning of another part of the sentence.

   The persons were shown the movie, Titanic, in seminar hall. (wrong)

   The persons were shown the movie Titanic in seminar hall.

5. **Pronouns and antecedent disagreement:** Rule: Pronouns must match their antecedents (the words they substitute for)in number case and gender.; indefinite pronouns have their own rules (Body-, -one, -thing,

each, either, and neither always take singular form; any , all, more, most, some, and none can be either singular or plural)

6. **Missing comma in a series:** Rule: Donot skip the final comma when listing three or more items. (standard rule is to put comma but sometime exempted)

7. **Subject and verb disagreement:** Rule: A subject must agree with its verb in number and person.

8. **Wrong tense or verb form:** Rule: verb should match time or order indicated by the context. And use correct form of irregular verbs.

9. **Sentence fragment**: Rule: adding a period does not complete a sentence.

10. **Unnecessary shift in pronouns**: Rule: Donot shift to second person after starting with the third person.

11. **Unnecessary shift in tense:** Rule: stick to one tense.

12. **Missing or misplaced apostrophe:** Rule: Add 's to show possession (but not after a plural ending in –s.)

13. **Comma Splice:** Rule: Do not use a comma by itself to combine sentences

14. **Wrong or missing prepositions:** Rule: Prepositions can be tricky; review usage guides and look up words you are not sure about. (between/ among, beside/besides etc.)

15. **Wrong or missing verb ending;** Rule: When revising, make sure you haven't left off any verb or chosen the wrong verb.

16. **Missing comma with a non-restrictive element:** Rule; add comma to set apart elements that do not restrict or define another part of the sentence.

17. **Wrong word:** Rule: donot depend on spell checker or auto correction (accept/except, affect/effect, then/than)

18. **Missing comma in compound sentence:** Rule: If you combine two complete sentences, you must separate them with a comma and coordinating conjunction (FANBOYS).

19. **Vague pronoun reference:** Rule: Donot use a pronoun unless it is clear to the reader what it stands for (this, that, these, those; who, whom, whose, which etc.)

20. **Missing comma after introductory element:** Rule: Place a comma after anything that comes before your main clause. (comma after therefore, however, moreover etc.)

(Dear learners see the examples of each type of error and learn the correct use. You can refer further readings for learning the examples.)

Helpful Online Resources for exercise:

- Passive and active voice handout: http://twp.duke.edu/uploads/media_items /passive-active.original.pdf

- Using first person effectively: http://twp.duke.edu/uploads/media_items /first-person.original.pdf

- Citations reference from Duke Libraries website: http://library.duke.edu/ research/citing/index.html

- Proofreading for common grammatical mistakes:

- http://bcs.bedfordstmartins.com/smhandbook6e/Player/MainFrame.aspx?tas k=handbook&taskid=3

- Numerous strategies for revising: http://twp.duke.edu/writing-studio/ resources/ academic-writing/revising

- Reading aloud: http://twp.duke.edu/uploads/media_items/reading-aloud.original.pdf

In the list of most common mistakes we can add:

Do not use the

- Vague or casual words or language. (a bit of, sort of etc.)
- Acronyms
- Slang (kids, guys)
- Idioms, (A1,)
- Contractions,(don't, isn't)
- Phrasal verbs (take away)

The objective of the AW is always more formal and objective.

Reading aloud is always a very good and simple technique to identify the language error. You can even trace any deviation from required cohesion and coherence by just reading aloud.

Types of Styles

The styles: By the style we refer to organization of an article. It includes formatting, referencing, citation and all other specifications. These are generally prescribed by various different agencies or well-known associations. But as long as you are strictly following a journal's specific guidelines it does not matter much, whether you know about the specific style or not.

Some important style their associations, their links and the subject coverage is listed for you in the lecture's pdf. These styles include APA, MLA, CMS, IEEE, Vancouver, Harvard, ACS AIP, CSE etc.

Style	Association/ Organization	Subject	URL of Style Guide
APA	*American Psychological Association*	**Psychology, Education, Sciences**	**www.apastyle.org/**
MLA	*Modern Language Association*	**English, Humanities**	**www.mla.org/style_faq1**
CMS	*Chicago Manual Style, University of Chicago Press*	**Medical, biological, life sciences**	**www.chicagomanualofstyle.org**
IEEE	*Institute of Electrical and Electronics Engineers*	Engineering	http://standards.ieee.org/guides /style/
Vancouver	International Committee of Medical Journal Editors	Medical, Biomedical, pharmacy, life sciences	http://www.icmje.org/index.html
Harvard	Harvard University		https://library.harvard.edu/writin g-guide#style-guide
ACS	*American Chemical Society*	Chemistry, life sciences, Pharmacy	http://pubs.acs.org/page/books/ styleguide/index.html
AIP	American Institute of Physics	Physics, Astronomy	http://kmh-lanl.hansonhub.com/AIP_Style_4 thed.pdf
CSE	*Council of Science Editors*	Biology, Natural Sciences	http://www.resourcenter.net/Scr ipts/4Disapi07.dll/4DCGI/store/S SFTOC.html
AMS	*American Mathematical Society*	Mathematics	ftp://ftp.ams.org/pub/author-info/documentation/handbk.pdf
AMA	*American Medical Association*	Medical, Biomedical, Pharmacy,	http://www.amamanualofstyle.c om/oso/public/index.html
AMS	*American Meteorological Society*	Geology, meteorology	http://www.ametsoc.org/pubs/A uthorsguide/html_vs/

The Big Three: APA, MLA, and CMS

Three main Style Schools, which are used to in formatting academic papers are APA, MLA and CMS.

- **APA style**: These are the official guidelines by the **American Psychological Association** (VI edition). This is the most suitable for social sciences, sociology, psychology, medicine, or social work.

- **MLA style**: The **Modern Language Association** provides guidelines for writing in humanities. Therefore, the artists, English and drama/ theatre students use this style MLA for a long time.

- **CMS style**: These guidelines are *Chicago Manual of Style* (16th edition). CMS style is used in the humanities, particularly with literature, history and the arts students.

They have many things in common like margins and spacing. But their attributing of references to source materials is the main distinguishing factor. E.g. For APA it is "references" on the other hand for MLA it is "works cited".–

It is a very small but very important difference but it may affect the fate of manuscript.

(http://www.collegescholarships.org/mla-apa-cms-styles.htm)

What is a citation and citation style?

A citation is a way of giving credit to individuals for their creative and intellectual works that you utilized to support your research. It can also be used to locate particular sources and combat plagiarism. Typically, a citation can include the author's name, date, location of the publishing company, journal title, or DOI (Digital Object Identifier).

A citation style dictates the information necessary for a citation and how the information is ordered, as well as punctuation and other formatting.

Let's have a look on some basic and important features of the big three different writing styles.

Uniform Requirements for Manuscripts Submitted to Biomedical Journals (the Vancouver Document) is at <http://www.icmje.org/index.html

The Big three: APA, MLA and CMS comparison of guidelines

TOPIC	APA	MLA**	CMS***
Association	American Psychological Association (6th edn)	Modern Language Association	*Chicago Manual of Style (16th edition) by University of Chicago Press*
Subject	Psychology, Education, Sciences	English, Humanities	*History, Medical, biological, life sciences*

Contd...

Verb Tense	**Use past tense or present perfect tense** for literature reviews and descriptions of procedures if the discussions are of past events; use past tense to describe results; use present tense to discuss implications of results and present conclusions(65–66)	Although the manual does not explicitly state a preferred tense, typically you should **use present tense** when discussing literature:	**Use present tense to refer to "timeless facts"** and to describe fictional works' plots(236)
Headings	**Upto five levels:** 1. Centered, boldface, upper case and lower case headings 2. Left-aligned, boldface, upper case and lowercase heading 3. Indented, boldface, lowercase heading with period 4. Indented, boldface, italicized, lowercase heading with period 5. Indented, italicized, lowercase heading with period(62–63)	**Double space between the end of one section and the heading label for the next, and between the heading and the section that follows; do not follow heading with punctuation; headings need not be bolded or italicized**(117)	**Set headings flush-left on a new line; each level of heading must be differentiated by type style and size; do not usefull capital**s(61)
Title Page	**Title page required:** **Title length<12words, upper and lowercase letters, centered, upper half of page (23);** Include author's name and institutional affiliation(23); include author note(23); **Running head** also included (229); Title page is page1,noted in the page header(229)	**No title page required;** on page one of the paper, type your name, your instructor's name, the course number, and the date on separate lines flush left and double-spaced; beneath that, type the title, centered in regular font(116–117)	**Title page should include full title of work with subtitle** (if any), the name of the author (author's affiliations typically not included), editor, or translator, and the name and location of the publisher; do not include date of publication and do not include a page number on this page(10)

Contd...

Page Numbers	Number all pages consecutively starting with title page Using Arabic numerals(229);include page number in runninghead:shortversionof title,followedby5spaces andpage#,maximumof50ch aractersincludingspaces and punctuation(229)	Number all pages consecutively in upper right-hand Corner of page, one half inch from the top and flush right; precede page numbers with your last name (117)	Each page of manuscript must be numbered consecutively(66–67)
Margins	One in chonall sides; lines should be flush-left; maximum Line length is 6.5inches(229)	One inchonall sides; do not justify right margin (116)	All text should be flush-left; margins of atleast 1 inch on all four sides of text(60)
Line Spacing	Double-space(229)	Double-space(116)	Double-space(59)
Paragraph Indentation	Indent first line of every paragraph and first line of every foot note ½ inch (5 spaces) from left margin (229)	Indent the first word of a paragraph ½ inch from left margin; indent set off quotations 1 inch from left margin(116)	Indent the first line of each paragraph using your Computer TAB key(60)
In-text Citation Emphasis	On the date of work created;	On authorship	Author or date, depending on source origin
Biblio-graphy Page	The reference list should begin on a new page and should be titled References, with the word centered on the page; double space reference entries and use a hanging indent; alphabetizeen tries by author last name(37)	The bibliography page should begin on a newpage and should be titled Works Cited with the words centered on the page; double space reference entries and use a hanging indent; alphabetizeen tries by author last name(230–231)	Full bibliography pages are not required but are recommended (684); the bibliography should bean alphabetical list of all sources used for research(685);it should follow the end of the paper and begin on a new page(685);acceptable titles include "Bibliography," "Works Cited," or "Literature Cited"(687)

Contd...

In-text Citations	When quoting, provide the author, year, and specific page number or paragraph number; date can follow author in an introductory phrase, or all citation elements can beat the end of the sentence (170–171): Robbins(2003)notes....(p.54 1). OR "..."(Robbins, 2003,p.541).	When quoting, provide the author and the page number; they may be separated or cited together(214–218): Town send notes"..."(613). OR"..."(Townsend 613).	When quoting, **use superscript numbers following the quotation to indicate a footnote or end note(665);a note should list the author, title, and publication facts** (661); foot notes appear at the foot of the page(671); end notes are grouped at the end, following the text and any appendices and preceding the bibliography (673)

"APA, MLA and Chicago comparison of Guidelines adapted from https://www.naropa.edu/documents/programs/jks/naropa-writing-center/citation-comparison.pdf"

Other important standards which are prescribed by any style include

Diction & Syntax (Word Choice & Arrangement)

Voice, economy of expression, clarity, person, scientific, etc.

Punctuation

Use the Colons, Semicolons and Quotes as per the style.

Abbreviation

Use standard abbreviations or as prescribed by the style.

- *Avoid abbreviations that are not already commonly used (see table for some common abbreviations).*
- *In most cases, write out the complete term or phrase with the abbreviation following in parentheses to continue using the abbreviation throughout the rest of the document.*
- *Once an abbreviation is used, it must continue to be used throughout.*
- *Avoid beginning a sentence with an abbreviation, especially those that begin with a lower-case letter.*
- *Do not use spaces within an abbreviation (e.g., U.S.)*

Numbers, Mathematics, and Statistics

- **For number, mathematics and statistics use formal abbreviations and symbols. Use standard unit e.g. SI or MKS unit throughout.**

Citations Managers

How references are being cited, is one of the most important distinguishing factor among different styles. For citation management, online citation managers are an easy way to keep track of all of the references. Mendeley, EndNote, and Zotero are some of these useful tools. These are tools for importing citations from various sources and article databases that can automatically integrate them into your research paper and bibliography.

Mendeley

Various Universities have an institutional subscription to this reference manager. The Mendely can help you organize your research, collaborate with others online, and discover the latest research.

End Note

EndNote is the standardized bibliography tool adopted by various universities and research organizations. EndNote enables users to search online bibliographic databases, organize references and create and format instant bibliographies. It also contains an integration interface with Microsoft Word, as well as a traveling reference library that follows the document for easy collaboration with others.

Zotero

Zotero is a free application that collects, manages, and cites research sources. It connects with your web browser to download sources.

Take Away Message

Dear learners in this chapter we learned the **basic language tips for the different sections of articles.** We learned how to ensure **cohesion and coherence** throughout the manuscript. We learned about the most common errors like apostrophe confusion, missing comma, unnecessary comma, subject verb disagreement, using improper pronoun etc. Always remember to avoid slang, contractions, acronyms, idioms and phrasal verbs. We learned about different schools of styles like APA, MLA, CMS etc. Chose the most relevant one and study the style guidelines.

Further Readings

- Passive and active voice handout:
  http://twp.duke.edu/uploads/media_items/passive-active.original.pdf
- Using first person effectively:
  http://twp.duke.edu/uploads/media_items/first-person.original.pdf
- Citations reference from Duke Libraries website:
  http://library.duke.edu/research/citing/index.html
- Proofreading for common grammatical mistakes:
- http://bcs.bedfordstmartins.com/smhandbook6e/Player/MainFrame.aspx?tas k=handbook&taskid=3
- Numerous strategies for revising: http://twp.duke.edu/writing-studio/resources/academic-writing/revising
- Reading aloud: http://twp.duke.edu/uploads/media_items/reading-aloud.original.pdf
- The basics of APA style: https://www.apastyle.org/learn/tutorials/basics-tutorialhttp://www.apastyle.org/products/asc-landing-page.aspx

References

- The 20 Most Common Grammar and Mechanical Errors, https://www.youtube.com/watch?v=VB8u_K0I-a8
- Lunsford, Andrea A. and Karen J. Lunsford. "''Mistakes are a Fact of Life: A National Comparative Study." CCC 59 (2008) 781-806.
- American Psychological Association. (2009). Manuscript structure and content. In Publication manual of the American Psychological Association (pp. 21-60). Washington, DC
- American Psychological Association. (2009a). The mechanics of style. In Publication manual of the American Psychological Association (pp. 87-124). Washington, DC: Author.
- American Psychological Association. (2009b). Writing clearly and concisely. In Publication manual of the American Psychological Association (pp. 61-86). Washington, DC: Author.

- www.easybib.com/guides/citation-guides/apa-format/
- http://www.icmje.org/index.html
- https://www.naropa.edu/documents/programs/jks/naropa-writing-center/citation-comparison.pdf
- Gibaldi, Joseph. MLA Handbook for Writers of Research Papers. New York: Modern Language Association of America, 2009. Print.
- Publication Manual of the American Psychological Association. (2010) Washington, DC: American Psychological Association.
- The University of Chicago Press Staff, ed. The Chicago Manual of Style. 16th ed. Chicago: University of Chicago, 2010. 1 Lunsford, Andrea A., Paul Kei Matsuda, and Christine M. Tardy. Easy Writer: A Pocket Reference, 4th ed. Boston: Bedford/St. Martin's, 2010. Print.

CHAPTER 7

Plagiarism (Introduction)

Dr Mona Semalty
H.N.B Garhwal University (A Central University)
Srinagar Garhwal-246174

Learning Outcome

After completing this chapter, you will be able to know

- Concept /importance
- Different types of plagiarism
- Consequences of plagiarism

Lesson Plan

- Introduction /definition of plagiarism
- Different types of plagiarism
- Few case studies regarding the same

Plagiarism is academic/literary theft (intentionally or unintentionally). Plagiarism occurs when someone uses other person's language, ideas, or any other type of text material, figure and graph which do not belong to common original knowledge without its acknowledgment.

"According to the Merriam-Webster online dictionary, to "plagiarize" means:

(http://www.plagiarism.org/article/what-is-plagiarism)

- "to steal and pass off (the ideas or words of another) as one's own
- to use (another's production) without crediting the source
- to commit literary theft
- to present as new and original an idea or product derived from an existing source"

Things to be done -

- You should not take or reproduce any theory, idea or any other study material of another person without acknowledgment.
- You must acknowledge the source weather in each case weather directly quoting, borrowing facts, paraphrasing of using other persons idea.

(https://www.urkund.com/en)

It is very important for the students / teachers / academician to understand plagiarism, its types and its consequences if so happens. Plagiarism has always existed but students have many more ways of plagiarizing without detection. Nowadays it's very easy for the student to download the information and use it partially or fully in their own work.

We must understand that academic writing demands zero tolerance towards plagiarism and the fact that it is a criminal offence to steal the work of another person. Also, it should be well understood that there are serious consequences of plagiarism. If you are faculty member, you may be suspended from the job/ concerned designation and you will not be allowed to be a supervisor for PhD student. And if you are a student your registration as PhD student may be cancelled and overall it depends on how seriously you have committed this offence.

(https://www.brad.ac.uk/library/help/plagiarism/frequently-asked-questions-about-plagiarism/)

Q.1 Plagiarism is wrong, or right?

Ans. It's wrong because it is a literary theft and academic misconduct.

Q.2 Can we copy from internet?

Ans. No, we cannot copy anything from internet no matter whether it belongs to a book/ journal/or from a web source. All you needs to do is give due credit to the source from where ever you have taken.

Q.3 How to cite when there is no name of anyone?

Ans. Sometimes, it may happen that there is no name of any author. But there must be something which can be cited. You might have taken it from a web source at least you can cite that in references.

Q.4 What happens if two or more persons get a same idea at the same time and they are unaware of it. Will they be accused of plagiarism?

It's very uncertain that two or more persons get a same idea at the same time. You will definitely find something that can be differentiated. Here you can discuss how little both the ideas are different.

Q.5 How to do citation? When everyone knows it.

Ans. It is a matter of common knowledge. And if you are dealing with subject specific common knowledge, you should have little backup information that can be cited or that can be used as a reference.

Q.6 How to cite secondary reference?

If you find something written in text book by citing some other reference. This is called secondary reference. In such a case you can cite both the references.

Why do we plagiarize?

There are different reasons that why a teacher or student do plagiarize. A teacher can plagiarize, may be, to take undue advantage in publication/writing projects/books etc. But there are plenty of reasons that why a student plagiarizes the things. Let's see, what are the reasons of plagiarism by a student.

Study Pressure

It is one of the reasons why do students plagiarize in academic writing. Students are busy with lot many things like assignments /presentations /term end examinations and along with that they also have to write their synopsis/papers so may be this is one of the reason they find it easy to cut /copy and paste.

Disorganization

Disorganization occur when lot many thing going in your mind, what to do? what not to do? and what to do first? What to do later?. Somewhere, in between, you may thing of just copying from internet or from any other source.

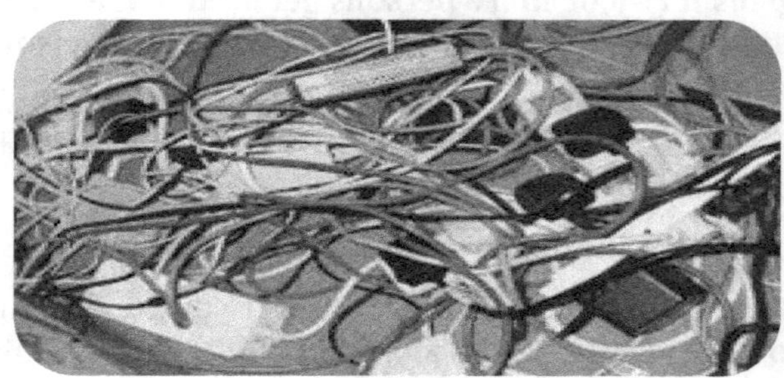

Poor Study Habits

When students do not take much interest in their research/work. They just take it very casually and they try to find any easy way of completing their assignment or whatever work they have been assigned.

English as a foreign language

Being nonnative speaker, students may not be comfortable in writing in English.

Lack of Strict Academic Discipline

Due to lack of understanding the implications of plagiarism, sometimes students as well as even teachers are not aware of consequences of plagiarism. It can sometimes may lead to plagiarism.

Careless Attitude

Careless attitude in teaching, learning and research may also attract plagiarism practice.

Types of Plagiarism

There are four broad way of classifying types of plagiarism.

Direct Plagiarism

Direct plagiarism is word-to-word transcription of a paragraph without the use of quotation marks or attribution. That means passing it completely or partially as his or her own work without attribution is direct plagiarism. Using other person's work is unethical, academic misconduct, and it makes grounds for disciplinary actions, if caught, as consequences of plagiarism may be very dangerous.

Mosaic Plagiarism (Patch writing)

Mosaic Plagiarism is borrowing phrases from any source without using quotation marks, or trying to find synonyms for the author's language keeping in view that the centric meaning of the text should not change. It is also called "patch writing". This kind of paraphrasing, whether you do it intentionally or unintentionally is an academic misconduct. And it is punishable even if you footnote your source.

Self-Plagiarism

Self-plagiarism is the case when a student reuses his or her own previous work, or mixes parts of previous works, without permission from each and every co-author involved.

For example, Self-plagiarism is when you are incorporating a part of your paper written at your graduation level into a paper you are going to write in your post-graduation. Self-plagiarism is about taking repeated advantage of one single work…it is like using one research work for taking two degrees or in other way you cannot copy figures /table or some text from your previous publication for the new manuscript. Technically you have to take the permission for reproducing the content…. Otherwise, it would be treated as self-plagiarism. **Accidental Plagiarism**

Accidental plagiarism occurs when you are busy doing literature work…and you just neglects to cite the sources, or misquotes their sources, or unintentionally paraphrases a source by using similar words, groups of words without attribution. Students should learn how to cite and should practice careful noting of the sources you have used (should develop note taking habit).

It means, be careful in noting down the tiny things you do, for example, you have taken few lines from a book and the book is misplaced…….then there may be problem in referencing and you could have avoided this, if you have already noted the reference of the book.

Other ways of classifying the types of plagiarism-

- Intentional

- Unintentional

To understand it more practically let us discuss few case studies of plagiarism and its consequences. All the case studies have been reproduced (and hence under double quote) from news papers and online news blogs.

1. **"CSIR scientist having fabricated /fake data in seven papers were retracted by the journals."(16 July, 2014):** "Seven papers of a CSIR scientist were retracted having fake data. Among them first three papers were published in PLoS ONE papers have Dr. Fazlurrahman Khan, Research Associate, as the first author and Dr. Swaranjit Singh Cameotra from the Environmental Biotechnology and Microbial Biochemistry Laboratory as the corresponding author. They were published online on April 17, October 1 and October 8 respectively. The other four papers that were too retracted also had Dr. Khan as the first author and Dr. Cameotra as the corresponding author. And they were published in the Journal of Hazardous Materials (on June 15, 2013), in Chemosphere (in November last year) and the other two in the Journal of Petroleum and Environmental Biotechnology (on August 21 and November 29, 2013)".

2. **"A doctoral thesis rejected on charge of Plagiarism" (20 May, 2012):** "The University of Madras has rejected a research scholar's doctoral thesis on charges of plagiarism and has banned the student from re-registering for the degree at the university".

3. **"Two professors found guilty of plagiarism"(04 March, 2014):** "Two professors from Zoology department working at an Ahmednagar-based college affiliated to the University of Pune, have been stripped off their status as PhD guides and two increments have been stopped, after they were found guilty of plagiarism. The professors Dr M Arif Shaikh and Dr Balraj Khobragade had allegedly lifted some contents from a research paper published in 2012 by a professor attached to Pune's premier Agharkar Research Institute's professor and used them in their own research papers. A departmental inquiry was ordered in 2012 by the Board of Colleges and University Development (BCUD) following a Professor & Head, Division of Animal Sciences at Agharkar Research Institute approached the varsity and registered a complaint against the two professors. After he found out

that some contents of his research paper was lifted by the two professors and they were used in their own research papers".

4. **"Coauthor of retracted stem cell papers commits suicide" (05 August 2014):** "The controversy over a high-profile stem cell finding by Japanese and Boston scientists that was retracted last month took a tragic turn Tuesday when a Japanese scientist who coauthored the work committed suicide, Japanese news reports said. According to The Japan Times, Yoshiki Sasai, 52, hanged himself at the research center where he worked and left two suicide notes and three other notes. One was addressed to Haruko Obokata, the young scientist who led the work and has been accused of scientific misconduct by Japan's RIKEN institution, where she is now trying to repeat the experiment".

5. **"Management Development Institute suspends professor for plagiarism" (Aug 22, 2013):** "The HRD ministry's allegation that Management Development Institute (MDI) associate Professor Amit Kapoor had plagiarized from a secret Expenditure Finance Committee note and later claimed copyright over it has resulted in widespread action".

"Kapoor has been put under suspension by MDI and Harvard University has promised to carry out an investigation and take appropriate action. Harvard comes into the picture as Kapoor was till recently the honorary chairperson of the Institute of Competitiveness, India, which has been recognized by the Institute of Strategy and Competitiveness at Harvard Business School. In fact, Kapoor was also given the Competitiveness Hall of Fame award by Harvard University. HRD ministry brought Kapoor's plagiarism to the notice of Michael Porter, professor of Harvard Business School and an authority on competitiveness, requesting him that the Indian affiliate should be disassociated as well as the Competitiveness Hall of Fame award and any other recognition to Kapoor be withdrawn. Meanwhile, sources in MDI said Kapoor has explained that most of the so-called secret documents were shared with him by the Planning Commission as it wanted him to prepare an index of competitiveness".

6. **"A former university vice chancellor has been accused of plagiarism" (February 25, 2015):** "Former Delhi University vice chancellor was arrested and sent to Tihar jail on Tuesday after a fellow professor complained of plagiarism and illegally using a chemical substance from the university's science lab. His fellow professor had accused him of plagiarizing his paper on biotechnology".

Kerala University: "A professor of psychiatry at an American institute has recently alleged that the Pro Vice Chancellor (PVC) of the University of Kerala had plagiarized from his work. The matter came to light earlier this month. Veeramanikantan has been accused of plagiarising around 42 percent of his doctoral thesis from the work of Paul Lehrer, a Professor of Psychiatry at the Robert Wood Johnson medical school, New Jersey, USA.

"In 2002 Lehrer had published a research paper titled "Psychology Aspects of Asthma" which was published in The Journal of Consulting and Clinical Psychology, which PVC is said to have copied from.PVC however, has denied the charges saying that he had quoted Lehrer's work and had been "dulely acknowledged" in the plagiarismcharge".

7. **"R. A. Mashelkar resigns from patent panel following plagiarism charge" Saturday, March 17, 2007:** "Dr. Mashelkar confirmed to The Hindu that certain lines used in their report's conclusion had been taken "verbatim" from a November 2005 paper that was authored by Shamnad Basheer, a doctoral student and an Associate at the Oxford Intellectual Property Research Centre, University of Oxford. (Feb 22, 2007)."

Indian Call for Preserving Academic Integrity: UGC has come up with following initiatives for preserving academic integrity.

- University Grants Commission (promotion of academic integrity and prevention of plagiarism in higher educational institutions) regulations, 2018 new delhi, the 23rd july, 2018, https://www.ugc.ac.in/pdfnews/ 7771545_academic-integrity-regulation2018.pdf

- Self plagiarism defined by ugc:self-plagiarism: https://www.ugc.ac.in /pdfnews/2284767_self-plagiarism001.pdf

- A new subject research and publication ethics introduced (https://www.ugc.ac.in/pdfnews/9836633_research-and-publicationethics.pdf)

- Guidance document "good academic research practices"; sept. 2020, https://www.ugc.ac.in/e-book/ugc_garp_2020_good%20academic%20research%20practices.pdf

Please go through these documents carefully and just be honest in research and publications to preserve your Academic Integrity. In following chapters we will discuss almost all the issues of Research and Publication Ethics.

Further Reading

- University Grants Commission (Promotion of Academic Integrity and Prevention of Plagiarism in Higher Educational Institutions) Regulations,

2018 New Delhi, the 23rd July, 2018, https://www.ugc.ac.in/pdfnews/7771545_academic-integrity-Regulation2018.pdf

- Self plagiarism: https://www.ugc.ac.in/pdfnews/2284767_self-plagiarism001.pdf

- Guidance Document "Good Academic Research Practices"; Sept. 2020, https://www.ugc.ac.in/e-book/UGC_GARP_2020_Good%20Academic%20Research%20Practices.pdf

- Plagiarism - Why students do it and how you can help, https://www.youtube.com/ watch?v=oCT7iamerdo

- Academic Integrity – Plagiarism, https://www.youtube.com/watch?v=MDFHd_31e_o

- http://library.csusm.edu/plagiarism/whatis/what_is_common.htm

- http://library.csusm.edu/plagiarism/howtoavoid/how_avoid_common.htm.

- https://www.bowdoin.edu/studentaffairs/academic-honesty/common-types.shtml

- https://abacus.bates.edu/cbb/examples/citations/index.html

- https://www.bostonglobe.com/news/science/2014/08/05

- https://www.ox.ac.uk/students/academic/guidance/skills/plagiarism?wssl=1

- https://www.turnitin.com/blog/is-plagiarism-just-bad-manners

References

- https://www.brad.ac.uk/library/help/plagiarism/frequently-asked-questions-about-plagiarism/

- http://www.plagiarism.org/article/what-is-plagiarism

- Roberts J. Plagiarism, Self-Plagiarism, and Text Recycling. Headache. 2018 Mar;58(3):361-363. doi: 10.1111/head.13276.

- Singh Harkanwal Preet, Mahendra Ashish, Yadav Bhupender, Singh Harpreet, Arora Nitin, Arora Monika. A comprehensive analysis of articles retracted between 2004 and 2013 from biomedical literature - a call for reforms. Journal of Traditional and Complementary Medicine. 2014;4:136–139.

- Kumar PM, Priya NS, Musalaiah S, Nagasree M.Knowing and avoiding plagiarism during scientific writing. Ann Med Health Sci Res. 2014 Sep; 4(Suppl 3):S193-8.
- https://www.bowdoin.edu/studentaffairs/academic-honesty/common-types.shtml
- https://timesofindia.indiatimes.com/home/education/news/Management-Development-Institute-suspends-professor-for-plagiarism/articleshow/21966512.cms)
- www.thenewsminute.com/article/former-university-vice-chancellor-has-been-accused-plagiarism-so-whats-new-about-it-22336

CHAPTER 8

Tools for Detection of Plagiarism

Dr Mona Semalty
H.N.B Garhwal University (A Central University)
Srinagar Garhwal-246174

Learning Outcome

After completing this chapter, you will be able to know

- What is the basic concept of Plagiarism detection?
- What are different mechanisms of plagiarism detection
- What are different factors of selection of software
- How do we detect the plagiarism

Lesson Plan

- Introduction to plagiarism detection
- Different mechanisms of plagiarism detection
- Factors affecting selection of software
- Plagiarism detection tools

Introduction

Plagiarism detecting tools involve different tools (software used to detect plagiarism).These tools can detect plagiarism in text material and there are number of software available to check plagiarism and the functioning of these software depends on certain mechanism. There are different mechanism for different types of software.

This is very important to know that similarity check can never be an automated process. Manual interventions are always needed to interpret the exact meaning of textual material (Or any phrase) in particular context. Even artificial intelligence has its own limitation in such tasks.

Now, how does any software detect the similarity or plagiarism?

In simple terms, it can be done by the comparison of submitted literature by the author with available literature on local or global level. Available literature includes printed as well as digital editions of books, book chapters, articles, blogs, newspapers, tweets, reports etc.

The similarity detection process can be done by-

- Fingerprinting
- Term occurrence analysis
  - o Substring matching
  - o Bag of words analysis
- Citation based plagiarism analysis
  - o Citation Pattern analysis
- Stylometry

Fingerprinting

Steps involved in finger printing (Fig. 8.1)

- Selects sets of multiple substrings in the document
- Sets are represented as fingerprints (Elements of fingerprints are called minutiae)
- Fingerprints of submitted documents are compared with the fingerprints of reference documents available in databases.
- It takes a lot of time which can be resolved by comparing a subset of minutiae

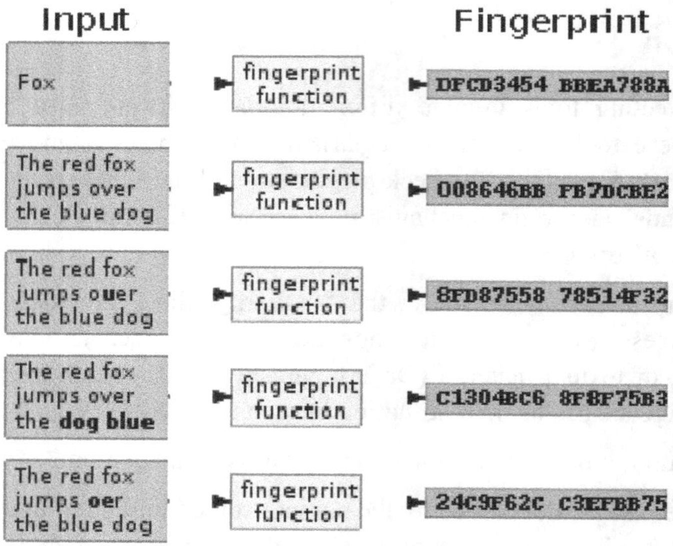

Fig. 8.1 Fingerprinting process

Term Occurrence Analysis

String Matching:

Steps involved in string matching (Fig. 8.2)

- Checks for verbatim text overlaps
- Uses suffix document models
- Requires high computational efficiency
- Expensive

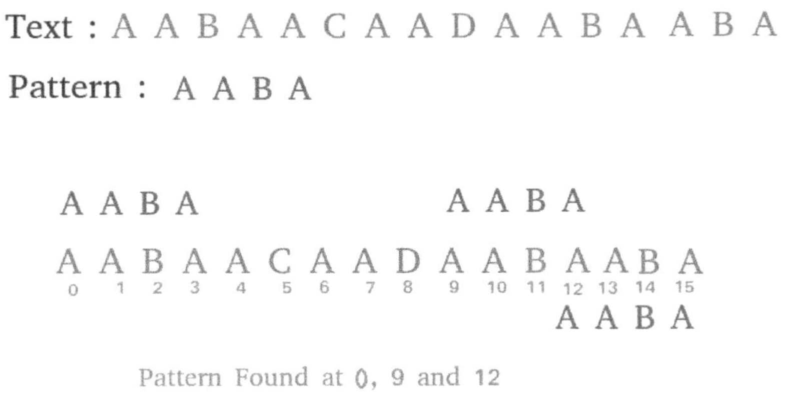

Text : A A B A A C A A D A A B A A B A

Pattern : A A B A

A A B A   A A B A

A A B A A C A A D A A B A A B A
0   1   2   3   4   5   6   7   8   9   10 11 12 13 14 15
                                                    A A B A

Pattern Found at 0, 9 and 12

Fig. 8.2 String matching process

Bag of words: Steps involved in bag of words (Fig. 8.3)

- Document is dividedin different parts
- Pair wise similarity computation is performed
- Based on traditional concept of information retrieval known as vector space model

Article ID	biolog	biopsi	biolab	biotin	almost	cancer-surviv	cancer-stage	Article Class
00001	12	1	2	10	0	1	4	breast-cancer
00002	10	1	0	3	0	6	1	breast-cancer
00014	4	1	1	1	0	28	0	breast-cancer
00063	4	0	0	0	0	18	7	breast-cancer
00319	0	1	0	9	0	20	1	breast-cancer
00847	7	2	0	14	0	11	5	breast-cancer
03042	3	1	3	1	0	19	8	lung-cancer
05267	4	4	2	6	0	14	11	lung-cancer
05970	8	0	4	9	0	9	17	lung-cancer
30261	1	0	0	11	0	21	1	prostate-cancer
41191	9	0	5	14	0	11	1	prostate-cancer
52038	6	1	1	17	0	19	0	prostate-cancer
73851	1	1	8	17	0	17	3	prostate-cancer

doi:10.1371/journal.pone.0162721.t001

Fig. 8.3 Bag of words process

Citation Analysis

- Analyses citation similarity (Fig. 8.4)
- Does not rely on textual similarity
- Citations should be necessary part of the document to be tested
- Newer concept

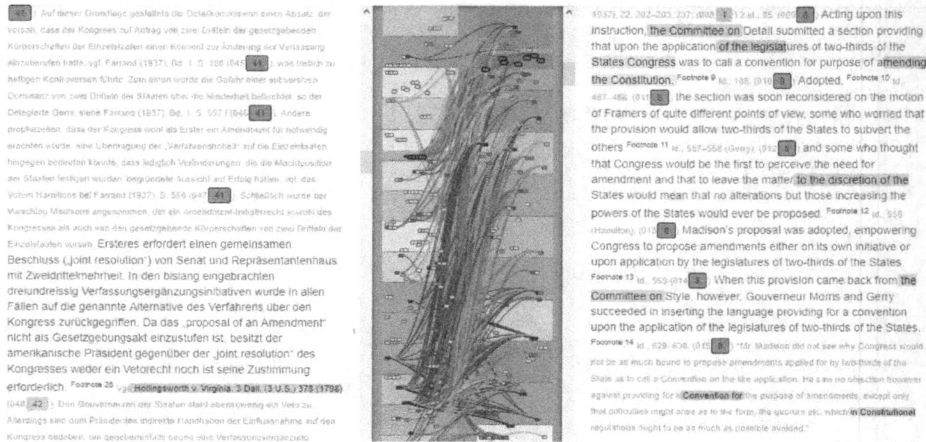

Fig. 8.4 Citation analysis

Stylometry

- Quantifies an author's attribution
- Matches for unique writing style of author (Fig. 8.5)
- Stylometric models are used for comparison

Fig. 8.5 Stylometry: example

Different tools of plagiarism detection -

- iThenticate
- Plagiarism Checker X
- Turnitin
- Viper
- Grammarly
- CitePlag
- Plagiarisma.net
- ProWritingAid
- MOSS (Measure of Software Similarity)
- DupliChecker
- PaperRater
- Copyleaks
- Search Engine Reports
- PlagTracker
- Plagium
- Prepostseo

Factors affecting the popularity of any plagiarism detection software-

- **Price** is the most important factor as there are few tools which are available freely but they have limited database and word limit to be searched. The best and fast software are costly and because of that their preference is limited until they have institutional subscription.

- **Scope of search** of a plagiarism tool depends on its database coverage. If the user is going to pay the price for such service, he/she will certainly go for the tools which have best coverage.

- **Analysis time:** If any plagiarism detection service has wider coverage of search and a big database, generally it will take a lot of time. To minimize the time better computational devices will be required which will increase the price of service.

- **Type of algorithm being used:** For a publisher, citation based plagiarism detection will be preferable than any other type. While for an author either fingerprinting or substring matching algorithm is more

Plagiarism Detection Software: PREPOSTSEO

Plagiarism detection can be done using PREPOSTSEO which is available online without any cost. It also gives an additional feature as google chrome extension so if you are using google chrome as web browser you can add this tool to your extension bar and it will ease your access. It has a word limit of 1000 words per search. You can just paste your text in the box or you can upload your file in .txt/.doc/.docx/pdf format. It is suitable to understand the process of plagiarism detection, but it is not suitable for thesis/dissertation/articles (Fig. 8.6).

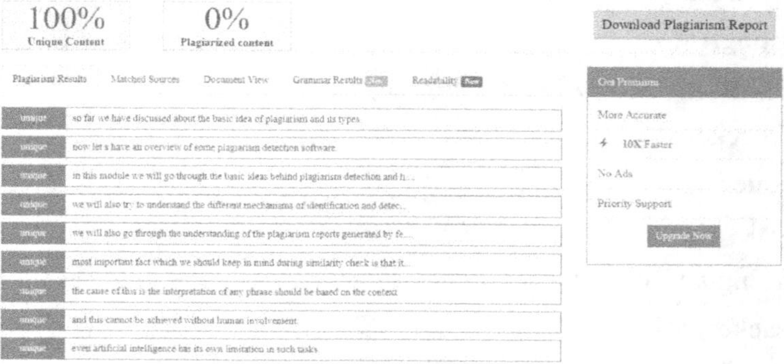

Fig. 8.6 Plagiarism detection PREPOSTSEO

Plagiarism Checker X

This tool is a PC-based application and it requires internet while generating the similarity report (Fig. 8.7). It accepts paragraph, large and small, .txt, .doc, .docx, .pdf files while uploading. It has wider coverage in comparison with PREPOSTSEO tool or other freely available software.

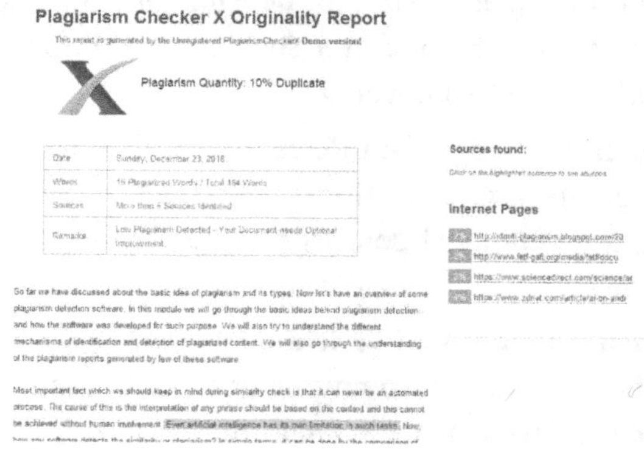

Fig. 8.7 Plagiarism detection withPlagiarism Checker X

DupliChecker

DupliChecker is web based freemium application for plagiarism detection with 1000 words limitation in its free version. Another option available with this service is uploading a .doc or .txt file. It is very easy to handle and suitable for small write ups. Similarity report of this tool is also very easy to understand (Fig. 8.8).

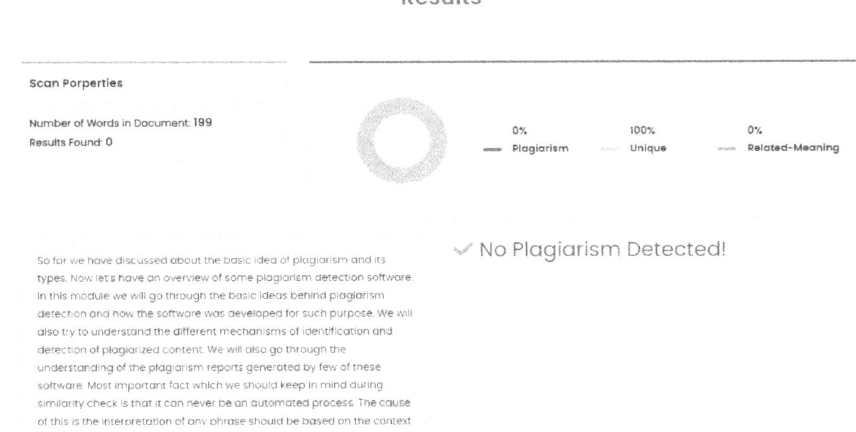

Fig. 8.8 Plagiarism detection with DupliChecker

Turnitin

IT is one of the tool which has best coverage of databases and fastest service till date. It is used by top universities and academic professionals worldwide. A similarity report of turnitin gives you the report on plagiarism of your documents as well as possible grammatical mistakes/suggestions. You can just paste your text in the box or you can upload your file in .txt/.doc/.docx/.pdf format. It is a web-based tool and subscription is mandatory. The report includes various highlighting colors for plagiarized content like:

- Blue (no matching words)
- Green (one matching word - 24% similarity index)
- Yellow (25-49% similarity)
- Orange (50-74% similarity)
- Red (75-100% similarity)

One can just know the degree of plagiarism by looking at the highlighting color of the text in the report. Every software works on some different/ unique mechanism (Fig. 8.9).

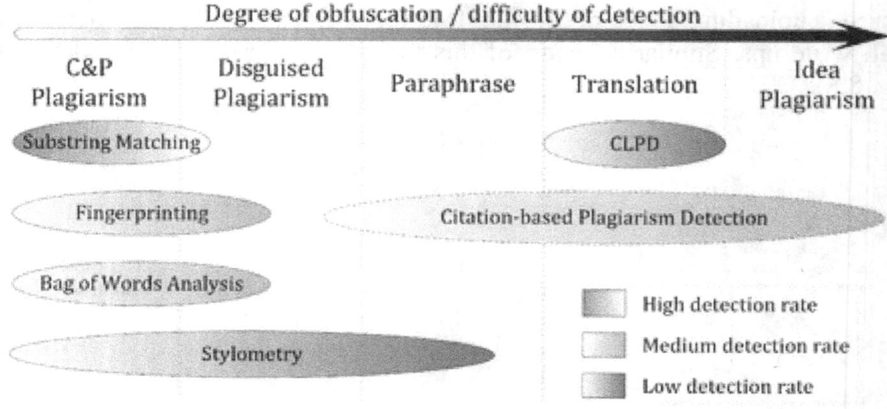

Fig. 8.9 Different mechanism by different plagiarism detection software

Using Urkund (now renamed as Ouriginal)

Now we will discuss using Urkund/Ouriginal (Fig. 8.10), which is one of the most widely used plagiarism detection software. This is the home page of Urkund/Ouriginal (https://www.urkund.com/ now changed to https://www.ouriginal.com/). Please refer new website.

Fig. 8.10 Home page of Urkund

Getting Urkund/Ouriginal ID: You have to get the Urkund/Ouriginal id from your institutional library, if you are a faculty member. Once you get registered in Urkund, you will get an email with the activation link.

You click on the activation link for starting your Urkund/Ouriginal access. You will also be given a Urkund email id which you can share with your students or you can use it for uploading the files to be checked.

In the mail itself a link to Urkund/Ouriginal quick start guide is given. Download the guide and go through it. This is 3-4 pages long quick start guide with all the necessary details about using Urkund.

Using the Urkund/Ouriginal account

Now, after activation go to the login page, enter your username and password (Fig. 8.11).

Fig. 8.11 Entering into Urkund/Ouriginal

The dashboard of your own Urkund/Ouriginal account will appear (Fig. 8.12). The dashboard will show all your previous work of plagiarism check.

Fig. 8.12 Dashboard of your own Urkund/Ouriginal account

Upload Document

Go to upload Document (Fig. 8.13). Select your analysis address (Urkund email id).

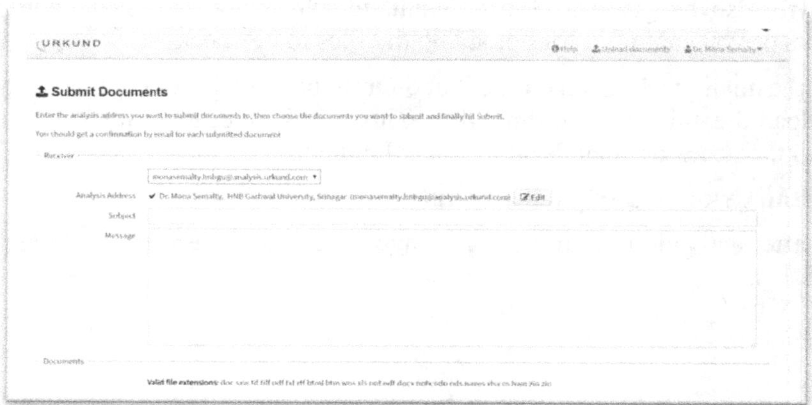

Fig. 8.13 Uploading File for plagiarism checking in Urkund/Ouriginal

Then upload the file. Either drag the file to the document section. Or click to select multiple files. Then submit document.

The submission confirmation will be sent to your email id. Depending on the length of the document, the system will take time to give you the report.

Studying and getting Report in pdf

Once the system finishes the plagiarism check it send the intimation in email with the link of report. The report can be seen in the dashboard (Fig. 14). Click the file. See the plagiarism. All the plagiarized content will be shown in color in the report with information of source. All the sources are listed on the top right panel. Uncheck the creative common content (provided you have attributed the source). This will reduce the plagiarism level.

Now you can click on the export to get the report downloaded in pdf format. You can use this report for submission to desired platform.

This was all about using Urkund/Ouriginal as the plagiarism detection tool. Please do refer the new website https://www.ouriginal.com/. You can also explore other tools for your academic writing.

Fig. 8.14 Generated Plagiarism Report in Urkund/Ouriginal

Further Reading

- Plagiarism - Why students do it and how you can help, https://www.youtube.com/watch?v=oCT7iamerdo

- Academic Integrity – Plagiarism, https://www.youtube.com/ watch?v=MDFHd_31e_o6

- How to avoid plagiarism and turnitin, https://www.youtube.com/ watch?v=RrPU_bFBeF8

- https://antiplagiarism.net/blogs/avoid-plagiarism-tool/

References

- Gipp, Bela (2014), Citation-based plagiarism detection, Springer Vieweg Research, ISBN 978-3-658-06393-1

- https://www.um.edu.mt/__data/assets/pdf_file/0018/261324/avdplagiarism. pdf

- https://elearningindustry.com/top-10-free-plagiarism-detection-tools-for-teachers

- https://lutow.acim56.info/a1252/

CHAPTER 9

Avoiding Plagiarism

Dr Mona Semalty
H.N.B Garhwal University (A Central University)
Srinagar Garhwal-246174

Learning Outcomes

After learning this chapter, you will be able to know

- How to prevent and resolve plagiarism effectively

Lesson Plan

- Types of strategies for avoiding plagiarism
- Preventing plagiarism
- Correcting plagiarism
- Avoiding text plagiarism
  - Direct Quotation
  - Paraphrasing
  - Summarizing
- Avoiding image plagiarism

Introduction

Plagiarism can be avoided simply by giving due credit to the source from wherever you have taken. So, we can say that the single key is "JUST BE HONEST".

Give credit, attribute, cite, and acknowledge the source honestly. And ask for permission for use of documents, figure and graph if you want to use the same in your work. Respect IPR and other people's ideas in academic exchanges. Ask for permission for Copyright for use of graphs for preventing plagiarism.

For example, if you want to define table "table is a piece of furniture with a flat top and one or more legs, providing a level surface for eating, writing, or working at". So, this is the basic definition of table. Now if you want to define table avoiding plagiarism you have reframe it keeping in view that the central idea should not change.

First write-down each and everything you do (noting down the source) from the beginning to end. Anticipate the time points of academic writing where the plagiarism may remain unattended.

Prevention Strategy

- Refer good quality, reliable resources in literature survey
- Keep record of each source
- Arrange/ Organize the literature for future use

Refer Good Quality, Reliable Resources in Literature Survey

A good quality resources may include a book /journal/or it may be a web source. So, here you have to understand and learn what are the good and bad journals. The journals should have good indexing like in Scopus, PubMed etc. For example, in case of arts and humanities where less numbers of journals are available and it's difficult to get good journal of your target area. You can go for those journals which are indexed at least in expanded science citation index or indexed by other subject related data base.

Do not go for predatory journals which contain lot of plagiarized material and referring those articles will be another plagiarism. So, go for original sources.

Keep Record of Each Source

In literature survey, we often forget to note down the source of some text and other materials you have taken from. This can be sorted out by -

- Developing good note taking and research habits.
- Note down and record all the bibliographic information
- Put the reference on to the top of copy when taking Xerox

Things to keep in mind while noting down references

- *URLs: Ensure to take down full URLs; they may be removed or modified, so the date on which it was accessed must be noted.*

- *Journal article: as per the referencing style and doi must be noted*

- *Books: Different Edition may be little different in contents, so ensure putting edition along with other information*

- *Check the Copyright information of web resource; © or creative common or public domain, and then decide what to take and how much*

Arrange/Organize the Literature for Future Use

- Adopt a way which suit you the most(you can arrange them by name /topic /date wise you can spiral bound hard copies for detailed studies)so that you will be able to locate it easily

- Do not forget to distinguish between direct quotes (word to word copy) and your own words.

- Adopt color coding, symbols, underlining, font change to distinguish direct quotations from paraphrases and summaries.

"Anything that won't sell I don't want to invent. Its sale is proof of utility, and utility is success."　　　　　　　　　　- Thomas Edison [direct quote]

Edison said that he doesn't want to invent a thing which doesn't sell because sale is the proof of utility and success (Edison).　　[Paraphrasing]

The successful research must be the need based one (Edison). [Summarizing]

Some More Points

- Put quotation marks
- Complete reframing of text is required when paraphrasing. Don't just simply rearrange sentences or replace a few words in a sentence.
- Check your rewritten version against the original version
- Form your own ideas and opinions about different issues,
- Compare voices/ides of authors and deduce your own idea/voice
- Correcting plagiarism
- Run the plagiarism checker
- See the similarity report
- Focus on the portion which shows high percent of copying first.
- The percent copying from a single source must be reduced on priority
- The results and discussion portion must be free from any plagiarism on priority. (you can understand, why?)
- Introduction is generally giving background, so chances of plagiarism are high, so focus on the correction

Plagiarism Avoiding/Correcting Strategies

Avoiding Text Plagiarism

- *Direct Quotation*
- *Paraphrasing*
- *Summarizing*

Avoiding Image Plagiarism

- *Draw yourself the same with different colors*
- *Modify and include your idea*
- *Never copy result figures*

Avoiding Text Plagiarism

Text plagiarism can be avoided by effective use of language for the better understanding and clarity of content. We must give due credit to the source. Next, we must respect the intellectual property right of scholars honestly. Care must be taken to cite all the references listed as and when required precisely as per the style recommended.

The three methods of avoiding plagiarism may be understood with the example (Fig. 9.1).

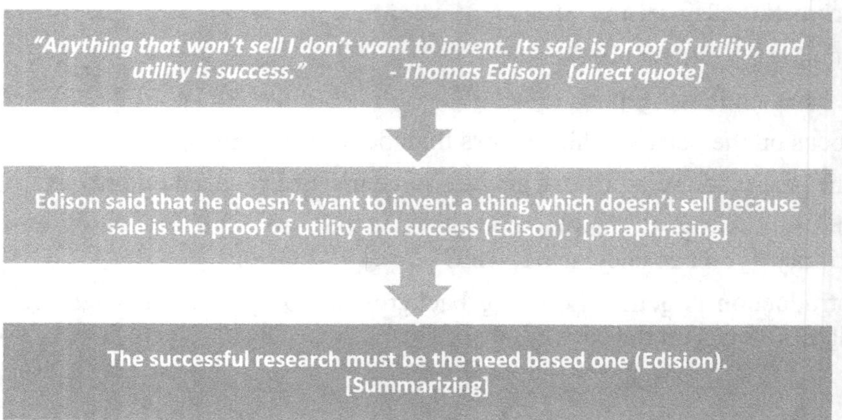

Fig. 9.1 The three methods of avoiding plagiarism: Example

Direct Quotation

Direct Quotation is a quote is a word, sentence, or sentences that a writer copies exactly from a source. One can quote directly by providing a reference. Direct quote also involves verbatim copying (for copying official definition and unavoidable quotes). But it's always better to paraphrase /summarize rather than directly quoting.

Let us see what the conditions are where one has to use quotes only -

- Official/ legal language

- Unique and cannot be paraphrased without changing the meaning.

- Already clear and condensed.

- Creating a particular effect (poetry, inspiring quote,) and addressing special context and author's opinion/ prestige/ aura

BUT NEVER OVERUSE THE DIRECT QUOTES

Now there are few things which must be taken care while using direct Quotes

- *Copy verbatim, without altering/modifying*

- *If it is longer than 3 lines (39 words),*

- *Use reference as usual,*

- *Do not enclose in quotation marks and*

- *Indent the lines by 5 spaces.*

An ellipsis (…) should be used to indicate the omission of word in between the quote. Square brackets [thus] are used to indicate some added text by you in quotes for clarification or grammatical correctness

Use of Single V/s Double quotation marks

You should use double quotation marks when you quote material from a source. And if you are also quoting passages from that source that were quoted in the original source, use single quotation marks to indicate that the original source contained the quotation.

Punctuating Quotations

"In the system of punctuation used in the United States, periods and commas go inside quotation marks except when you use in-text citations. In those cases, periods and commas go outside the quotation marks".

Paraphrasing

What is paraphrasing?

A restatement in your own words of someone else's ideas is called paraphrasing

Paraphrasing is not just changing a few words of the original sentences it requires complete reframing of sentences which have your own voice and your own understanding (about the author's idea). Correct and accurate paraphrasing requires highly-developed writing skills. So you have change both the words and the sentence structure of the original, without changing the contextual meaning. Never forget to attribute the source and citation is required.

Paraphrasing Tools

One of the useful techniques for paraphrasing is to rearrange the sentence without losing its meaning. And it can be done by describing, expanding or explaining the content as per your convenient. You can also go for breaking the long sentences in to smaller to get easy access of the content.

Modify the order content& structure of sentences.

It is done by following these strategies.

- Use synonyms
- Use different forms of words
- Change the voice or perspective
- Modify the order & structure of sentences
- Do not look at the original content while writing

Use Synonyms

- Words or expressions which have similar meanings

  E.g. use- utility, make-build etc.
- Best resource for finding synonyms: thesaurus
- Be alert to exact meaning word, the contextual meaning may be different sometimes.

Changing words should never be the sole method of paraphrasing; must be complemented with other strategies

Use different forms of words

Change the word from an adjective to a noun or from a noun to a verb. This change necessitates changes in sentence structure and organization. For example, we must know the official rule before implementation. The knowledge of official rule is must before the implementation.

Change the voice or perspective

Change voice from active to passive or passive to active. We must know the official rule before implementation. The knowledge of official rule is required before the implementation. But avoid unnecessary voice conversion, the voice must be simple and easy to follow. Do not overuse passive voice unnecessarily

Summarizing

"Reducing the source text to its main points."

This avoids overuse of direct quotations and paraphrasing large sections of the original text. Focus is on own understanding and presenting in your own words. But you still need to acknowledge the source of the information, add your own comments to show your analysis and interpretation of the work.

Referencing

Learn proper referencing method/ style even down to the positioning of commas and full-stops

Remember various common writing or citation styles we discussed remember the BIG THREE "CMS, MLA, APA"

Adopt the style as per your subject area and recommendation of your department / institute/ journal.

Avoiding image plagiarism

- Draw yourself the same with different colors ; use Google image/ www.lucidchart.com or other tools for recreating figures
- Use and search free to reuse figures
- Search Wikipedia/Wikimedia and other free to reuse resources
- Use the advanced google search/ filter for free to reuse figures
- Modify and include your idea

Ask for permission to reproduce even it is your own figure in any other journal until you own the copyright. Never copy result figures e.g. Japanese researcher reproduced stem cell research figure in result section in a paper in Nature; punished when caught.

Let's summarize some key points

- Always acknowledge, attribute and cite reference honestly

- Use direct quote, paraphrasing and summarizing along with proper citation of reference for avoiding plagiarism
- Learn the proper referencing style
- Avoid the plagiarism in image by using advanced filters

Further Reading

- https://services.unimelb.edu.au/__data/assets/pdf_file/0004/821668/5297-Avoiding-PlagiarismWEB.pdf
- White paper the ethics of self-plagiarism, https://www.ithenticate.com/hs-fs/hub/92785/file-5414624-pdf/media/ith-selfplagiarism-whitepaper.pdf
- https://antiplagiarism.net/blogs/avoid-plagiarism-tool/
- Plagiarism - Why students do it and how you can help, https://www.youtube.com/watch?v=oCT7iamerdo
- Academic Integrity – Plagiarism, https://www.youtube.com/watch?v=MDFHd_31e_o

References

- Correct and accurate paraphrasing, http://www.academicintegrity.uoguelph.ca/
- http://isites.harvard.edu/icb/icb.do?keyword=k70847&tabgroupid=icb.tabgroup108986
- Kumar PM, Priya NS, Musalaiah S, Nagasree M, Knowing and avoiding plagiarism during scientific writing. Ann Med Health Sci Res. 2014 Sep;4(Suppl 3):S193-8. doi: 10.4103/2141-9248.141957.
- https://www.wikihow.com/Use-Simple-Words-in-Technical-Writing
- Debnath J.Plagiarism: A silent epidemic in scientific writing - Reasons, recognition and remedies.Med J Armed Forces India. 2016 Apr; 72(2):164-7. Epub 2016 Apr 16.

CHAPTER 10

Journal Metrics-I

Dr Ajay Semalty
H.N.B. Garhwal University (A Central University)
Srinagar Garhwal-246174

In previous chapters, we covered the concept of academic writing, basics, English in academic writing and plagiarism. Now we will discuss the Journal and author metrics in this week. In this chapter we will discuss the journal metrics.

Learning Outcome

After completing this chapter, you will be able to know.

- Concept of journal metrics
- Distinguishing journals on the basis of metrics
- Targeting journal on the basis of these metrics
- Impact Factor as a journal metrics

Lesson Plan

Here we will discuss:

- What are Journal metrics?
- Types of metrics: Journal and author level metrics
- Impact Factor as the journal metrics

You have done your research work and now you want to plan a paper. However, as far as the timing is concerned we will also discuss when you should plan the paper. Technically you should not wait for the completion of your entire research work for planning a paper.

Planning a Paper

When you are planning a paper, what should be the first step.

You will have to define

- What to publish (Content, extent, type): will deal with this later...
- Where to publish

Correlating what with where….

Correlating what you want to publish with where you want to publish is very important aspect.

Whenever you go for buying and choosing the things you compare the things. You compare the things every time you go for targeting a car to own, a pair of shoes to buy, or to choose a college for taking admission. You decide the things on the basis of some desired features/ merits/ factors. Isn't it! The same holds true for targeting a journal and deciding.

Deciding Where to Publish

Our decision of publishing a paper in a journal is governed by some factors. For deciding where to publish, we check 3 factors of a journal

- Impact
- Speed
- Reach

The Journal metrics answers these three factors. You need some checkpoints to know that this journal is suitable for you. And these three factors: impact, speed and reach decide the suitability of journal for you (Fig. 10.1).

Fig. 10.1 Factors for targeting journals

Let's define "Journal Metrics"

"Journal metrics provide extra insight into three aspects of our journals – impact, speed and reach – and help authors select a journal when submitting an article for publication."
- Elsevier

So "Impact speed and reach" are the key factors. Please do remember that you may have somewhat different targeting in your mind. Sometimes, you are ready to wait to publish in a good impact; sometimes you are not ready to wait. So you will rather focus a journal with good speed. And sometime you need to reach more and more people rather than focusing impact and speed. Or some time the combination of these three aspects.

Impact Means

Let us discuss, what does the impact means in academic writing? For example you have written and published an article and someone else refers your work and cites your article. That showed that your article had an impact on this and in the same way, if lots of article cite your work it shows that your article had a great impact. Now, what the Bibliometrics (journal/authormetrics) does that shown impact can be expanded to the journal or researcher as proof of their impact as well. From there, the impact can be spread even further. Your impact as a researcher spreads to a group like a faculty or the University. So, you can see that one cite from an article shows the impact to the entire linked system. This impact is not just the number, as some researcher feel. Their research is more than that. It is not like that a derivative work is trying to demolishing the previous work. Actually, what a citation proves is the IMPACT that an article has made and thus become more important/useful. These bibliometrics are more powerful and useful tool in health/ biomedical and Sciences. Whereas, the other faculties like architecture, law, business and engineering has their own hurdles which they need to cross. Engineering faculties tend to publish a lot in conferences which are harder to track and therefore, harder to be cited. The arts' faculties love books and the books have exact cites problems due to irregular publishing schedules making it harder to track. Law on the other hand has a different problem. Although they publish heavily in journals much of their work is aimed to judges and politicians rather than the other academics. The architecture people have their unique problem in the sense that their lot of work have an artistic component which is hard to track and cite. Business have their own problems. Mainly because the bibliometrics is more focussed to sciences and these are just entering in their domain now.

(Introduction to Bibliometrics, UTS Library, *https://www.youtube.com/ watch?v=yBHneMkeUHA*)

So, impact means

- What kind of aura/ prestige/ impact a journal possess
- The quality supported with sound statistical indicators
- Impact gives the confidence in the authors for their selection of journal.

Speed

What does the speed mean?

Authors need to publish their work on time specially in fast moving research areas.

"Early rejection is far better than the very late acceptance"

Many of you will agree with me. Sometimes we wait for a very long time of 6-7 months and then we come to know that our paper is rejected. At that moment the time delay is more painful than the rejection. In contrast I like the journals which give rejection even in 48 hrs to one week. Or clear the authors if the article is suitable for further peer reviewing/ processing. No issue, it is good to not to remain in dark or the uncertainty and plan the article for new journal again without losing much more time.

So for that, see, "The frequency of the journal": semi annual, quarterly, bimonthly, monthly, fortnightly or weekly. Yu have to check the frequency of the journal for assessing the speed of publication. Two major factors you can check

"average number of weeks taken by the peer review process"

"average number of weeks taken for first decision/final decision/ online/offline publication"

Let's take a look on this **Figure 10.2** which shows the journal metrics with respect to the review speed. As we can see, for a journal submission to first decision and submission to final decision is shown in weeks for year 2014 to 2017. It is consistently increasing time between submission to first decision and submission to final decision. The review speed idea can be seen by this and assess when you may expect the first decision. Alternatively, whether you can wait that time for the final decision or not, after the first decision. Your decision may vary depending on these things.

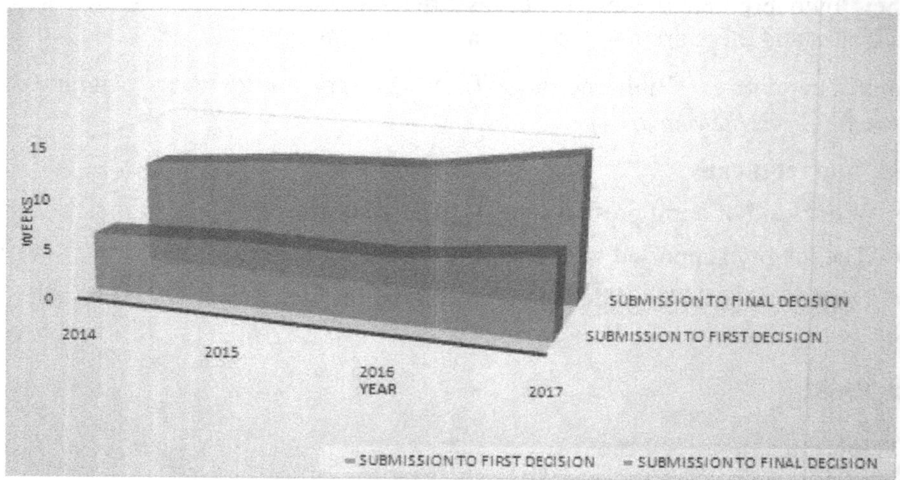

Fig. 10.2 Journal Metrics: review speed

Online Article Publication Time

There are two parameters for this

- Time taken from manuscript **acceptance to the first appearance** of the article online (with DOI).

- Time taken from manuscript **acceptance to the final appearance** online of the fully paginated article.

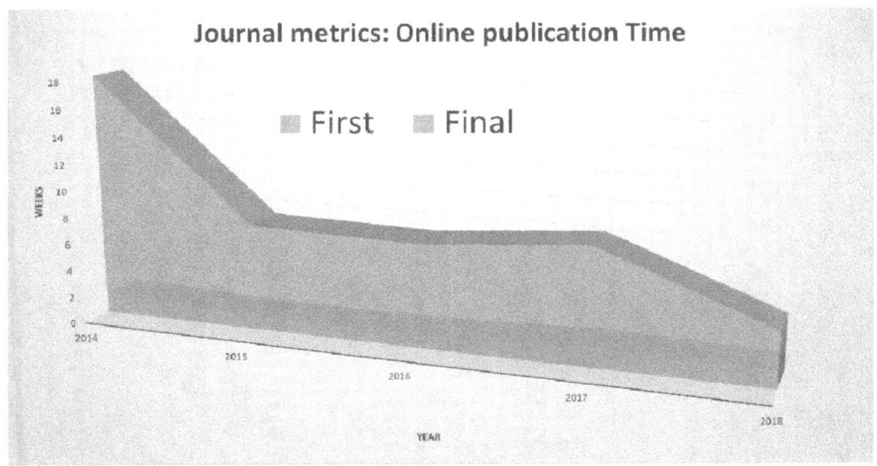

Fig. 10.3 Journal Metrics: online publication time

The figure 10.3 shows the change in online publication time over the years for a particular journal. Average time taken for first appearance to final paginated version of article is shown here. It can be seen that over the years the time taken to first and final appearance has decreased for this journal. Where it was about 16 weeks from first to final appearance time in 2014, it reduced to just 4 weeks in 2018.The technological advancement, online automated system, availability of huge number of articles for finalizing an issue etc. may be the reasons of this reduction in publication time. But overall this is a plus point for a journal and prospective authors.

So, as compared to previous years overall publication time is decreased for the journal and hence this factor may be able to attract more authors. The availability of this type of journal metrics on the journal's page is very useful.

"Time is invaluable / priceless". Everyone wants to publish the paper as early as possible. A researcher better understands its importance when the thesis submission is pending just for one publication. You might remember many examples around you among Ph D scholars whose Ph D submission was withheld just due to lack of publication on time.

Reach

The third important factor is reach.

- Reach of journals to readers and authors: the popularity
- Wide geographical area: the global presence
- The number of downloads countrywise & the number of Corresponding author countrywide for last 5 yr are the indicators of reach of the journal

Fig. 10.4 Geographical reach of journal

Source: https://journalinsights.elsevier.com/journals/0888-613X/authors

This graph (Fig. 10.4) indicates the number of primary corresponding authors at the country/region level in the last five years. We can see that for this particular journal France is holding second rank with 61 contributing authors for the journal. This metrics is clearly showing the global contribution of authors to the journal countrywide and indicating the effective reach of the journal.

Summary

We learned about the three factors: impact, speedand reach, on the basis of which we want to target our manuscript to a suitable journal. These three factors decide, on which metrics we will focus. Let's see these metrics in detail one by one.

In continuation to the standard journal metrics let's see why the metrics are needed firstly.

A standard/ metrics is needed:

- To assess the relative suitability of the journals

- To rank the journal on merit (Impact, Speed, Reach)

- To have the benchmark of competitiveness (we will be in condition to decide the competitiveness only after seeing particular metrics) of journals

- To have the benchmark of competitiveness of authors: For assessing the researchers for grant application, job/ promotion, awards etc.

- To attract the more and more potential authors and the relevant research work

- To save time in targeting the most suitable journal by the authors: Authors can not study the several articles of several journals and spend his valuable time just for choosing a suitable journal. The metrics save the time for the researcher by helping him/her for targeting the most suitable journal.

The metrics may be of two types Journal metrics and author metrics. And these may be further sub classified as shown in Table 10.1.

Table 10.1 Types of Metrics

Metrics		IF
	Journal metrics	ISI Ranking
		Citescore metrics
		Eigen factor
		Article Influence
		Immediacy index
		SNIP
		SJR
		Altmetrics
		H5
	Author Metrics	H index
		I10 index
		G index

Journal Level Metrics

Impact Factor (IF)

It is the most Important basis of selection of journal. It is a measure of the reputation of a journal.

"Impact factor is a measure of the frequency with which the "average article" in a journal has been cited in a particular year."

A journal's Impact factor is presented by a digit followed by year. e.g. on the home page of Expert Opinion on Drug Delivery journal (https://www.tandfonline.com/action/journalInformation?show=journalMetrics &journalCode=iedd20) 2019 Impact Factor: 4.838 2019). Many more other informations like usage (downloads, views) and speed/ acceptance rate are also provided on the same page. All these may be helpful in targeting a journal.

Impact factor is a dynamic index which changes on the basis of citations and number of papers every year.

How calculation is done?: IF

- "In any given year, the impact factor of a journal is the number of citations, received in that year, of articles published in that journal during the two preceding years, divided by the total number of articles published in that journal during the two preceding years." (https://en.wikipedia.org/wiki/Impact_factor)

$$IF\ (Year\ X) = \frac{\text{The number of citations in year} \times \text{to articles published in a journal in years } X-1 \text{ and } X-2}{\text{The number of articles published in that journal in year } X-1 \text{ and } X-2}$$

For example a journal's 2017 impact factor calculation will be:

IF(2017) = (Citations in yr 2017from articles of yr2016 + Citations in yr 2017 from articles ofyr2015)/ Published papers 2016 + Published papers 2015

IF(2017) = (1450 + 1550) / (150 + 260) = 4.878

But remember, you need not to calculate these things yourself. These are the automated calculations and record which are kept/ tracked and given by several agencies. And next question is who these agencies are? Who does award IF to journals?

WHO? : IF

- Thomson Reuters/ Journal Citation Reports® (JCR®)/ Clarivate Analytics: If provided by JCR is mostly recognized across the subjects, countries and institutions.
- The Institute for Scientific Information ISI
- Scopus-SJR, CitesScore: a different set of journal and author metrics is given by Scopus (will be discussed later).

Indexing Agencies

- Science Citation Index® (SCI®) and ESCI
- Social Sciences Citation Index® (SSCI®).
- PubMed: NCBI
- Web of Science
- SCOPUS
- EMBASE
- Chemical Abstracts
- ISI
- Index Copernicus
- Sci factor
- ICI
- Google Scholar (the most wide coverage)

In general, higher the number of well-known indexing agencies by which the journal is indexed the more potential is the journal for targeting the manuscript.

These indexing agencies index the journal articles and once these journals are indexed in these indexing services, agencies like Clarivate keep track of the citations and publications and then provide IF time to time. For example: JCR releases the IF in the month of July. For example in 2018 July, the IF of 2017 was released.

Note: Self-declaration of IF by journals is not valid. These agencies are globally recognized for analysing and releasing the IF. Self-declaration is giving award to ourself just and is an unethical practice..

IF depends on

- No. of papers published and citations
- Presence in various Indexing services

- Frequency of publication of journal
- Online nature and wide dissemination (OA)
- Publisher's policies of self-citation
- Nature of journals (Review only/ Interdisciplinary)
- Miscellaneous factors: Manuscript/presentation/editorial quality

Summary

We learned what is impact factor? "Impact factor is a measure of the frequency with which the "average article" in a journal has been cited in a particular year". We learned how IF is calculated, presented; who provides the IF. We learned about the indexing agencies.

After going through the basics of IF let us see the factors on which the IF is dependent on.

IF Factors depend on

- **No. of papers published and citations:** By definition IF depend on both the number and citations directly.

- **Presence in various Indexing services:** More indexing, higher chances of citations and hence the IF

- **Frequency of publication of journal:** It is very important factor. A journal published fortnightly publishes more number of papers and hence has good chances of higher citations as compared to quarterly, bimonthly and monthly journal. That's why chances of getting more citations and improved IF is higher in these cases.

- **Online nature and wide dissemination (OA):** Being online is itself a prerequisite for getting IF. But there are journals which are very prestigious but donot have IF. (specially in cases of arts and humanity journals). OA journals provide free full text and hence have more readership, higher chances of citation and hence chances of getting IF is good.

- **Publisher's policies of self citation:** Some journal pose conditions on authors to cite some the papers of their own journal/publisher, this increases their citatioins and hence the chances of improving IF

- **Nature of journals (Review only/ Interdisciplinary):** Review and interdisciplinary journals have wide coverage and readership. Review articles are almost always cited at higher rate than that of a research paper with new data. Every article's introduction part almost always cite state of art and/or relevant review papers rather than describing the background in details. So, almost in every discipline, the review journals have higher IF.

e.g. in Pharma sciences: Advanced Drug Delivery Reviews, Expert Opinion on Drug Delivery. You can check your field of research ....

Miscellaneous factors: Manuscript/presentation/editorial quality:

- **Manuscript Quality**: Better the quality of manuscript more will be the citations.

- **Presentation quality** of manuscript makes it easy to follow by the readers and researchers. A nicely presented manuscript is likely to be cited more.

- **Editorial quality:** Many journals have started adding introductory video and data set with manuscript as additional resources. This increases their presentation quality and reach. They put these article's intro video or abstract video onto social media like you tube/ facebook and this result in more readership and hence the citations are expected to be increased.

Impact Factor in Academic Recognition

The Academic Performance Indicators (API) as prescribed by latest Regulation (2018) University Grants Commission of India for Career Advancement Scheme and for assessing bio data of an academician for University Teaching jobs.

It allows **8 points** for each research paper of a peer reviewed/ UGC listed journal (for single author)

This score for paper in the journal is augmented on the basis of impact factor:

- paper in refereed journal without impact factor – by 5 points
- paper with impact factor less than 1 - by 10 points;
- papers with impact factor between 1 and 2 by 15 points;
- papers with impact factor between 2 and 5 by 20 points;
- papers with impact factor between 5 and 10 by 25 points;
- papers with impact factor above 10 by 30 points.

Upto two authors 50 % each; more than 2 authors 70/30 (main, corresponding/ rest of the authors)

Therefore, higher the impact factor of the journal in which you have published higher will be your API score.

The international agencies and universities also take the help of various journal metrics and author metrics to assess the performance and suitability of a candidate. It may be the total IF of an author's publications or it may be the number of citations or the number of SCI indexed journal articles together with the citations. So, we can say that IF does play a vital role in assessing the expertise and academic performance of a researcher and faculty member.

Therefore, please be careful and not fall in the trap of predatory journals who wrongly claim the high impact factor.

If you want to verify the journal's IF, either you should have the list of IF issued by the agencies or simply type the name of journal in google search bar, if it would have the actual IF the google would show in the box otherwise not.... (However, it may not show you the current year's IF).

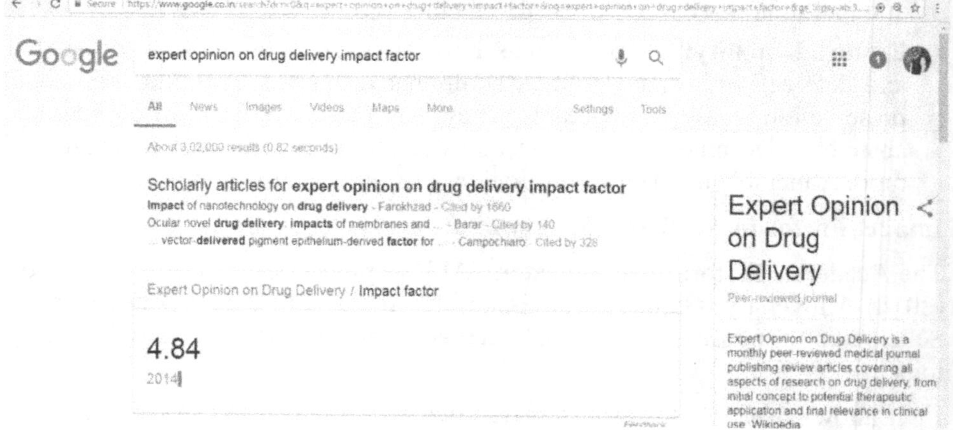

Fig. 10.5 Google Scholar in identifying Impact factor (raw method)

Other Approaches

Go to the URLs shown here...

http://www.citefactor.org/journal-impact-factor-list-2014_I.html

http://lms.uop.edu.jo/lms/mod/data/view.php?d=29&advanced=0&paging&page=1 (ISI Thomson Reuters Impact Factor List 2017)

Or search individual publisher's website like Bentham, Elsevier, Springer, chose the journal from their list and then open the Main page-- IF will be shown if exists with year and name of agency.

But for other journals "Don't rely upon the IF shown in their home/main page"; first check whether the journal is indexed in SCI or ESCI by checking the website of SCI. This measure is suitable for Science journals. But for Arts and humanities journals, SCI indexing and IF is very rare to find. These journals are not generally indexed by the indexing services dedicated to Sciences, biomedical etc. And in some cases journals are not there in English and if exist, are not online. So, In India, UGC constituted a committee Consortium for Academic and Research Ethics (CARE) to refine and strengthen research publication in all the disciplines. In the sciences, technology engineering, biomedical and allied subjects the web of science and Scopus indexed journals have been excluded from the analysis. For rest of the journals the committee is given the task to analyze and list the approved journals including the subjects the Social Sciences, arts and humanities specially.

So, you can have the idea that without IF or without any indexing the journals need the analysis, certification or approval for considering them in API.

IF: Limitations

- "The IF is not perfect, but to be fair, every metric has its flaws."
- Just a single highly cited article can improve the IF of a journal.
- Being a Journal level metrics doesn't discriminate between good/bad author / article.
- The worst article in a journal has the same IF as the best article in that same journal.
- IF is a slow process and does not always depicts true picture.
- Difference in frequency and subject of the journal do not allow meaningful comparison of the journals of different subject.
- Inclusion of self-citation

The Journal Impact Factor quartile

Journal Citation Ranking and Quartile Scores (Q1-Q4) Based on Impact Factor (IF) data, the Journal Citation Reports published by Thomson Reuters provides yearly rankings of science and social science journals, in the subject categories relevant for the journal (in fact, there may be more than one).

It is the quotient of a journal's rank in category (X) and the total number of journals in the category (Y), so that (X / Y) = Percentile Rank Z.

$$Q1: 0.0 < Z \leq 0.25$$

$$Q2: 0.25 < Z \leq 0.5$$

$$Q3: 0.5 < Z \leq 0.75$$

$$Q4: 0.75 < Z$$

Each subject category of journals is divided into four quartiles: Q1, Q2, Q3, Q4. Q1 is occupied by the top 25% of journals in the list; Q2 is occupied by journals in the 25 to 50% group; Q3 is occupied by journals in the 50 to 75% group and Q4 is occupied by journals in the 75 to 100% group. Those who are publishing in Q1 journals are given more weightage than Q2, Q3 and Q4 of the same category. Therefore, even if your paper is published in lower impact factor journal of Q1 as compared to other person's article (of same field) published in higher impact factor journal of Quartile 4, the weightage of your paper is greater.

Finding the Journal quartile:

Step 1: Enter the publication title into search on the Journal Citation Reports' platform and go to its page. 2.

Step 2: Find the button <Get Full Report> and go to the page with more information about journal.

Step 3: See the division "Rank" pointing to the next page with data about journal's quartile.

Note: *InCites displays the best quartile for journals that appear in multiple Web of Science Research Areas. When a research area is specified, the quartile for that particular journal and research area is displayed.*

Summarizing IF

You can publish your work in good journals with impact factor even without paying even a single penny.

Alternatively, target journals with good indexing does not matter if currently they are not having IF. Because these journals have the good chances of getting IF in near future. E.g. Current Drug Discovery Technologies (PubMed and SCOPUS indexed journal without impact factor till 2020, but with Scopus Citescore 2.2. It is very much possible that it gets IF this year only…

Just search and target a good journal and give the justice to your hard work….

Take Away Message

- Metrics are needed for benchmarking of journals and authors
- "Journal metrics provide extra insight into three aspects of our journals – impact, speed and reach – and help authors select a journal when submitting an article for publication."
- IF is one of the most important journal metrics.

- It's a dynamic index and depends on number of citations and number of papers published for last two years.
- Indexing agencies track these and IF is provided on the basis.
- Always chose good impact journal or Journal with good indexing.

Further Reading

- Cross JO, Impact factors – The Basics, the e resource management Handbook, https://www.uksg.org/sites/uksg.org/files/19-Cross-H76M463XL884HK78.pdf
- The Impact of Impact Factors, http://www.springer.com/gp/partners/society-zone-issues/the-impact-of-impact-factors/4592
- Bergstrom CT and West J, Comparing Impact Factor and Scopus CiteScore, http://eigenfactor.org/projects/posts/citescore.php

References

- Impact factor, https://en.wikipedia.org/wiki/Impact_factor
- https://journalinsights.elsevier.com/journals/0888-613X/authors
- https://en.wikipedia.org/wiki/Impact_factor
- http://www.citefactor.org/journal-impact-factor-list-2014_I.html
- http://lms.uop.edu.jo/lms/mod/data/view.php?d=29&advanced=0&paging&page=1
- Research Indices - I: Impact Factor, https://www.youtube.com/watch?v=nPLnJqLEknY
- Introduction to Bibliometrics, UTS Library, https://www.youtube.com/watch?v=yBHneMkeUHA
- http://help.incites.clarivate.com/inCites2Live/indicatorsGroup/aboutHandbook/usingCitationIndicatorsWisely/jifQuartile.html

CHAPTER 11

Journal Metrics-II

Dr Ajay Semalty
H.N.B. Garhwal University (A Central University)
Srinagar Garhwal-246174

We have discussed the basic concept of journal metrics and started the same with Impact Factor in the last chapter. Now we will discuss some more Journal metrics in this chapter.

Learning Outcome

After completing this chapter, you will be able to know

- Some of the journal metrics and their importance in targeting of a Journal for your manuscript

Lesson Plan

Journal metrics

- 5 Yr IF
- ISI ranking
- CiteScore Metrics
- Eigenfactor
- Article Influence
- Immediacy index
- SNIP
- SJR
- Altmetrics
- H5

We discussed that bibliometrics or journal metrics depend on the three major factors: IMPACT, SPEED and REACH. After discussing the Impact factor, as the first, and foremost journal metrics, let us discuss some more journal metrics.

5 Yr Impact Factor

The 5-Year IF extends the 2-year window of the regular IF. This normalizes the time frame limitation of standard two yr calculation based Impact Factor. It is suitable for slow moving areas of research where the citation is slow and need more time.

"A base of five years may be more appropriate for journals in certain fields because the body of citations may not be large enough to make

reasonable comparisons, or it may take longer than two years to publish and distribute leading to a longer period before others cite the work."

- Elsevier

5 Yr IF (Year X) = (Total No of articles cited for X-1, X-2, X-3, X-4 and X-5 Yr)/ (Total number of published articles in X-1, X-2, X-3, X-4 and X-5 years)

Pros and cons of this IF are more or less same as that of regular IF. This gives the holistic approach and flexibility with respect to time frame and some time the actual picture comes with 5Yr IF rather than the regular 2 Yr IF. And you can see in most of the journals both the IF are generally mentioned in their Home Page of the journal.

ISI Ranking

Ranking of journals on the basis of IF. While ranking, the comparison of ranking must be either for 2 yr IF or 5 Yr IF. Donot compare the 2 Yr IF ranking of a journal with 5 Yr IF ranking of other journal. **"A rank should always be in context to the subject category."** Within a subject category, you will have to compare the journals within a same subject category and type of IF. So, the Journals are ranked within a subject category on the basis of their IF.

Metrics by Scopus: CiteScore™ metrics

Scopus provides a set of metrics: CiteScore metrics. CiteScore™ metrics are a new benchmarks of measuring journal citation impact. "CiteScore metrics are comprehensive, transparent, current and free metrics calculated using data from Scopus®."

It is a group of eight indicators

- CiteScore
- CiteScore Percentile
- CiteScore Tracker
- CiteScore Quartiles,
- CiteScore Rank,
- Citation Count,
- Document Count,
- Percentage Cited

Metrics by Scopus: CiteScore, CiteScore Percentile: Let us discuss some of the most important Scopus metrics.

CiteScore

"CiteScore calculates the average number of citations received in a calendar year by all items published in that journal in the preceding three years. "*

- https://journalmetrics.scopus.com/index.php/Faqs

"CiteScore counts all articles published in a journal, where the IF only counts research articles and reviews, and excludes news articles, editorials, book reviews and the like, from the metric's denominator." (doi:10.1093 /eurheartj/ehx446)

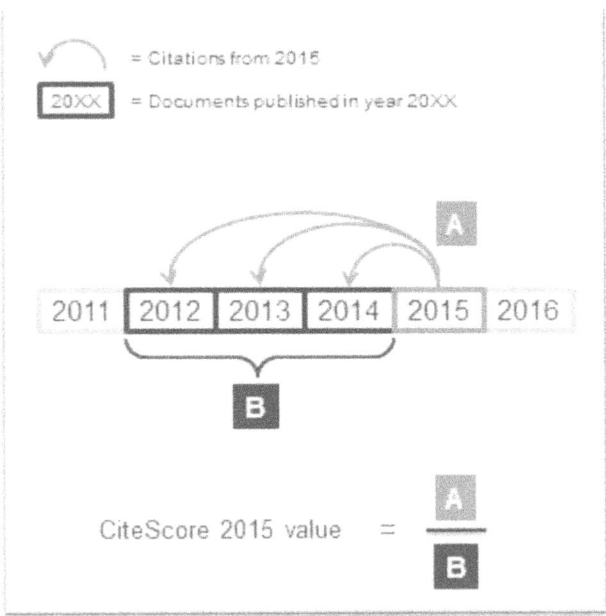

Fig. 11.1 Cite score Calculation (Source: https://www.elsevier.com)

Let's see how CiteScore of year 2015 can be calculated (Fig. 11.1). Average number of citations for last three years (Yr 2014, 2013 and 2012) divide by total number of articles of last three years. So CiteScore includes all the items of the journal, not just the research and review articles.

CiteScore Percentile is similar to "ISI ranking of IF". CiteScore Percentile gives the journal's standing in the field on the basis of CiteScore rank of Scopus indexed journals. So, it is a Scopus subject ranking of the journal (Fig. 11.2).

CiteScore Tracker is current year CiteScore.

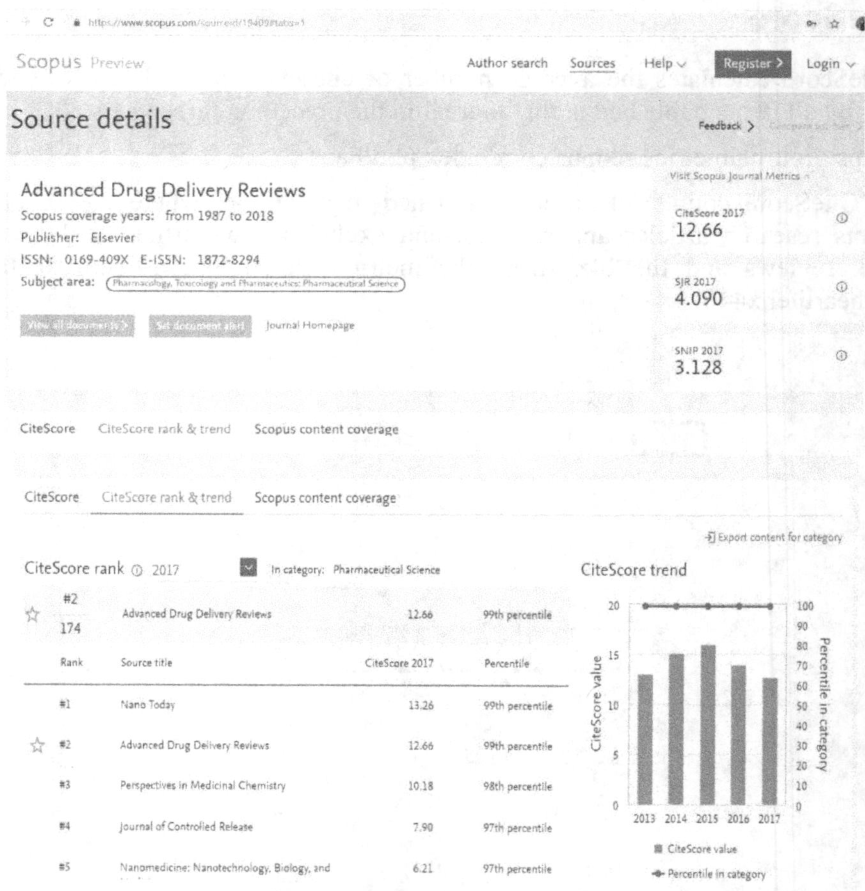

Fig. 11.2 Journal Metrics by Scopus
(Source: www.scopus.com)

In the landing page or the home page of Scopus indexed journal these metrics are freely available. CiteScore, CiteScore rank and trend, Scopus content coverage is always mentioned there. As you can see in this example of Scopus metrics of a well-known journal of Pharmaceutical Sciences: Advanced Drug Delivery Reviews. It is the second in category and its CiteScore is 12.66 with CiteScore percentile of 99. A title will receive a CiteScore rank, CiteScore and CiteScore percentile if it is indexed in Scopus. You can go through rest of the elements of scopus CiteScore Metrics.

Limitations

Like all other metrics it also have some limitations. The Scopus metrics is not suitable for journals which publish a lot of non-research material. E.g. The Journals like Nature, Science and the New England Journal of Medicine publish a lot of non-research material (like news, blogs, letter to the editor etc.), and the CiteScore metric punishes them for it. These Non-research articles are widely read, but not well cited, resulting in a lower CiteScore ranking for these journals.

Moreover, the CiteScore metric has also been reported to favour Elsevier's own journals over the others.

Summary

- Among the journal metrics, 5 yr IF is more flexible in giving time for citation unlike IF which gives only 2 year window.

- ISI is the journal IF rank in a subject category.

- CiteScore™ metrics is a set of 8 metrics by Scopus as Journal metrics.

- Cite Score is 3 Yr IF of scopus indexed journals. Covers all items not just Research and reviews

- CiteScore Percentile is similar to "ISI ranking of IF"

- CiteScore Tracker is current year CiteScore

- Please do learn about rest of the indicators by Scopus…

Eigenfactor®

The Eigenfactor® Project is an academic research project co-founded in January 2007 by Carl Bergstrom and Jevin West, and sponsored by the *West Lab* at the Information School and the *Bergstrom Lab* in the Department of Biology at the *University of Washington*.

It is a rating of the total importance of a scientific journal. It is a count of the number of citations in the target year to articles from the previous 5 years.

Eigenfactor®

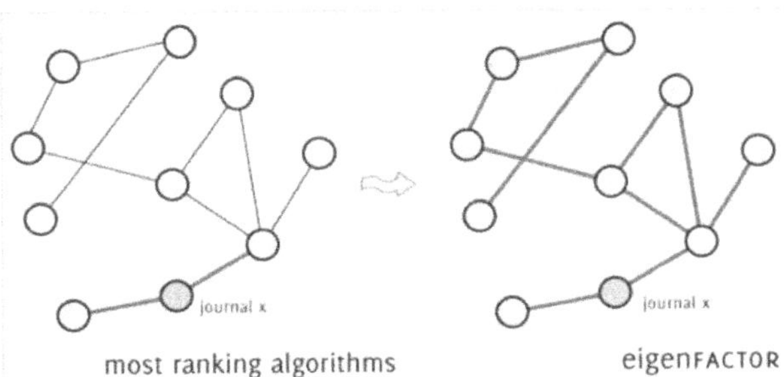

Fig. 11.3 Concept of Eigenfactor(Source: http://www.eigenfactor.org)

- Based on the idea: "Citations from highly ranked journals are more important and influential than those from lower ranked journals."
- It is more robust than the Impact Factor metric.
- "Scholarly references join journals together in a vast network of citations".
- But its algorithms (Fig. 11.3) use the structure of the entire network (instead of purely local citation information) to evaluate the importance of each journal."
- Excludes self-citations
- freely available

The graph shows how the Eigenfactor and Article influence are shown for a journal.

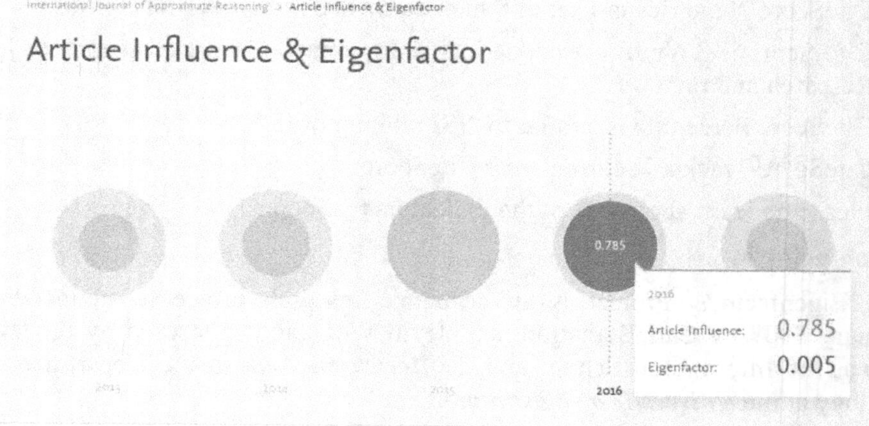

Fig. 11.4 Concept of Article Influence & Eigenfactor.

Article Influence

- The *Article Influence* determines the average influence of a journal's articles over the first five years after publication (Fig. 11.4 & 11.5).

  AI is roughly analogous to the *5-Year Journal Impact Factor*.

$$\text{Article influence} = \frac{0.01 \times \text{Eigenfactor score}}{X}$$

$$\text{Where } X = \frac{5\text{-year Journal Article Count}}{5\text{-year Article Count from All Journal}}$$

- If AI >1.00 → All articles of the journal have above-average influence.
- This metric exclude self-citations.

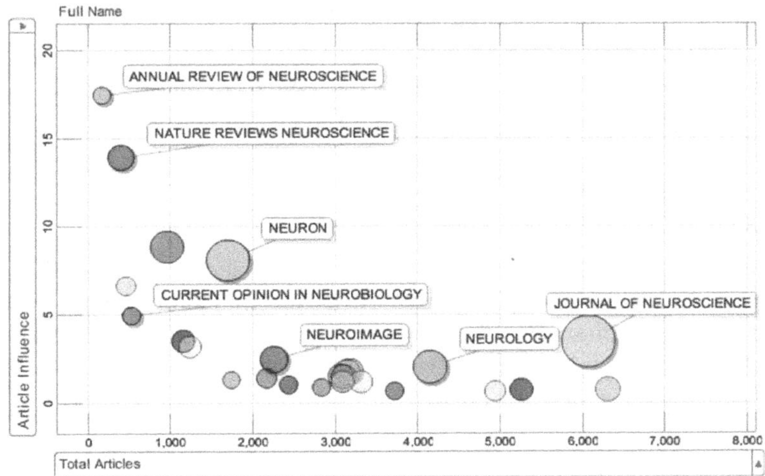

Fig. 11.5 Article Influence in neuroscience journals

(*Source:* The Eigenfactor TM metrics, December 2008, The Journal of Neuroscience:
The Official Journal of the Society for Neuroscience 28(45):11433-4,
DOI: 10.1523/JNEUROSCI.0003-08.2008; http:// www.eigenfactor.org/bubble/neuro/); "Article
Influence Scores and total articles published for the top 25 journals by Eigenfactor score in the
field of Neurosciences. Several prominent journals, including The Journal of Neuroscience, are
labeled. The volume of each circle reflects the Eigenfactor score of the corresponding journal"

Limitation

- It assigns journals to a single category, making it more difficult to compare across disciplines.
- It isn't much different than raw citation counts

Immediacy index

It assesses how immediately a citation is received for an article of journal.

- 1-year IF that looks at a much shorter time period,
- The **journal Immediacy Index** indicates how quickly articles in a journal are cited.
- It is derived by taking citations in year X to articles published in year X divided by the total number of articles published in year X.
- It aims to show the importance of article leading to early/urgent citation.

Limitation

Associated with the limitations as that of IF e.g. Reviews are cited promptly without any significant reason.

Immediacy index may not be giving the clear picture in case of some subjects while in case of some subjects it is very relevant and significant indicator. So the importance of this index varies from subject to subject. The difference in applicability of the index vary from subject to subject.

Summary

- Eigenfactoris freely available 5 yr IF of a journal. While the Article influence is an advanced eigenfactor based indicator. They give more weightage to the citations of highly ranked journals of the subject, exclude self-citations and rank the journals in a particular single subject category.

- Immediacy index is a kind of one year IF and indicates the how quickly a journal article is cited and hereby important for fast moving research fields.

After discussing various journal metrics like IF, EF, AI, Immediacy index let us discuss next metric that is SNIP.

Source-Normalized Impact per Paper (SNIP)

- The SNIP Measures the impact of a paper within a subject field
- It provides contextual citation impact

Fig. 11.6 The Ramman, UNESCO declared World Heritage Dance form of District Chamoli, Uttarakhand, India and Pandav Nritya dance form

Let us discuss the contextual citation with the example of Impact of a paper on Ramman, Pandav Nritya"(Fig. 11.6). A person has done the research on the Ramman, which is a UNESCO-declared World Heritage Holy Dance form of District Chamoli, Uttarakhand, India. Very few people know about that so citations of this kind of people will be very less. So the citation may be less for these kind of papers but the impact of such a paper is very high for those who are interested or for the relevant subject field. You do the work on the common dance forms or the hot topics you will be cited normally. But the research on Ramman or Pandav Nritya (another rare dance form from the famous Epic Mahabharat practiced in High Himalayas of Uttarakhand) like specialized or untouched area will draw the contextual citation and these citation are normalized as per the source in way such that the article's impact cannot be under weighed just because it was cited less.

How SNIP is calculated

$$SNIP = IPP/DCP$$

- IPP is Impact per paper: average number of citations received in a particular year (X) by papers published in the journal during the three preceding years (X-1, X-2, X-3)

- DCP is Database Citation Potential: citation potential in the subject field it covers

- SNIP enables direct comparison of sources in different subject fields.

- Basic journals tend to show higher citation potentials than applied or clinical journals.

- It helps authors to identify the most suited and performing journals of their subject field. And this is very important for arts, sociology, humanities where the bibliometrics do not work properly or not available for the journals or not giving the clear picture.

Limitation

Accurate categorization of journal is needed in a subject and it is very typical for interdisciplinary (ID) subject journals. We cannot categorize every journal in a specific subject category specifically when the journals are having wider scope in coverage, nowadays.

SCImago Journal Rank (SJR)

- It is essentially a variation on the Eigenfactor and a prestige metric based on the idea that all citations are not created equal.

- In this metrics the source of the citation is weighted based on that source's importance.

- The weighting is calculated using a three-year window of measurement and uses the Scopus database.

- Citations issued by more important journals will be more valuable than those issued by less important ones.

So, the important journals will be those which in turn receive many citations from other important journals. It is just like getting an award from a well-known agency/ government is more meaningful and valuable than that of getting award from an unknown NGO/foundation. So, SJR awards the citations from important journal higher as compared to the citations from less important journals.

Altmetrics

- *A shortcut for the phrase "alternative metrics," meaning alternatives to citation metrics.*

- Essentially attention metrics. It attempts to measure the amount of attention and public conversation stimulated by a given article.

- Variety of inputs used: Depending on who is doing the measurement and how they have chosen to weigh each input relatively, a variety of inputs are used (Fig. 11.7).

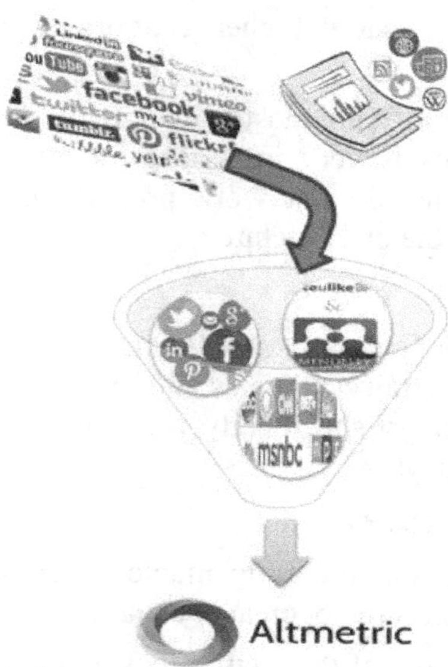

Fig. 11.7 Concept of Altmetrics

- Typical inputs: activity on social networks and social bookmarking sites, mainstream media and blog coverage, and any comments on the article.

- It is valuable to those working in controversial fields; institutions, librarians, journal editors etc. Institutions and researchers can have an idea what kind of projects are doing well, which faculty member or the research group is performing well; For librarians it may help them to give idea about most accessed journals; and journal editors can have the idea which journal is performing well and getting more readers and authors.

- Some altmetrics are given by most of the journals for their individual articles; most downloaded, most viewed etc.

Limitation of Altmetrics

Altmetrics is completely social media dependent tracking of items of any type. It does not provide the scholarly merit/impact of the article/item.

- Articles with high citation counts may have low Altmetric scores and vice versa.

- Measurements of attention, not measurements of quality.

- A negative/ wrong or controversial work may have very high altmetrics.

- Even attention measurement is not sure, if seen on internet, for how much time... (Studies have shown that most page views on the internet last less than 15 s,)

- Faking and manipulating Altmetrics is easy. It is easy to have fake ids and manipulation of metrics is possible.

H5 index

When at least h articles in a publication are cited at least h times each, it is called as H5 index. Please remember H5 index is journal metrics while the H index is an author index which we will discuss in next chapter.

H5 index is used to rank journals using articles published during the previous 5 years. But the size of the journal/ number of papers in a single issue may alter/ distort the correlation with IF.

For example, a huge journal such as PLOS ONE, with an IF around 3, has a higher h5 Index than Nature Neuroscience whose IF is above 16. So, the mass of the article may change the scenario with respect to this index.

Take Away Message

- Journal metrics like IF, 5Yr IF, Scopus CiteScore metrics, Eigenfactor, Article influence, SNIP, SJR, Altmetrics and H5 index are the well-known journal metrics useful for researchers, faculty, employers, institutions etc.

- Eigenfactor, Article influence are freely available and they provide 5 yr IF subject specific ranking.

- While the SNIP is source normalized 3 yr IF metrics which enables direct comparison of sources in different subject fields and is very useful for arts and humanities. SJR is a 3 year window based ranking of Scopus indexed journal and gives the importance to the citations received from high ranked journals. Altmetrics is social media based metrics which focuses on the attention not the quality. H5 is number of article cited number of times for 5 years for a journal.

- There is no perfect metric for research evaluation. Pros and cons are associated with each metric.But still we need to rely on some metrics for targeting our manuscript to the most suitable journal. It saves our time, efforts and provide quality solution.

Further Reading

- Scopus Tutorial: CiteScore metrics in Scopus, https://www.youtube.com/watch?v=zOZ852yJ9bw

- Cross JO, Impact factors – The Basics, the e resource management Handbook, https://www.uksg.org/sites/uksg.org/files/19-Cross-H76M463XL884HK78.pdf

- SJR & SNIP versus IMPACT FACTOR, https://youtu.be/GGDB1y3SUyA

- https://www.metrics-toolkit.org/

- Altmetrics explained in under 2 minutes, https://www.youtube.com/watch?v=-E5BIf4h5DQ

- Measuring research impact, https://library.leeds.ac.uk/info/1406/research_support/17/measuring_research_impact

- Bergstrom CT and West J, Comparing Impact Factor and Scopus CiteScore, http://eigenfactor.org/projects/posts/citescore.php

References

- https://journalinsights.elsevier.com/journals/0888-613X/authors
- https://journalinsights.elsevier.com/journals/0959-6526/citescore
- https://journalmetrics.scopus.com/index.php/Faqs
- The EigenfactorTM metrics, December 2008The Journal of Neuroscience : The Official Journal of the Society for Neuroscience 28(45):11433-4, DOI: 10.1523/JNEUROSCI.0003-08.2008
- Crotty D, Journal Metrics, Article III; OtherMetrics: beyond the Impact Factor, Cardiopulse,doi:10.1093/eurheartj/ehx446
- Crotty D, Journal Metrics, Article IV; AltMetrics, Cardiopulse doi:10.1093/eurheartj/ehx447
- http://www.scopus.com
- http://www.elsevier.com
- Bergstrom CT and West J, Comparing Impact Factor and Scopus CiteScore, http://eigenfactor.org/projects/posts/citescore.php

CHAPTER 12

Author Metrics-I

Dr Ajay Semalty
H.N.B. Garhwal University (A Central University)
Srinagar Garhwal-246174

We have discussed the basic concept of journal metrics various journal metrics like IF, 5Yr IF, AI, CiteScore, Eigenfactor, ISI ranking, Altmetrics etc., in the previous chapters. Now we will discuss Author level metrics in this chapter.

Learning Outcome	Lesson Plan
After completing this chapter, you will be able to know	Author metrics
• H Index as the Author metrics and its importance	• H index • Identifying h index from different platforms

Author Metrics

- H index
- I10 index
- G index

When we are working in a particular area of research, we often need to follow the most experiences researcher of that field. It becomes essential almost every time.

- Among a group of researcher you want to have a quick decision about their level of expertise for appointing project fellow, faculty member etc. At that time do you think you will have time to read all the articles by the all researchers who have submitted their candidature.

- Author indices provide the solution

h-Index

Why we do the research?

What is the benchmark of success for a researcher?

"Anything that won't sell I don't want to invent. Its sale is proof of utility, and utility is success."
- Thomas Edison

So we do the research…

…….To produce knowledge/ product/service → Productivity

…….To disseminate our knowledge in research fraternity → Impact

- So a researcher is recognized for his productivity (e.g. number of papers) and impact (e.g. number of citations).

- The **h-index** (sometimes called the *Hirsch index* or *Hirsch number*) is an author-level metric that attempts to measure both the productivity and citation impact of the publications of a scientist or scholar. (Wikipedia)

- The h index measures simultaneously the quality and quantity of scientific output.

- So h index was the solution given by a physicist of University of California, San Diego Jorge E. Hirsch in 2005, as a tool for determining theoretical physicists' relative quality.

- Later, it is used for assessing **productivity and impact** of a researcher/ research group/ institution.

- It's an index which increases on the basis of citations and number of papers continuously with the passage of time.

Like Impact factor it is also a dynamic index. But it increases only (does not decrease). Let us discuss how h index is calculated.

How ?

- Jorge E. Hirsch proposed the h-**index** or Hirsch **index** in 2005 as an author level metrics.

- "**h- index** is the number of papers (of a person/ group or institution) with citation number equal or greater than h" (Hirsch, 2005).

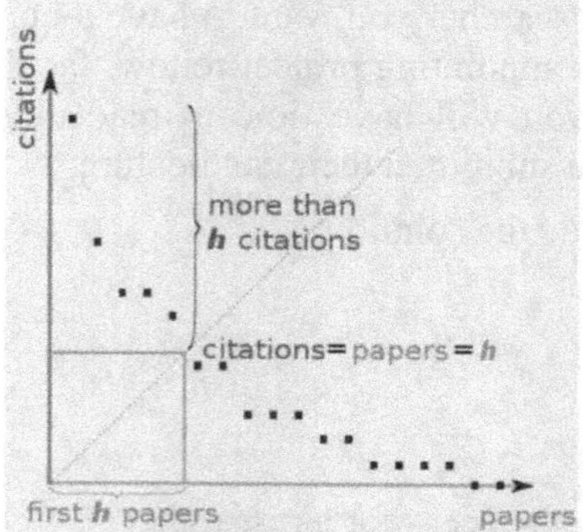

Fig. 12.1 h Index by Hirsch

The definition of the index is that a scholar with an index of *h* has published h papers each of which has been cited in other papers at least h times. Thus, the h-index reflects both the number of publications and the number of citations per publication.

You take the example of this graph given by Hirsch himself in his work. What he did? In this graph citations and papers were plotted (citations in y axis and number of papers in x axis). You can see the first "h" papers which were cited more than "h" times is called h index of that author. When the citation equals to the papers that is h.

We can say that if a researcher's h index is 20, it means that out of his/her publications each 20 papers are cited 20 or more times.

Hope you have got the essence of h index. You have got the idea.

WHO? :*h*-index

Now, who provides the h index. There are the data bases scopus web of science and most commonly google scholar. These are the three most important sources of h index.

Automatic calculators (Citation databases)

- Web of Science
- Scopus
- Google scholar

These databases calculate the h index automatically. But in these databases books, foreign language journals, patents and offline journals are not indexed / covered.

Manual Calculation

The *h*-index can also be manually determined using citation databases like we did using graph. You need not to calculate manually.

*** As "h" index are given by various agencies, when mentioning the h index in your CV or in grant applications do not forget to mention the source database of the same. Which data base you are quoting must be mentioned e.g. h index (WOS/Scopus/ Google scholar). Depending on the databases h index may vary. In general, the h index given by google scholar is the highest for any scholar. It is lower than the Google scholar, for web of science and SCOPUS h index. Because they have selected indexing. Google scholar covers almost all the online resources but Web of Science and Scopus have got the limited indexing.**

Summary

H index is the most important author metrics. It indicates qualitative and quantitative productivity and impact of the researcher. H index is the given by number of papers and number of citations. Actually it is the number of papers cited number of times (atleast). And it varies depending on the database in which h index is mentioned. Use the h index for your CV and for identifying the expertise and for evaluating the expertise of author or researcher. By this way you will follow the right person for your research. In the next section we will see how we can identify the h index from various databases.

In the previous section we discussed what is h index. In brief with definition number of papers cited atleast that much number of times is the h index. And who provides the h index. In this section we will be dealing with locating the h index in different databases Scopus, Web of Science and Google scholar. We will begin with locating the h index in the SCOPUS database.

Identifying the h Index from Scopus

Now, we will learn using the Scopus database for locating the h index.

First step is to type the URL of SCOPUS (https://www.scopus.com). Enter into the SCOPUS database (Fig. 12.2). This the user interface of SCOPUS. If you have the institutional ID/subscription of SCOPUS you can add your login and password and can enjoy the database with many more features. As we are not having the institutional subscription of the SCOPUS so we are just going to the free lookup of Scopus. For that we will go for the author search and this is the information about SCOPUS.

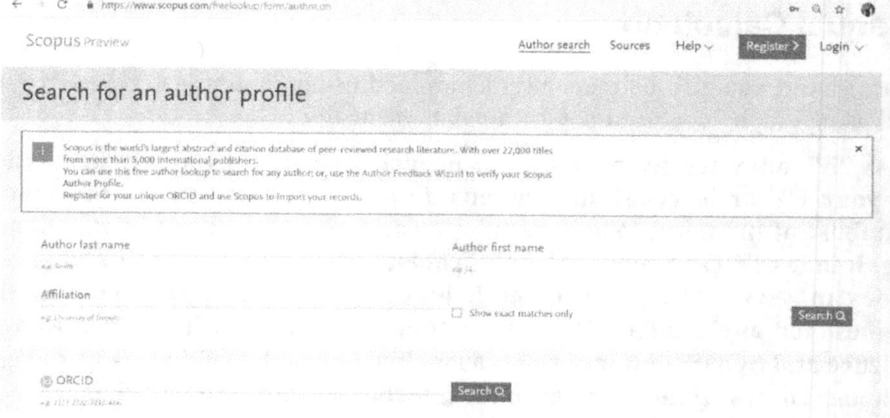

Fig. 12.2 Using the Scopus database for locating the h index

"Scopus is the world's largest abstract and citation database of peer-reviewed research literature. With over 22,000 titles from more than 5,000 international publishers. You can use this free author lookup to search for any author; or, use the Author Feedback Wizard to verify your Scopus Author Profile. Register for your unique ORCID and use Scopus to import your records."

We will discuss the ORCID in next chapter.

Now, what next?

These are the fields which you have to utilize for searching the author. Author's Last name, author's first name, affiliation if you know. I have entered Loftssona very famous scientist who is known for his cyclodextrin research, initial of first his name, affiliation if you know, his ORCID, if you know and then click search.

Now, this is the result given by Scopus. As you can see, the two author results we found with the name Loftsson last name and first name T. You can refine your results with respect to source title, affiliation, city, country and with respect to document count from high to low or author affiliation etc.

In this case we got the two entries with the name. We know the author is from University of Iceland and he is having 268 documents and h index 55. So, this is the direct h index. However, this is the new development with Scopus as far as i know. Previously in Scopus free lookup in this page particularly the h index was not shown directly. Nowadays it is shown in this page only. Otherwise you had to click on the name of the author then only the h index was visible. Now, we will clicked it thereafter we will have the author details, ORCID of author, subject area document and citation trend we will be observing with a particular graph. We can see the h index for this author Loftsson h index is 55, document by author 268, total citation 14610 by 8192 documents. You can view the h graph if you are having the subscription, you can see the other name formats of the author, subject area you can view in which he has contributed, document and citation trend we can see. We can see that in 2017 the total citations were 1305, in 2018 total documents published were 14 and the citations were 1304. While his endeavour started as per Scopus in 2009 with 6 documents which now has reached to in 2019 with citation 587 and documents 6.

What else we can get from this user interface or author profile page. Total documents, total citations, total papers we can have the idea from this chart. Now we can scroll down to get the list of all the publications by him. And if we have identified the author to follow we need not go any other source. If the article is in open access we can freely download them. Or we will have to go the Scopus indexed subscription based journals.

Let us got to author search again, take another example. Let us enter the name of great scientist Einstein A. Search.

Then we landed in to a page with 68 authors result. Now we will have to focus on the affiliation of the author. We know that the Einstein was affiliated to the Princeton. So we will scroll down first we will see whether there are multiple authors or not. If there are multiple authors we will goto the affiliation then click on the institute of advanced studies entry for getting the author detail. In case of Einstein we can see there are 102 document which are cited by 199915 documents, 37 coauthors were there (the number shall change with the passage of time). We can scroll down to have the entire list of Einstein's publications.

So, with the help of affiliation we come to know that this is the genuine author detail. We can see that Einstein is having h index 40. Though he has ceased working still the h index is growing. That is the point we discussed tat h index always keep on moving with upward trend. Documents by the author are 102 and total citations are 22746 (the number shall change with the passage of time). We can have the entire list of papers.

We have discussed locating h index from Scopus database. Now let us move to locating the h index from Google scholar.

Locating the h index in Google scholar

Enter in to google scholar. Type the name of the scholar or the scientist which you want to search for example I again enter the name of great Albert Einstein (Fig. 12.3).

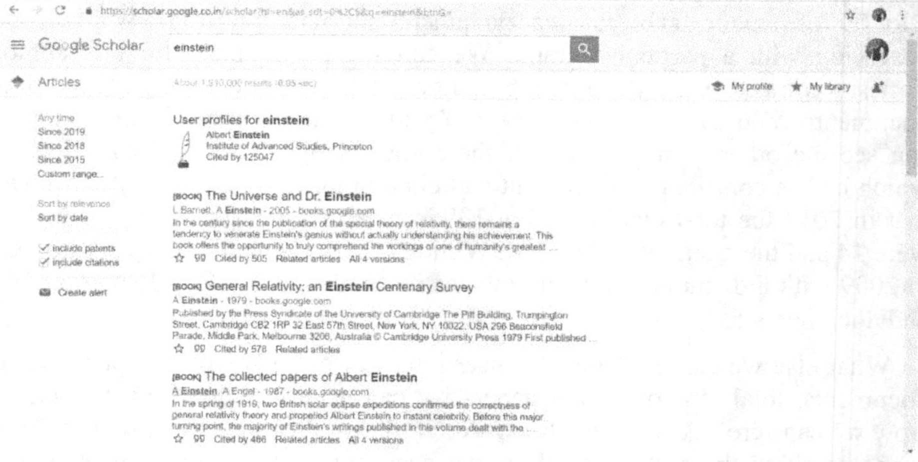

Fig. 12.3 Using the Google scholar for locating the h index

Now see the difference. As you can see her all the publications by the Einstein, User profile of Einstein we can enter into (Fig. 12.4). In his use profile, there is no verified email. If you want to follow a particular author you can click on follow. In this case you will get an alert and information from google whenever the author publish a new work.

Fig. 12.4 Google scholar profile of Albert Eistein

However, in case of Einstein, you can see by the year also in this order and if you click on the year it will be in chronologically reverse order. In 2011, one paper is there which is cited by 4 articles. Actually, these are the reprints of Einstein's paper that's why these are shown in year 2011. In 2011 it was cited by 275 times and so on with the language difference is also shown in scholar platform. You can go through the list of publications of great author and the researcher in his author profile. Now let us focus on this very important information which are endeavours of the great Einstein that Total Citations are 1,24,348, with the h index 114. This is the difference while google scholar is showing very high h index as compared to Scopus with i10 index of 381.

What are the citation? As per google scholar, if you click on it you will get the definition. It is the number of citations to all publications. The second column has the recent version of this metric which is the number of new citations in last 5 year to all publication. This is basically the h5 index for the particular author.

Now, you can also have the definition of H index. H index is the largest number h such that h publications have at least h citations. The second column has the recent version of this metric which is the largest number h such that h publications have atleast new citations in the last 5 years. This is the h 5 index. And this is very peculiar to google scholar.

I10 index only shown by google scholar. I10 index is the number of publications with atleast10 citations. The second column has the recent version of this metric which is the number publications that have received atleast 10 new citations in the last 5 years (Fig. 12.5).

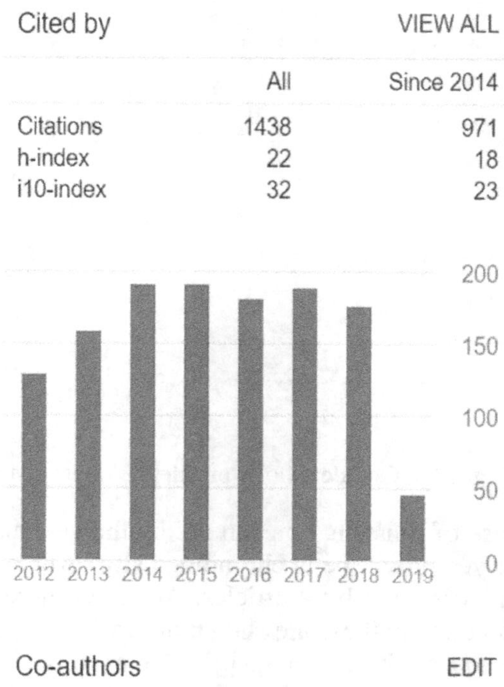

Fig. 12.5 Google scholar profile of Albert Eistein showing h index, i10 index and citations (accessed on June 09, 2018)

Google scholar sees all these citations, h index and i10 index in the 5 year window for giving the clearer picture in the slow moving areas of research. You can also focus on the bar diagram in which the year wise citations of the author's papers are given. Say for example in year 2015 total citations was 6768, in year 2014 it was 6083, in 2018 it was 6634, in 2019 it has reached to, so far it is 1927.

So, you can have the clear picture in this bar diagram in which you will get the year wise citations per year since 1990.

Here, we finished the understanding and locating the h index in Scopus and Google scholar.

Summary

We studied h index in this chapter. H index is an author level metrics. It is the number of papers cited atleast that much number of times. H index may be located from the three major databases: Scopus, Web of Science and Google scholar. We discussed how we can locate them. Use the h index in your CV when you are applying for some positions or promotions. It will give you an easy recognition and easy way for evaluating your expertise in that particular field.

And try to search the h index of your mentor, supervisor and the experts of your field. Identify the experts of your field so that you can follow in your research. The h index is the unbiased indicator of proficiency of the expertise in the research field. So, dear learners use this h index for your career and for overall research.

In the next chapter we will be covering the further details in h index and other author level metrics.

Further Reading

- Web of Science: Citation Report,
  https://www.youtube.com/watch?v=7qssEKTHQII
- Costas R, Bordons M, The h-index: Advantages, limitations and its relation with other bibliometric indicators at the micro levelJournal of Informetrics 1 (2007) 193–203. doi:10.1016/j.joi.2007.02.001
- Bergstrom CT and West J, Comparing Impact Factor and Scopus CiteScore, http://eigenfactor.org/projects/posts/citescore.php

References

- https://journalinsights.elsevier.com/journals/0888-613X/authors
- https://journalmetrics.scopus.com/index.php/Faqs
- The EigenfactorTM metrics, December 2008The Journal of Neuroscience : The Official Journal of the Society for Neuroscience 28(45):11433-4, DOI: 10.1523/JNEUROSCI.0003-08.2008
- Crotty D, Journal Metrics, Article III; OtherMetrics: beyond the Impact Factor, Cardiopulse,doi:10.1093/eurheartj/ehx446

- Crotty D, Journal Metrics, Article IV; AltMetrics, Cardiopulse doi:10.1093/eurheartj/ehx447
- http://www.scopus.com
- http://www.elsevier.com

CHAPTER 13

Author Metrics-II

Dr Ajay Semalty
H.N.B. Garhwal University (A Central University)
Srinagar Garhwal-246174

We have discussed the various journal metrics various journal metrics like IF, 5Yr IF, AI, CiteScore, Eigenfactor, ISI ranking, Altmetrics etc., and h index as one of the author metrics in the previous chapters. Now we will discuss some more important aspects of h index and rest of the Author level metrics in this chapter.

Learning Outcome

After completing this chapter you will be able tolearn

- About H Index and some more Author level metrics

- Author ids

Lesson Plan

- Factors affecting h index
- Importance of h index across the discipline
- i10 index
- G index
- Author ids
- Miscellaneous metrics

Factors affecting h Index

Let us begin the author level metrics. In continuation to our discussion of h index let us discuss the factors affecting h index, the factors on which h index is dependent on.

- **No. of papers published and citations:** Higher the number of papers and citations higher would be the h index.

- **Subject and Field of research:** Science researchers and allied disciplines generally have higher h index than researchers of Arts and humanities. And within a same subject some researchers working on hot topics are cited more than others.

- **Presence of articles in impact and accessible journals:** The easy access and broad readership of a journal also increases citations and hence the h index.

- **Stage of career:** Obviously, a senior researcher would have higher h index than new entrants or young researchers. Because he/she have published more number of papers as compared to the fresh entrants of the area.

- **Miscellaneous factors:** Type of papers (methodological papers/ review papers), innovative research papers etc.: The pioneer research work or review papers are cited more. Like Hirsch's 2005 paper being pioneer, is cited highly and hence plays a vital role in improving h index.

h-index in Academic Recognition

In general higher the h index greater the researcher (in terms of productivity and citation impact).

It's a direct and simple measure of comparing the potential of scientists of same field.

In biomedical sciences:

h index < no. of years of experience → Average/ Below Average Researcher

h index = no. of years of experience → Successful Researcher

h index > no. of years of experience → Outstanding/ Unique Researcher

h index give easy access to the potential or proficiency of the researcher.

Hirsch's Observations: Career levels

- *h*-index: Disciplines and career levels

If we see the H index for the scientist after 20 years of experience we can predict the level of scientist (Table 13.1).

Table 13.1 H index for the scientist after 20 years of experience

Hirsch (Scientist Performance after 20 years)	h index	Level of Scientist
	20	Successful
	40	Outstanding
	60	Truly unique

Source: Wikipedia

h-index: Disciplines and career levels

Hirsch observed that for physicist, h index may be the indicator of level of scientist (Table 13.1). An h index of 12 indicates advancement to tenure; H index 18 indicates full professorship; h index 15-20 indicates Fellowship in the American Physical Society; h index 45 or above indicates the level of scientist as Membership in the United States National Academy of Sciences (Table 13.2). So, the level of scientist is well indicated by the h index.

Table 13.2 h-index: Disciplines and career levels

Observer/ subject	H index	Inference
Hirsch (for physicist)	12	Advancement to tenure (Associate Professor) at major research universities
	18	Full professorship
	15–20	Fellowship in the American Physical Society
	45 or higher	Membership in the United States National Academy of Sciences
London School of Economics (Social Sciences)	2.8 (in law),	A full professor in the social sciences had an h-index about twice that of a lecturer or a senior lecturer, though the difference was the smallest in geography
	3.4 (in political science),	
	3.7 (in sociology),	
	6.5 (in geography),	
	7.6 (in economics)	

While London school of economics observed that the average index of faculty members of different discipline were significantly different. It can be see that the h index was quite low in social sciences for as compared to the researchers of sciences. It was 2.8 in law; 3.4 in political science; 3.7 in sociology; 6.5 in geography and 7.6 in economics. It was also evident that a full professor in the social sciences had an h-index about twice that of a lecturer or a senior lecturer, though the difference was the smallest in geography (Source: Wikipedia).

Limitations

- The h-index doesn't work well across fields, as publication and citation practices differ

- Databases coverage differ in publication sources and time range.

- Can be used only for comparing the same career stage and discipline.

- H index increases only and does not elicit the current impact/ potential of researcher. It invariably increases even if author has stopped research/publications. (see h index of Einstein …)

- It favours older scientists

Check your own h index or the h index of the mentors of your field (Fig. 13.1). You can have the idea of the journals for communicating your manuscripts and to have the list of articles along with links published by the researcher in one platform.

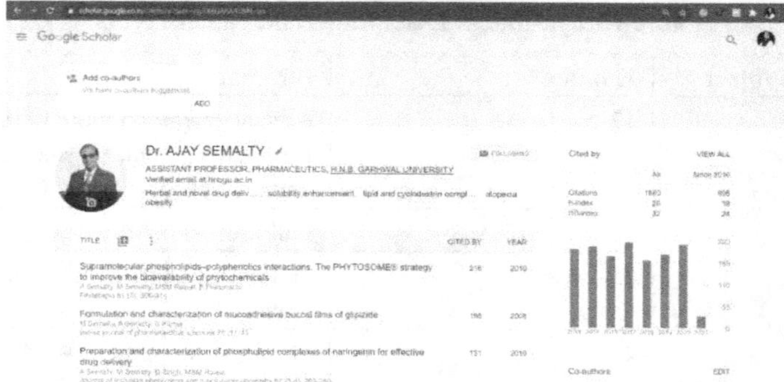

Fig. 13.1 Google scholar profile page showing h index of Course Coordinator & Author

Summary (h Index)

- Get known by h index (Add it to your CV)
- Higher h index = more productivity and impact
- h index is the major benchmark used by the employers for selection/recruitment and/ or assessment of Researchers.
- Higher h index is also an indicator of credibility of authors and it improves the confidence of readers and researchers following the authors.

i10 index

- i10 index refers to the number of paper with 10 or more citations.
- It is h10 index with reference to authors…
- It is shown only in google scholar profile of authors

Search your own h index and I 10 index (Fig. 13.2)

Fig. 13.2 h index, i10 index and citations of Course Coordinator& Author

G index

- The g-index is an index for quantifying productivity in science, based on publication record (an author-level metric).
- Suggested by Leo Egghe in 2006; Egghe observed the drawback of the H-index: it does not take into account the citation scores of the top most cited articles; The G-Index is an improvement of the H-index.
- The g-index gives more weight to highly-cited articles
- **"[Given a set of articles] ranked in decreasing order of the number of citations that they received, the g-index is the (unique) largest number such that the top g articles received (together) at least g^2 citations."**

Advantages of the G-Index:

- It considers the citations of top articles of authors.
- It helps to distinguish between the impact of two authors.
- The G-Index provides inflated values for providing credits on the basis of citations of each paper. More credit for highly-cited papers are given.

Disadvantages of the G-Index:

- Still the debate is going on about the superiority of G-Index over H-Index.
- But still G index is not be as widely accepted as H-Index.

Author/ researcher Ids

- Scopus Author Identifier
- Open Researcher and Contributor IDentifier - ORCID'
- Research Gate RG Score
- Misc. :Linkedin ID

Scopus Author Identifier

- Authors with publications indexed in Scopus are automatically assigned a Scopus Author ID.
- "It is an algorithm running within the Scopus database which automatically assigns papers to a unique author identifier using data such as addresses and subject areas."
- Check whether your paper are there completely or not? If not add your papers. Check if some other person's articles are there in your profile? If yes remove them.

- Also check for possibility of presence of your multiple profiles in Scopus. It is possible due to your multiple or old affiliations; for ladies due to change of sirname after marriage,

- Authors can send feedback to Scopus and request to merge the multiple accounts, for removing any errors and misattributions etc.

- The Scopus author preview can be checked freely

Finding Scopus Id

- Researchers, Faculty often need a Scopus Author identification number at time of applying for jobs, grants etc.

Find your IDnumber at Scopus:

Step I: Go to Scopus Free Look Up page (Fig. 13.3).

https://www.scopus.com/freelookup/form/author.uri

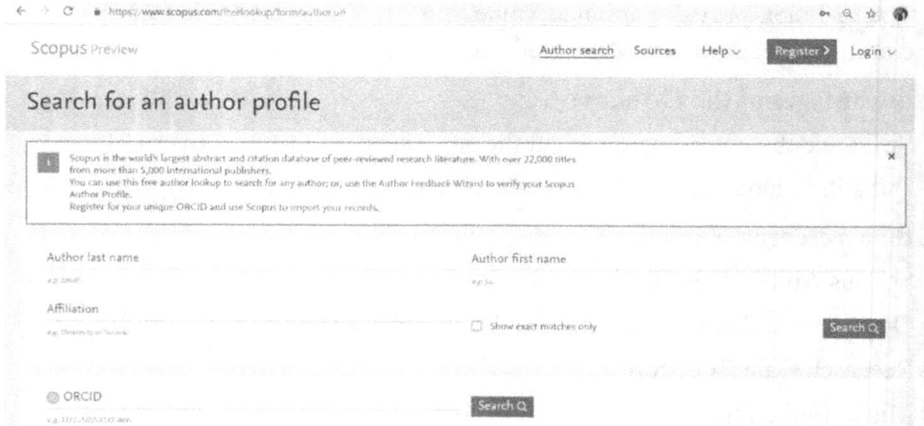

Fig. 13.3 Finding Scopus ID

Step II: Type in your last name, first name or initials to locate your record (if you know your ORCID id, just enter it)

Step III: Find your assigned number: Select the appropriate author record, and you can view your Author Identification record along with other details (Fig. 13.4).

You can see h index, documents count/ list by authors, total citations and year wise citations of your publications indexed in Scopus. But please remember the h index by Scopus is always less than that of google scholar.

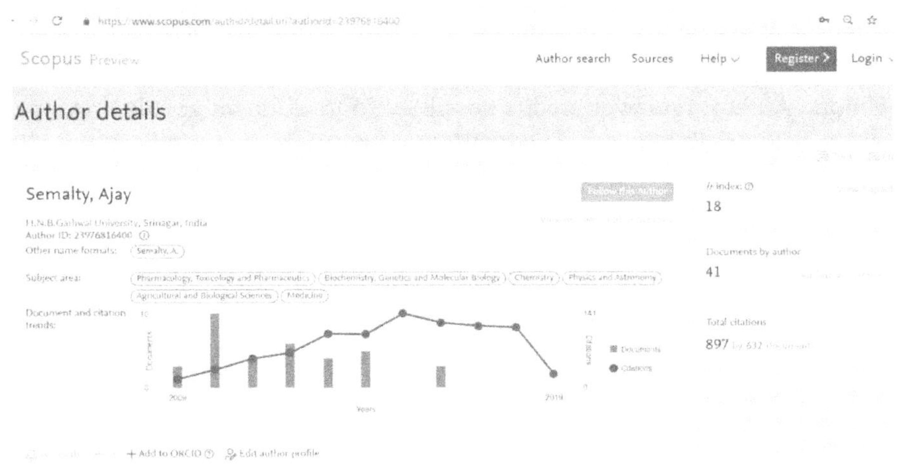

Fig. 13.4 Scopus profile showing Scopus Author id and other author metrics
(Accessed on July 2018)

Summary

- Apart from h index (Hirsch Index), I 10 index is another author metrics; technically it is H10 index; available only through google scholar

- Another Author metrics is G index, which focus on the highly cited articles for giving due credit.

- Author ids are needed for presenting at the time of grant/job application, promotions and for CV.

- Scopus author Id can be seen freely from Scopus free lookup page.

- You must get your scopus id and use it for your CV and as signature in your email id.

- It increases the impact of researcher.

After covering the Scopus ID let us discuss some other author ids.

Open Researcher and Contributor IDentifier - **ORCID**

- The ORCID iD is a nonproprietary alphanumeric code. This serves to uniquely identify authors and researchers.

- It is a platform-independent identifier which creates and maintain a registry of unique researcher identifiers.

- Offers a transparent method of linking research activities and outputs to these persons.

- This ID also avoids name change and name similarity problem in giving credits
- Unique Author/ reviewer credits possible: To track down genuine reviewers and to give credit to the reviewer ORCIDs are used.
- Fake authors/ reviewers can be avoided:

Getting ORCID

- Go to https://orcid.org/
- Register to get unique ORCID identifier. In Scopus, you need not to register. It is automaticall there if you have published in any of the Scopus indexed journals. But here you will have to register.
- Addyour information (Enhance your ORCID record with your professional information and link to your other identifiers such as Scopus or ResearcherID or LinkedIn)
- Use your ORCID ID:Include your ORCID identifier on your Webpage, when you submit publications, apply for grants, and in any research workflow to ensure you get credit for your work.

Research Gate RG Score

- Research Gate is a kind of scientific social media. If you are a researcher you might be using it frequently already. Research Gate (founded in 2008 by a group of scientists) is a social networking site for scientists and researchers to share research articles, and search collaborations.
- It also acts as discussion forum on research issues.
- Each author has a unique profile there.
- RG Score is a single number that is attached to a researcher's profile.
- "The RG Score measures scientific reputation based on how your work is received by your peers."
- RG Score focuses on authors and their interaction with peers.
- It is calculated once a week. RG Score is giving the current impact of the researcher on weekly basis.
- Many critic don't rely on RG score still due to several factors e.g. even the conference abstracts, and unpublished data, paper can be loaded as dataset, self triggered discussion may increase the score etc.
- RG Score is not totally dependent on your scholarly articles just the discussion may trigger the RG Score.

Miscellaneous metrics

So many new metrics are coming day by day with the advent of new publishing groups and new metrics from old publishing houses

Plum Analytics

This service aims to provide a more comprehensive measure of a researcher's scholarly impact by gathering data about usage of data sets, open access publications, presentations, blogs and other types of scholarly communication.

SciVal

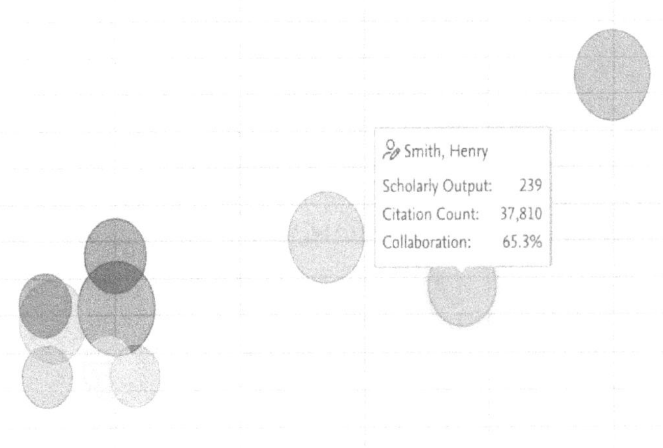

Fig. 13.5 SciVal: an institute/ country/researcher's performance indicator based on Scopus data

- Recently (in 2014) Elsevier introduces a new metrics SciVal®: an institute/country/researcher's performance indicator based on Scopus data (Fig. 13.5).
- "SciVal, is the tool by Elsevier Research Intelligence. It delivers research performance metrics, based on the Scopus database, for a large number of Universities and other institutions in > 200 countries."*

SciVal enables you to:

- Search for your Scopus profile
- Find your aspirational peers
- Select the metrics
- Benchmark yourself to demonstrate research excellence

Take Away Message

- Apart from h index, i10 index and G index are two other author metrics.
- I10 index is H10 index; number of papers cited at least 10 times
- G10 index is the modified form of h index to give credit to highly cited article.
- Scopus Id, ORCID and RG score are the author ids which you must utilize to showcase your credentials for jobs, grants, promotion etc.
- Many other metrics are there like SciVal which is very useful due to their pragmatic approach in eliciting the research performance of researcher, institute, country etc.

Further Reading

- Hirsch, J. 2005. An index to quantify an individual's scientific research output – PNAS, Available at http://www.pnas.org/content/102/46/16569.full.pdf
- Find your h index, https://www.youtube.com/watch?v=wmnqCge-h_M
- Indices of Research II : h-Index Part 1, https://www.youtube.com/watch?v=BAhPzxWVtVE
- ORCID and Scopus: Manage your author profile, https://www.youtube.com/watch?v=UJBq0f0Qnrs

References

- https://en.wikipedia.org/wiki/H-index
- Hirsch JE, An index to quantify an individual's scientific research output, Proc Natl Acad Sci U S A. 2005 Nov 15; 102(46): 16569–16572.
- Judit Bar-Ilan, Which h-index? — A comparison of WoS, Scopus and Google Scholar, Scientometrics, 2008; 74(2): 257- 71.
- Bornmann L, Daniel HD, What do we know about the h index? Journal of the Association for Information Science & Technology, 2007; 58(9): 1381–1385
- Glänzel W, On the H-index—a mathematical approach to a new measure of publication activity and citation impact. Scientometrics, 2006; 67: 315–321
- https://www.scival.com/
- https://journalmetrics.scopus.com/index.php/Faqs

- Crotty D, Journal Metrics, Article III; OtherMetrics: beyond the Impact Factor, Cardiopulse,doi:10.1093/eurheartj/ehx446
- http://www.scopus.com
- http://www.elsevier.com

CHAPTER 14

Literature Review-I
(Introduction)

Dr Ajay Semalty
H.N.B Garhwal University (A Central University)
Srinagar Garhwal-246174

In previous chapters we covered the basic concept of academic writing, basics, English in academic writing, plagiarism and journal/ author metrics. Now, we will discuss the Literature Review in this week.

Learning Outcome

After learning this chapter, you will be able to know

- Concept of literature review
- Purpose
- Types of Sources
- Basics of doing the quality Literature review

Lesson Plan

- What is literature review?
- Purpose
- Sources
- Process
- Planning

Initiating Research

When we are initiating the research the very first questions comes in our mind: from where to start? What should we do?

- From where to start?
- Reading & Thinking…. (but what to read and what to think?)
- The prospective topics must be read.
- The related body of knowledge must be gone through.

For this needs a Good homework and must have a good understanding the need/resources.

- What is literature review and what to search?
- Why to search?
- How to search?

Now, let us first discuss

What is Literature Review (LR)?

Literature review (LR) can be defined in many ways.

"Systematic and organized compilation and critical study of related body of knowledge is literature survey."

"A literature review surveys books, scholarly articles, and any other sources relevant to a particular issue, area of research, or theory, and by so doing, provides a description, summary, and critical evaluation of these works in relation to the research problem being investigated."

Literature reviews are designed to provide an overview of sources you have explored while researching a particular topic and to demonstrate to your readers how your research fits within a larger field of study.

It is the foundation of a sound and successful research. Until and unless you do a good literature review you cannot do a good research. The complete success of research depends upon the LR. On the basis of level of research, it may be the blend of summarizing and synthesizing.

Let us understand what is summarizing and what is synthesizing?

- **Summarizing** the historical background and recent development of the field of research.

- **Synthesizing:** Systematic reorganization and reshuffling of the information to develop knowledge/ reasoning/ problem definition.

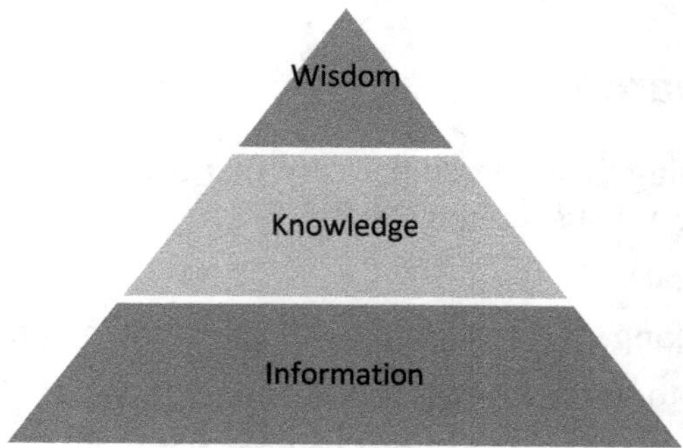

Fig. 14.1 Triangle of information, knowledge and wisdom.

It is all about gathering the information and information when summarized and synthesized leads to knowledge which upon further processing leads to wisdom (Fig. 14.1).

And donot forget! We are in 21st century where information is an inevitable and indispensable weapon. From where you get the information, quality of your information, how you collect, manage and use it, are needed to be learnt for the successful completion of any endeavour in any part of life. The success in each field of life depends on your efficiency of using the information effectively.

What?

What you have to search?

- Research/ review articles,
- Books,
- Thesis
- Conference papers
- News
- Blogs
- Databases
- Misc.

We will discuss these resources in detail in the next chapter.

Why?

Why we do it? What is the purpose of doing LR? However, definition itself shows the purpose. Let's discuss the purpose of LR.

- To prepare a handy guide on the topic/subject
- To understand the origin of idea and its development through the passage of time
- Identifying the research gap
- Defining research problem & establishing rational of the study
- Ensuring novelty
- Establishing aim and objectives of the study

To prepare a handy guide on the topic/subject

The body of knowledge of any field of research is always huge. Can you carry all the body of knowledge at a time? Is it possible? Can you present it with that hugeness? The hugeness will not be easy to follow. So, what is the way out? You require a condensed, power packed, compilation of the body of knowledge for ready reference and guide. It is just like preparing notes of a lecture.

Huge knowledge is there in the books and in your teacher's lecture. But what you do? You compile them in the form of notes so that you can readily refer them in the systematic manner. And this is the main purpose of the LR.

To understand the origin of idea

Without going through the literature search,

- Can we originate an idea?

- Can we know how it developed with time?

To understand the origin of idea and its development through the passage of time, Literature review is needed. The time factor is very important in studying a research of any time. It may change the entire need or the significance of a research sometimes e.g. research on pager technology. Before going to the research on pager technology have you gone through whether it is in there in the trend or not? Because with the passage of time the technology has already become outdated. So, see the timing. Whether the significance of the topic is time bared already or not or what is the potential? So, origin of idea depends on the time factor predominantly.

Identifying the research gap

LR identifies the research gap. The gap or the missing link between existing knowledge and expected outcome can be explored. Search, what is the limitation/ deficit of the current research? The proper identification of gap is the key of success of research. You will have to identify the gap, what is missing there and what can be done and for that we do the LR.

Defining research problem & rational

This is related with the research gap. Until you identify the research gap you cannot define the research problem. A good definition of problem is like choosing the right way. Wrong problem cannot lead to a right research. And for that LR are needed to establish the rationality: what & why? Means what you are doing and why you are doing it must be justified.

The properly organized information help in defining the problem and establishing the rationality of research. Can you plan a drug delivery system for buccal route of the drug which irritates the buccal mucosa? Where is the rational?

Ensuring novelty

Once research gap is identified ensuring novelty of research is important many times even at this stage. The novelty of the research may be ensured only through a well planned and executed LS. If you are planning and aiming for a breakthrough research or patentable research, it is the very first and very vital component. At this stage only, if it does not seems to be novel then you cannot expect it to be patentable at all.

Establishing aim and objectives of the study

The properly identified research problem may be planned only after the LS/LR. LS leads to establishing aim of the research. Due to body of knowledge available and studied the, aims may be divided in suitable tangible and result oriented objectives.

Section Summary

Literature survey is the foundation of a good research and paves the way to a successful research. "Systematic and organized compilation and critical study of related body of knowledge is literature survey."It is about summarizing and synthesising the idea of research, exploring research gaps, rationality, novelty; establishing problem definition and planning the aim and need based objectives of the research.

We discussed so far what is Literature Review (LR) and why we do it. Now let us discuss how?

How?

How we can do LR. This is the process flow of LR. It is a five-step process (Fig. 14.2).

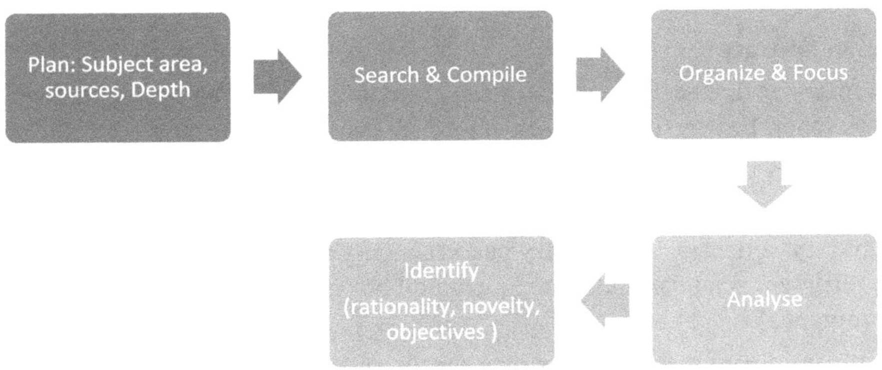

Fig. 14.2 Process of Literature Review

Let us discuss these one by one.

Plan

"A good planning is the half work done. In each work you plan well ahead even for very smaller things. So why not in research and LR. Plan with your team, supervisor, collaborator and the mentor regarding

- Broad subject area
- Tentative topics
- Type of sources to be searched
- Quantity of literature
- Level of historical background to be covered
- Converging strategy
- Synthesising/ analysing strategy

The outline/roadmap of LR must be drawn after proper planning and discussion

Tuning with your team and mentor is the key of successful research. And the planning is the base of LR and research. So, it must be done with good tuning and discussion with all the stakeholders.

In all, what we are going to plan? To which extent are we going to cover the background of the research? What thing? On which basis we are going to analyse the aspects of research? What would be the convergent strategy? We will discuss the convergent strategy in detail later.

But let us address the next question: where to search? What are the types of resources to search?

Types of Sources

- **Primary sources:** The original thesis or data collected and presented by a researcher. E.g.: research papers in journals, patents, Ph D thesis etc.
- **Secondary sources:** The original data and studies of an original study quoted or collected by another author e.g. review papers,
- **Tertiary sources:** The primary and secondary data collected and presented by different scientists/authors e.g. edited books, news, blogs, and other web resources.
- Each and every source has its own merits and demerits. Choose the source wisely. The selection of a type of source depend upon the subject, scope and level of the study.

"LR is developing Tertiary source with the help of primary and secondary sources."

- Types of sources

The sources may also be categorized in offline and online type.

- **Offline sources**

Local library resources: Books (enlist major reference books to be referred); Journals (The list of relevant journals in library); New papers, Official/ standard books (IP, regulatory guidelines etc.)

- **Online sources**

  Enlist and identify the quality journals, publishers, portals for reference. There are end number of resources in the online world. Virtual world is filled up with the information sources. There is the bombardment of information, ocean of information. You have to identify the quality information and process it into the form of knowledge. You evaluate the sources or take the help of your supervisor/ mentor in identifying the quality sources. There are end number of low quality journals or substandard portals which provide low quality information which are most of the times, unreliable. Until and unless you identify the quality sources you cannot make or establish a quality foundation for your forthcoming research. What you are going to do is based on the LR. So, refer the quality sources.

- e.g. Google scholar (use scholar not just google). Google is a search engine but the google scholar is the platform for searching scholarly articles only and avoiding junk information. It saves your time and efforts.

- Journals of Prestigious publishers (Elsevier, Springer, Oxford University Press, Macmilan, Medknow, Taylor& Francis, Bentham, Informa Healthcare, Kluwer, Lancet, Nature etc.), Pubmed (for biomedical, sciences and life sciences and allied fields), governmental data repositories, other OERs like Youtube, Wikipedia, OA journals, OERCommons etc. (We will discuss the online sources in a separate chapter in the last week of the course.)

The Most Important sources: SUPERVISOR

Your mentor/ supervisor is the most important and vital source. Have a good tune with your supervisor. Ask for help and guidance. It would be easier in your mind what to search, where to search and how to search after consultation and guidance of your supervisor.

"Collecting body of knowledge from the supervisors: Experience Survey"

Lot many practical things are not in books or scholarly articles. Many times, your supervisor tell the things which you could not deduce or know even by studying a lot. Experience survey make the things easier and save timing for your survey and overall research.We will discuss in a separate chapter also.

But you have to ask! And be in recipient mode. But the supervisor is also needed to keep himself or herself updated with the field of research. So, it is the duty of all the supervisor and team members to be updated and the new information and knowledge must be shared within the team and discussed.

Horizontal as well as vertical flow of information must be there for updating the body of knowledge in the team.

Section Summary

- Planning a literature survey must be done carefully. The outline of LS must be drawn in good tuning. The sources may be primary, secondary, tertiary or offline or online.

- You must search from quality sources.

- The most important source of LS is supervisor or mentor, this is called experience survey.

We learned what to plan and how to plan. Now let us discuss how to search

How to search?

- Start with the homework regarding the type of source to be searched and the depth of knowledge to be accessed with vital inputs of experience survey (supervisor's inputs). Student might not be in condition to judge what to refer, how to refer when to refer? So, ask your supervisor.

- Define the tentative breadth and depth of your LS on the basis of subject nature, topic, and the level of research. It may be condition like SINK OR SWIM. If you have not planned properly your LR will sink. So, plan effectively and come out of the ocean of the knowledge, swim through it.

- There are no thumb rules for length, and it varies depending on the subject, level and scope of the study. (remember the total length of Einstein's thesis)

- The length of a Ph D LS may be the entire length of the PG dissertation. Remember, recall, what was the total length of Ph D thesis of Einstein?

- In general for short research dissertations LS may be 5- 6 page long with 1500 to 2000 words; while for Ph D it generally range from 8 to 10000 words or more.

- It's a choice of yours to begin offline or directly start on line search.

Off line Literature survey

- **Begin with the Text and reference books:** Literature collection should be commenced from the theory part from books and then from the research and review papers of journals available in the departmental or institutional library.

- **Refer official/ regulatory books:** Related encyclopedias and pharmacopeia (for pharma research) should be referred for understanding the basic concepts. Referring and searching should be done with good planning. Encyclopedia may have huge knowledge, but you have to pinpoint.

- **Start with one end of a string:** A particular journal should be taken once, and all its available issues should be searched. Take your textbook, go to reference book then see articles, then go to major cross reference. Avoid sinking of your LR.

- **Record properly:** While searching, if any important article is found, note down its reference (Author, title, journal, year, volume, issue, page no.).

- **Compile:** When scanning of one journal is finished then collect the issues in which articles of interest have been identified and get the articles Xeroxed. Now move to the next journal in the same way. Hard copy of important articles must be studied exhaustively. Take them home, read, underline them. Hardcopy reading triggers the idea, discuss with your team and mentors and develop the body of knowledge.

- **Searching abstracts physically:** To refer compilation of abstracts, like Manual of Aromatic Plants Abstracts (MAPA), Indian Science Abstracts (ISA) is very helpful and timesaving. After a particular abstract of importance is identified from these, the same can be then traced from that journal's database. However, many of these abstracts are also available online in which you can search rather more easily. But you can begin offline in early stage.

- It is better to start with offline search so that you can have the first string in your hand and then you can keep on going to higher level of knowledge.

- **Old thesis/records:** The project reports or previous thesis may also be consulted. It may help in designing research which may be done in the Department/ institute itself. Though in India, Shodhganga has made the thesis available online, but still many universities and institutes have the thesis lying in the almirah, still unexplored and reviewed. Referring the own institutional theses may allow researcher to think about the practical aspect of planning the subject within the institutional available resources.

This may also help in getting the cross references (references cited in text or reference in an article) which may in turn be collected further.

Online search

- Being time consuming, the offline method is obsolete now and just limited to your own institutional library

- The online search is very fast and can save time.. Work which was done in 6 months can be done in 6 days now.

- Check from your librarian if you have institutional access of SCOPUS, Web of science etc. if yes explore it.

- If you do not have the access to these paid services, explore https://www.ncbi.nlm.nih.gov/pubmed (a free web resource from U.S. National Library of Medicine)

- or just use other search engines like google (use google scholar, not simply google)

Take Away Message

- We learned, what is Literature review? It is systematic compilation and organization of critical study of body of knowledge of the related field of research. It is the foundation of the successful research.

- It is summarizing and synthesizing the idea, identification of research gap, establishing rationality, exploring novelty, establishing problem definition, laying down the aim and objectives of the research.

- LR starts from effective planning. Length and breadth must be planned well ahead. Tentative outline must be planned wit supervisor and team members.

- The search must be done from quality sources. The sources may be primary, secondary, tertiary, off line and online. The most important source is Supervisor. It is the experience Survey.

- Online search saves time and improve the effective ness in compiling and organizing the body of knowledge.

Further Reading

- Anson, Chris M. and Robert A. Schwegler. The Longman Handbook for Writers and Readers. 6th edition. New York: Longman, 2010.

- Reviewing the Literature: A Short Guide for Research Students,

- https://uq.edu.au/student-services/pdf/learning/lit-reviews-for-rx-students-v7.pdf

- https://writingcenter.unc.edu/tips-and-tools/literature-reviews/

- IRM M L11 How to read a Research Paper, https://youtu.be/cT-UrjqGQYY

References

- https://www.kent.ac.uk/learning/resources/studyguides/literaturereviews.pdf

- https://www.journals.elsevier.com/
- https://www.ncbi.nlm.nih.gov/pubmed
- Scopus Content coverage guide, https://www.elsevier.com/?a=69451
- www.wikihow.com/Write-an-Article-Review#/Image:Write-an-Article-Review-Step-4-Version-3.jpg
- http://library.atmiya.net/research%20commons/researchguide/literature_review.php

Literature Review-II
(Process)

Dr Ajay Semalty
H.N.B Garhwal University (A Central University)
Srinagar Garhwal-246174

We will cover the process in this module. We were discussing the Literature Review, the foundation of a quality research. In continuation of previous module, we will discuss online review and process of literature review in this module.

Learning Outcome

After learning this module, you will be able to

- Know the complete process of Literature Review (LR)
- How to search effectively?
- Organize, compile, analyse Literature Review (LR)
- Identify the aim and objectives

Lesson Plan

- Online Literature Survey/Review(LR)
- Boolean search
- Organizing and compiling
- Analysing
- Identifying research gaps, rational, aims and objectives

Online Search

Using the proper key words is the first and foremost thing. Type appropriate key words in task (search) bar of website of SCOPUS, PubMed etc. The selection of key words is very important. This provides the list of articles, abstract, full text (if available free or free for institutional subscribers), and the link to journals associated with the key word given in task bar.

Select proper key words, try changing key words, keep searching using filters like review, research etc. It will improve the effectiveness of LR. Try different filters like review only, research only or both. The selection of the key words decide whether the quality of your review will **sink or swim**.

Alternate strategy for searching: Explore searching in subject specific journal or the publisher's website with proper key words, atleast abstract you will get. Even if you are not having the subscription of that journal atleast the abstract you will get.

Remember, many good journals provide the free full text. Search the reputed journals of your field of research, which provide free full text.

Use Boolean search: Universal tools for online searching

"Boolean search is a type of search allowing users to combine keywords with operators (or modifiers) such as AND, NOT and OR to further produce more relevant results."

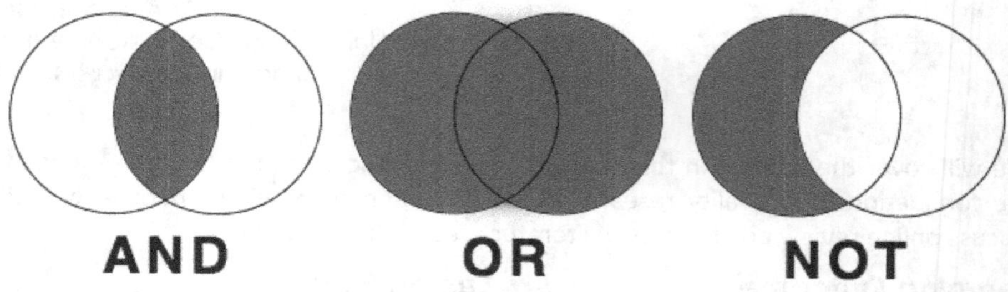

Fig. 15.1 Major Boolean Operators

These operators develop the **basis** of **logic database and mathematical** sets. These operators connect your search words and either **narrow** down or broaden the results.

Note that AND, NOT and OR will generally need to be in upper case/ capital letters when used as a search operator.

- **Quotes:** Quotes are used to search for an exact phrase. *Example: "network administrator"*

- **Parenthesis:** Modifiers can be combined to create a more advanced or complex search. *Example: network AND (administrator OR architect)*

- **AND:** It includes two search terms. *Example network AND administrator*

- **OR:** Broaden your search with multiple terms. *Example: "network administrator" OR "network manager"*

- **NOT:** Use to exclude a specific term. *Example: administrator NOT manager*

Other Common operators

- Phrase search:
- **Double quote** will search for fuzzy phrases
  - Will search for both singular and plurals
  - "Heart attack" will find heart attack and heart attacks both
- **Curly bracket** {} will search for the specific phrase
- {Heart attack} will search only for heart attack

- Wildcards
  - Asterisk (*) replaces characters
  - Tox* will search for toxic, toxin, toxicity etc.

Six Common Boolean operators

- AND
  Limit results
- OR
  One term OR another
- - or NOT
  Exclude a term from the search
- -site:
  Exclude a website from the search
- ~
  Synonyms of term
- ""
  Exact phrase

Use common/advanced filters

- Use filters like year, subject, type (review only, review, research only) etc. by choosing and applying the filters.
- Use the advanced filters
- Example: Google scholar: The most widely used database does not have much filters, but still useful for many basic LR.

Using google scholar

Using operators in Google scholar:

- Open Google scholar
- Type a key word
- E.g. Let us use the Boolean operator "cyclodextrin complex"
- Information given by scholar: list of articles with title, author and abstract,
- Related articles, versions available and link for pdf download is given.
- If freely available, article can be downloaded in single key.

This much you all know….

Let us explore some more things

Can you see the quote sign below each article detail, It is the key to get citation information "cite" that to in any of the writing styles you can choose.

Citations are given in 5 major styles in scholar (Fig. 15.2)and it can also be managed using BibTeX EndNote RefMan RefWorks.

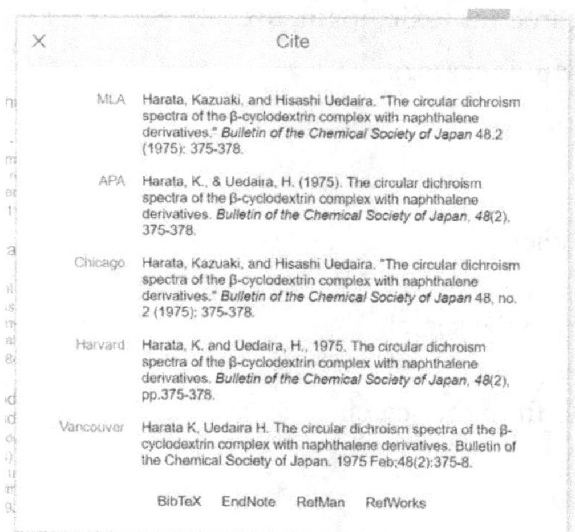

Fig. 15.2 Importing Citations from Google Scholar

We will discuss this part in next module in literature management.

See the number and details of articles which cited an article

You can apply the filters like

Time range (from 2009, from 2018 etc.) or customized the time range as per your need;

Type of article (include patent/ citations), by default these are included.

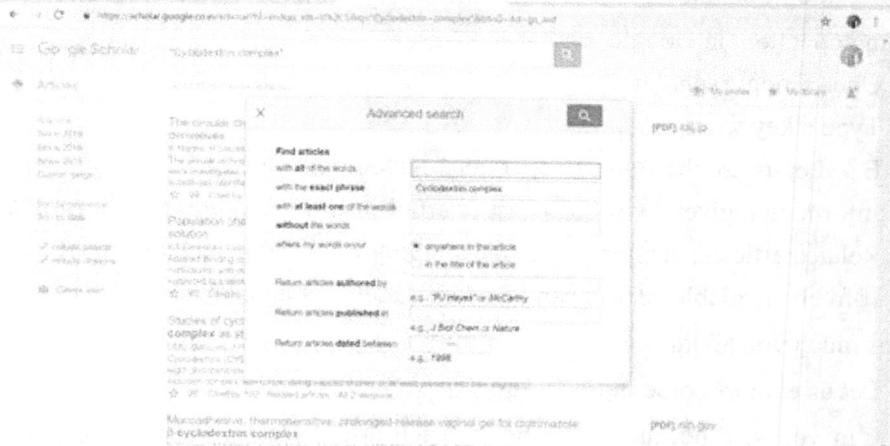

Fig. 15.3 Advanced search in Google scholar

Go to advanced search

- Use the choice of words/ phrases, location of words, author, publisher etc (Fig. 15.3).

Section Summary

- Online searching is time saving method of LR.

- We may use Boolean search operators for filtering our LR.

- Use AND, OR, NOT, quotes, Parenthesis as the most common operators.

- Google scholar is the most commonly used searching platform and use of filters and Boolean operators can make our work easier.

Key features of Literature Review

Let us discuss the key features of LR or LS.

- **Size (depending on the subject/ level of research):** Size means what should be the length and breadth of the content to be collected. It depends on the subject, type of research, scope of research and level of research. We can not expect a huge LR from a postgraduate dissertation work. Can you expect? No. So, we will have to keep the length and breadth defined. Otherwise it is an ocean of knowledge. Each and every topic is an ocean of knowledge. You can never be a Ph D of an entire subject. You can be Ph D of a particular topic, may be. The proper choice and effective LR methodology may allow you to swim across that ocean. Otherwise your LR will sink and your quality of research work will go down.

- **Quality:** Search/ refer the quality articles from quality sources/ journals/ publishers/ books. Wrong methods can not lead to right research. Substandard articles, if you will refer, you can not expect a good research. Refer the good quality source. It is utmost important. Quality articles only can lead to reproducible results. So, you must rely on the quality. Otherwise, if you will refer a wrong/ substandard article, you will follow it you will try to replicate the results of that article and what will happen? You will lose your time, money, resources and all your efforts. So rely on the quality articles, journals, books and other quality resources only. And that must be discussed in the planning stage itself with your team members and your supervisor.

- **Versatility:** Versatility means including all the types of resources not just one. Your LR must contain a good mix of review, research, patents, books, news, meta-analysis, regulatory guidelines, historical articles/ studies, geographical distribution of compiled studies etc. Geographical distribution of compiled studies means, I found an article on a topic from a Chinese author only and referred all the references from the same article. It may be the need of some time or case. So, I just skipped the geographical distribution of authors citation and skipped the chance of reviewing the topic in an unbiased form. It increases the understanding and quality of field of research.

- **Up-to-date**: The current literature at least references of last 5- 10 years must be there. The "uptodateness" is very important till the finish of work. That's why Ph D LR is an ongoing process till the thesis is bound. You started writing your thesis on 2017 last. You finished the LR in 2017 and you could submit the thesis on year 2019. So your references are only before of year 2016 or 2017 (or just a few references of year 2017). So what is lacking here that you are not having references of year 2018 or 2019 it. It might be an objectionable issue when it is coming before the reviewer because he or she will observe the current references.

- **Do not skip the historical or pioneer studies/** articles of the subject/topic. Though in one hand we are saying that only current references should be cited. But we must not avoid the historical or pioneer reference. For example, if am discussing or criticizing the h index, can I skip the paper of Hirsch of year 2005? the pioneer paper/ reference. I cannot skip citing that pioneer reference. So, you must cite the pioneer study but donot put the overdose of older references. Sometimes, we just try to fill the paper with older references. For example, you got a review paper of year 1998 which was very relevant to your work. You put all the cross references of that article which were very old, way back from 1946 to 1980. So, what you are doing in 2019 that you are stopping your literature review upto 1980 or 1990. It is not at all acceptable.

- **Completeness:** Complete the LR with respect to the scope of study (topics, subtopics, allied topics etc) with due focus.

- **Be critical:** Include the studies which may not favour your tentative hypothesis also. You must be self-critical. You planned or trying to hypothesize a thing. You are collecting the supporting evidence. But you must not skip the things which are against your hypothesis or any other contradictory/ controversial things. It will be an unbiased ground of LR.

- **Discuss with the supervisor/ mentor/ team members/ collaboraters** for any doubt/ help. There must be a good tune in between.

Ask for help

If nothing above is working, then donot hesitate in asking for help. For getting full text of a certain article if you donot have access you can explore

- Asking for help from authors/ co-authors (through email)
- or the persons with access of the full text of journal

- In social media platform like LinkedIn/ Facebook pages
- In scientific platforms like Research Gate

Section Summary

- Literature search is an important and vital part and you must be aware of the key features of the LR.
- Cover the LR in required size, quality, versatility, contemporary or current knowledge pool, completeness, criticalness and with support from your team supervisor and other stake holders.
- Next we will discuss how we can organize the LR

Organize & Focus

- Compiled the articles/ resources what next?
- **Organize the compiled material**
  - o Chronologically
  - o Topics and subtopics wise
  - o Publication wise: e.g. books, research, review patents etc.
  - o Progression of method/ technology/ trend
  - o Unanswered question/paradox

Any of the basis may be there for organizing the literature on the basis of your subject and field of study.

- **Save and store the material smartly** (folders/ subfolders, search the raw notes, record bibliography, URLs etc.) so that you can trace them at any point of research. For example: If a matter has been copied in word file from a website, the exact URL or website of source and the date on which it was accessed should be mentioned at the end of file. Many times online pages and its content may get changed or be removed. So, note the date of access. Recording each and every time when we are collecting is a very good habit. Sometimes, we have a good content/ literature but we donot have the source noted with us. Then it becomes a very typical condition. So, make a habit to record reference properly whenever you are collecting matter. Be alert and attentive when you are collecting or searching the material.
- Focus on convergent search

Convergent search

- Beginning from broad topic and taking them to a focal point by consecutively discussing less broader points, is the convergent search.
- The search cannot be random, endless or directionless. Giving the shape to the direction of search is very important.

- The **strategy** must be planned well in advance at the very first stage of planning a LR by discussing with the research team/ mentor.

- **Begin with the historical background,** continue with the developments with time. Limit the coverage to a certain development among multi-dimensional developments of the field of study. You can not go to every dimension of your research just focus on the target dimension.

- Search should begin for broader area of research and then should be taken systematically to the focused area.

- Structuring your literature review is like an inverted cone (Fig. 15.4).

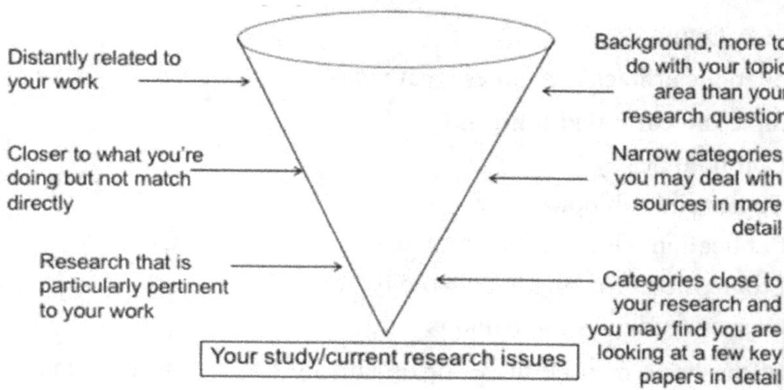

Fig. 15.4 Convergent search: Structuring literature review

At last you are having a very few key papers in your hands in the convergent search. And this goes pinpointing your research area and your aim and objectives.

At the end of the LR you will end up in one, two or three papers which you can say that these are the key papers in which direction you are going to work and you are going to follow for your work. You are going to make these key papers, the basis of your research work.

So, focusing of your efforts is called convergent search.

For example, if we want to do work on "antidiabetic formulation development of some plants"

We shall search in the following sequence:-

Diabetes as disease → medicinal plants investigated for antidiabetic activity → Herbal antidiabetic formulations → Plant survey of selected plants → survey of previous work done on antidiabetic activity on the selected plants and formulations.

Analysis

- Analyse whether the LR fulfils the criteria of a good LR like completeness, up-to-date, relevant, quality, size, focusing etc.
- Refer your outline and LS plan first. Check your plan and see how it was shaped and how it came up with convergent LS.
- You just cannot make a compilation of summaries of previous studies.
- Have a blend of summarizing and synthesizing the idea progressively.
- See the strength and weakness of studies as a critique
- Do not overlook contradictory, negative or paradox studies.
- Analyze the rationality and novelty of the tentative hypothesis and make it strong with supporting documents.

Check whether your plan is working or not?

Identify: Rational/aim/objectives

- Once you have analysed the literature systematically and technically, identify the research gap and highlight the same.
- Synthesize the hypothesis of the work
- Present the rational of the work and hypothesis
- Define the problem properly
- Plan the Aim and SMART objectives: "Specific, measurable, Attainable, Relevant and Time Bound"

Specific: Pinpointed topic. Focusing is prime property.

Measurable: The objective must have some benchmark to measure. Then only you can quantitatively compare them with other studies.

Attainable: The objective must be attainable. It should not be vague or too gross. The objectives should tangible.

Relevant: Need based relevancy is very important.

Time Bound: It should be completed in a Time frame or be possible to finish in certain defined time frame 3- 5 years.

Take Away Message

- Online LR saves time and plays a vital role in LR. Use of proper keywords and Boolean search help in effective and focussed LR and avoiding undesired/ irrelevant literature.
- Literature review must have all the key features like size, quality, versatility, correctness, completeness, criticalness, convergent etc.

- Literature review must have all the key features like size, quality, versatility, correctness, completeness, criticalness, convergent etc.

- Learning to organize and compile the LR on the basis of time frame, topic wise, publication wise etc is very much essential.

- Analyzing the LR critically leads to Identifying research gaps, rational and help in laying down the aim and objectives.

Further Reading

- Conducting a Systematic Literature Review, https://www.youtube.com/watch?v=WUErib-fXV0

- Using PICO to Structure Your Literature Search, https://www.youtube.com/watch?v=iOSWnQpVMjc

- Tips on Literature Review, https://www.youtube.com/watch?v=1lt7sS2G5MU

- Alvesson, M., & Sandberg, J. (2011). Generating research questions through problematization. Academy of Management Review. doi:10.5465/AMR.2011.59330882

- Reviewing the Literature: A Short Guide for Research Students, https://uq.edu.au/student-services/pdf/learning/lit-reviews-for-rx-students-v7.pdf

References

- Topic 0019: writing your literature review, https://snazlan.wordpress.com/2016/11/09/topic-0019-writing-your-literature-review/

CHAPTER 16

Literature Review-III
(Literature Databases)

Dr Ajay Semalty
H.N.B Garhwal University (A Central University)
Srinagar Garhwal-246174

We have discussed the basic concept and process of Literature Review (LR).We have discussed the basic steps required to do an effective and quality LR. Now, we will discuss various online literature databases available for literature review in this chapter.

Learning Outcome	Lesson Plan

After learning this chapter, you will be able to
 	
• Know and use the various online literature databases for LR
	• Major Databases
	  o Subscription based/free databases
	  o National/International Databases
	• Using these databases

Let us see the major classification of databases. These databases may be classified on the basis of subscription or free databases.

Major Databases

SUBSCRITION BASED

SCOPUS	-	MD*
WEB OF SCIENCE	-	MD
Social Science Citation Index (part of WOS)	-	SS*
EBSCO HOST	-	MD
Proquest	-	MD
IEEE Xplore	-	Engineering, CS*

Major Databases: FREE

PUBMED/ MEDLINE	-	MEDICAL, BIOMEDICAL
AJOL	-	MD
HUBMED	-	MEDICINE
GOOGLE SCHOLAR	-	MD
JSTOR: Journal Storage	-	SCIENCE
Microsoft academic	-	MD
Mendeley	-	MD -
DOAJ	-	MD
DOAB	-	MD
OAD	-	MD

DOAR (Directory of Open access repository) - MD

(Compiled from *https://en.wikipedia.org/wiki/
List_of_academic_databases_and_search_engines)*
MD: Multidisciplinary; SS: Social Sciences; CS: Computer Sciences

Scopus

Scopus is the subscription based one of the largest databases covering various subjects. It covers all types of peer reviewed resources like literature, scholarly article of journals, books, conference proceedings etc.

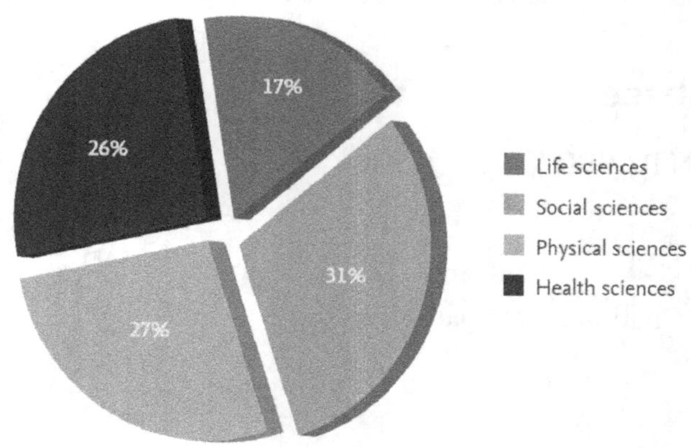

Fig. 16.1 Coverage of Scopus database

Scopus is a multidisciplinary database and it covers Science, technology, medical, biomedical, social sciences, humanities (Fig. 1). It is updated daily with 60 million records from more than 21500 journals by more than 360 publishers.

Citation detail is available for each article and Reference list with link for most of the articles.

What are the functions? These are… Searching, managing citations & my Scopus account

Let us take a sample topic*(Scopus - The Dream Research Database!, https://youtu.be/nkuHSASo-1U, CC)*

Sample topic: "What is the effect of e-cigarette on smoking behaviour?"

- Select the key words/phrases
  - "E-cigarette"
  - "Electronic cigarette"
  - "Electronic cigarettes"
  - "smoking behaviour"
- Scopus: Phrase searching and wild cards
- Phrase search:
- Double quote will search for fuzzy phrases
  - Will search for both singular and plurals
  - "Heart attack" will find heart attack and heart attacks both
- Curly bracket {} will search for the specific phrase
  - {Heart attack} will search only for heart attack
- Wildcards
  - Asterisk (*) replaces characters
  - Tox* will search for toxic, toxin, toxicity etc.
- Now go to document search
- Type the key words
- Use OR
- Click the type of document you want to search from drop down list. We selected article title, abstract, keywords.
- Click on PLUS sign to add more operators like AND
- Type next set of keywords "smoking behaviour"
- If required, Limit the search for date range, document type required

- Search yielded 48 articles, with latest on top, you can see the information, title, author, journal, abstract;

- The left navigation bar gives more choices or filters for language, year, author, subject area etc.

- Access the pdf/html of the article

Section Summary

Literature databases are available freely or with some subscription. Scopus, Web of Science, Proquest, IEEEx are the major subscription based databases. Pubmed, HUBMED, Google scholar, JSTOR, MS academic, Mendeley are the major free databases. Scopus is one of the largest multidisciplinary paid databases with various advanced search options.

Practice using Boolean search filter the articles as per your requirement practice the same in google scholar like free databases. You can use many more features like citation, analyse the search results etc. But being paid/ subscription based it is not freely available. However, if you want some full text ask for help from your friend who has the access in his/ her institute.

Web of Science (WoS)

- Web of Science was previously known as **Web of Knowledge.** It is a Multidisciplinary database (Fig. 16.2).

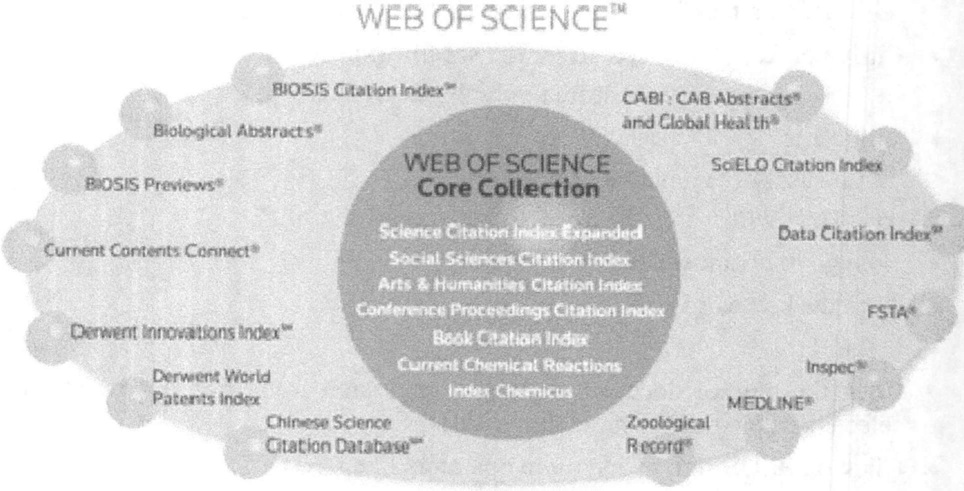

Fig. 16.2 Web of Science erstwhile Web of Knowledge: cluster of database

- WoS is a huge subscription-based scientific citation indexing service. It was originally produced by the Institute for Scientific Information (ISI) and now it is maintained by Clarivate Analytics.

- It searches conference papers, patent data, datasets and data studies, books, etc.

- Web of Science: Effective group of databases. WoS is a cluster of databases under WoS.

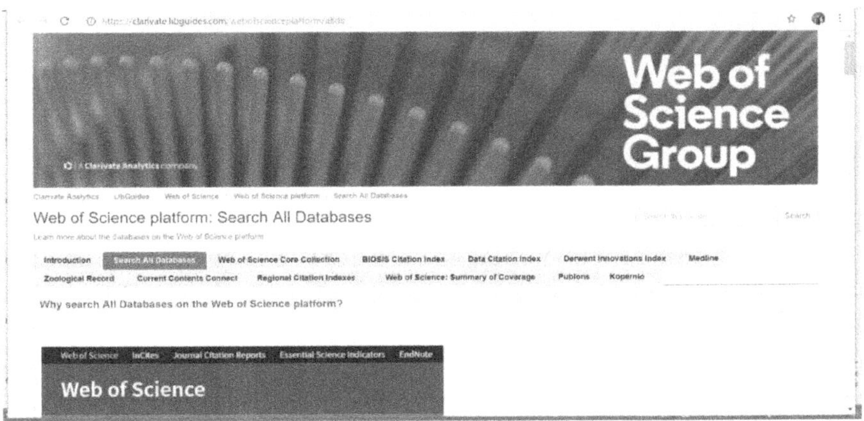

Fig. 16.3 Homepage of Web of Science

- **Searching WOS**

  Let me take you to the home page of WoS for getting the know-how of the uses of this database (Fig. 16.4).

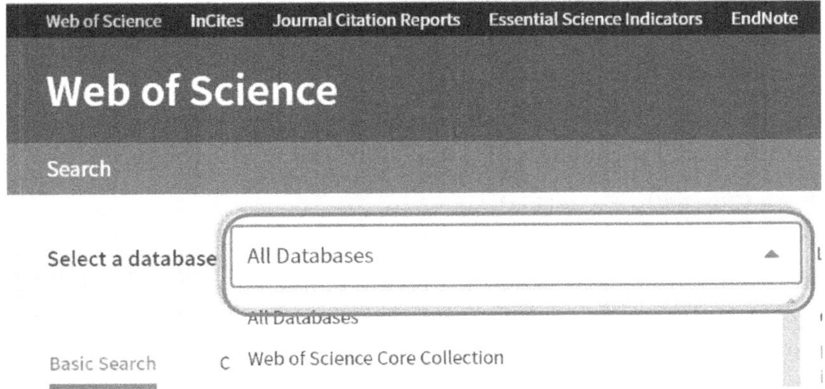

Fig. 16.4 Searching WOS: Search bar in Web of Science

This is the home page of WOS (Fig. 16.3). It has the default search window which searches all databases (Fig. 16.4). However, you can chose any specific database from the list. For example, if you are looking for Social science citation index, you can chose the core collection which includes this. You can either search the whole core collection or some part of it by selecting or deselecting on the basis of your need. Deselecting all other except the needed database shall narrow down your search.

So, you can start your search with simple topic search along with another fields on various basis.

For example you added two topics economics AND soil for search which yields 259 results from core collection.

You can refine your results on the basis filters provided like subject categories/ document types.

This was a brief introduction of WoS. The major problem is that it is not free. It needs the subscription. You can ask for the help from the friends or colleagues who are having the institutional subscription of WoS. You can send them the DOI/ title/link of the article and can request for the full text.

You can see the list of all the databases in this Wikipedia page (*https://en.wikipedia.org/wiki/List_of_academic_databases_and_search_engin es*). You select your subject specific database from the list and search the required content.

PubMed

PubMed is a free database covering citations and abstracts from the fields of medical, biomedical, health, Sciences (basic and applied including pharmaceutical sciences), life sciences, behavioural sciences, chemical sciences, and bioengineering.

This free web resource ismaintained by the National Centre for Biotechnology Information (NCBI) in U.S. National Library of Medicine.

Searching PubMed

In PubMed there is no need to use Boolean operators for searching the database just use the advance search option, the Boolean operators are in built there.

When you type some specific key words, it gives you a list of articles with most recent one on top (Fig. 16.5). You can sort the articles on the basis of best match, publication date, author etc.

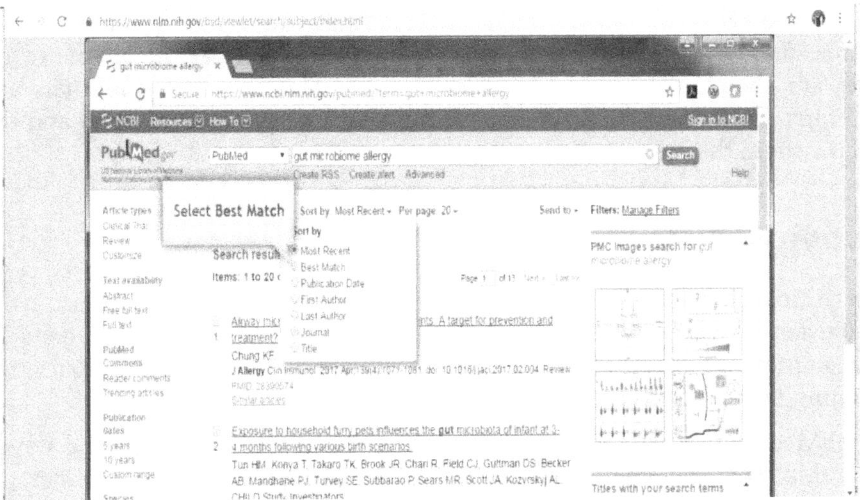

Fig. 16.5 Searching PubMed

- In the left navigation bar various other filters like article types, text availability, publication date etc. are available for refining the results.

- Use these filters for effective and focussed Literature review.

- Below each article a link to similar articles is there for focussing the search.

Other free databases

- Google scholar (already addressed): Easily and freely available with advanced filters.

- Directory of Open Access Journals: DOAJ (www.doaj.org)

- DOAB (Directory of Open Access Books)

- DOAR

Open Access articles in subscription-based Journals

Open Access Policy of journals allow free full text availability in subscription-based journals, for this the authors pay for putting their journal in open access. Funding agencies like NIH provide the grants to researchers for OA articles. It gives wide coverage, chances of better citation. It provide good reach to the article.

Many paid journals provide their one issue per year with free full text. Alternatively, many journals give place to OA articles in their every regular issue.NIH and other funding agencies give funds for open access publication of articles in subscription-based journals.

Do not exclude the subscription-based journals from your search. You may get free quality articles from the subscription-based journals also. One or two issues are always there which contain free full text from these journals also. Just enlist the journals of your field focus their OA policy and trend and then search with in the journal.

Section Summary

Web of Science is a subscription based huge multidisciplinary database which is a cluster of databases of various subjects. It covers all types of contents like book, conference proceedings etc. Various filters and features can be used for searching in the platform.

PubMed is free database by US national library of medicine and covers, medical, biomedical, chemical sciences, life sciences and allied disciplines.

Just enlist the journals of your field focus their OA policy and trend and then search with in the journal.

You can search your subject related database from the list we provided.

Indian Databases

Information and Library Network (INFLIBNET) Centre

Information and Library Network (INFLIBNET) Centre is an autonomous Inter-University Centre of the University Grants Commission (UGC) of India. It helps in modernizing university libraries in India and connecting them with high speed network. It provides quality learning and literature database for students, teachers and researchers.

This centre provides a huge variety of databases. For LR purpose I am only giving the databases relevant to LR only.

- Information and Library network centre (founded by UGC)
    o SHODHGANGOTRI
    o SHODHGANGA
    o eSHODHSINDHU
    o ICSSR DATA SERVICE
    o INDCAT
    o VIDWAN: EXPERT DATABASE
    o Refer: https://youtu.be/asdLiBbVoXI

SHODHGANGOTRI

- Repository of electronic version of approved synopsis submitted by research scholars to the universities for registering themselves for the Ph.D programme (Fig. 16.6).

- It reveals the trends and directions of research being conducted in Indian universities

- avoid duplication of research

- By many ways a synopsis can be searched, it helps in planning a research.

- Linked to full text available in SHODHGANGA

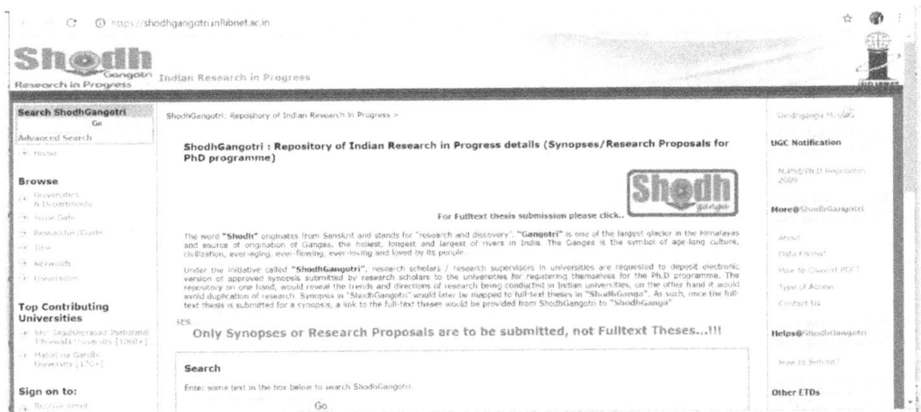

Fig. 16.6 SHODHGANGOTRI

SHODHGANGA

- Shodhganga is a reservoir of Indian Theses (Fig. 16.7 & 16.8)

- Available to the entire scholarly community in open access

- Plagiarism free content (URKUND)

- Advanced search: subject/ title/ scholar/ supervisor/ university wise

- The repository has the ability to capture, index, store, disseminate and preserve ETDs submitted by the researchers.

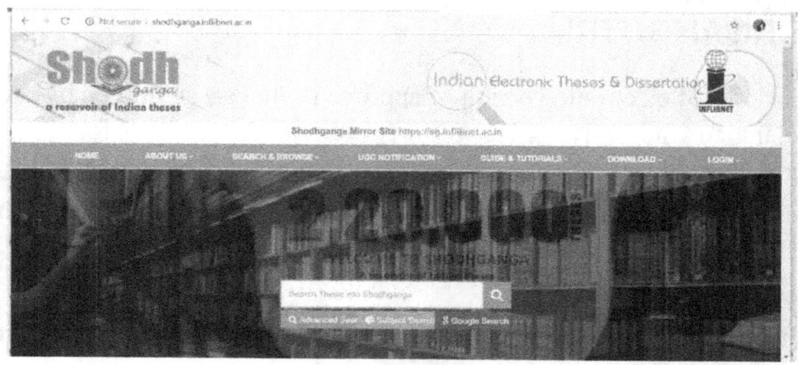

Fig. 16.7 SHODHGANGA

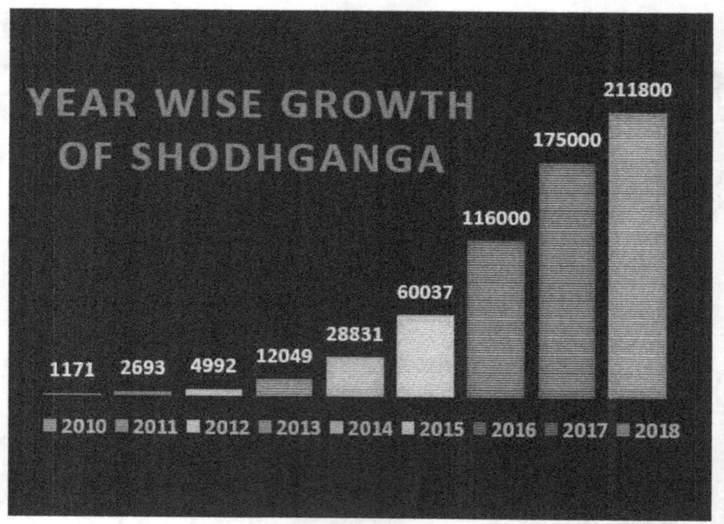

Fig. 16.8 Growth of Shodhganga (http://shodhganga.inflibnet.ac.in)

E-SHODHSINDHU

- Consortium for Higher Education Electronics Resources An initiative by MHRD (Fig. 16.9).

- "Provides access to e-resources (15,000+ full-text journals from 25 publishers, 5 bibliographic databases, 2 legal databases, 4 factual databases and one standards database) to Universities, Colleges and Centrally Funded Technical Institutions in INDIA."

Fig. 16.9 E-SHODHSINDHU by Inflibnet

ICSSR DATA SERVICE

The ICSSR granted a project entitled "Setting-up of Indian Social Science Data Repository for the ICSSR's Data Service" to the Centre (Fig. 16.10). The ICSSR signed a MoU with Ministry of Statistics and Programme Implementation, New Delhi for providing National Sample Survey (NSS) and Annual Survey of Industries (ASI) datasets which, in turn, are being handed over to INFLIBNET Centre for hosting onto the repository.

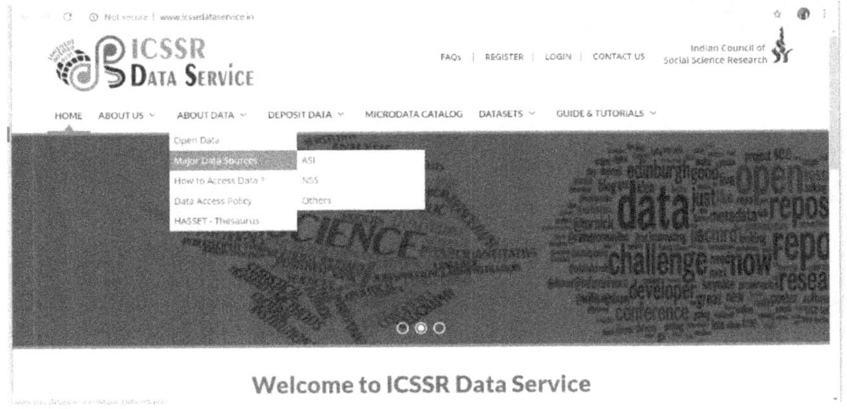

Fig. 16.10 ICSSR DATA SERVICE

National Sample Survey (NSS) is conducting nationwide integrated large scale sample surveys, employing scientific sampling methods, to generate data and statistical indicators on diverse socio-economic aspects.

Annual Survey of Industries (ASI) is principal source of industrial statistics in India

INDCAT

- IndCat is unified online catalogues of Books, Thesis and Journals available in major university libraries in India (Fig. 16.11).
- More than 1.25 crore bibliographic record of books.
- Simple and advanced search options.

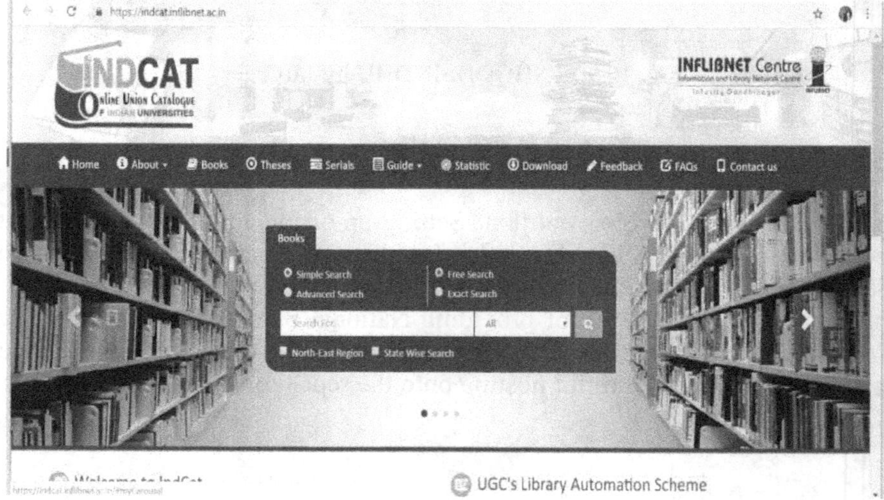

Fig. 16.11 INDCAT

VIDWAN

- It is a premier database of profiles of scientists / researchers and other faculty members working at leading academic institutions and other R & D organisation involved in teaching and research in India (Fig. 16.12 and 16.13).
- It provides important information about expert's background, contact address, experience, scholarly publications, skills and accomplishments, researcher identity, etc.

VIDWAN profile

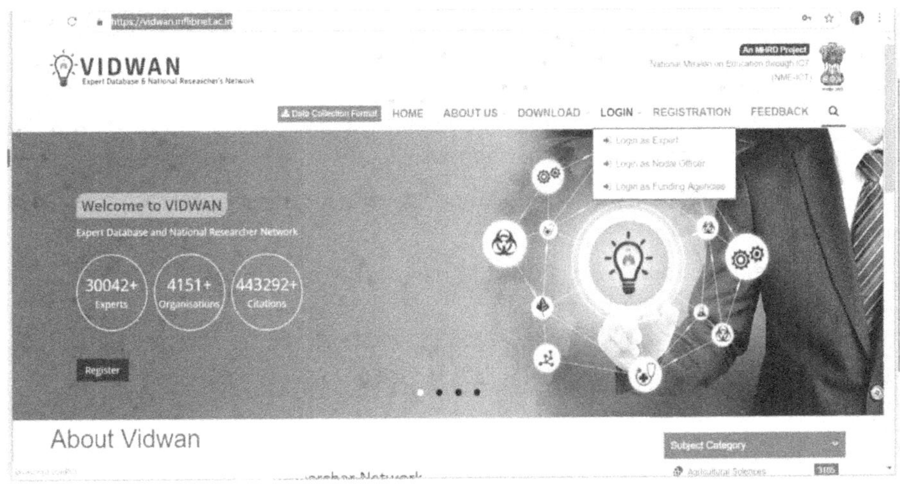

Fig. 16.12 VIDWAN database

Fig. 16.13 VIDWAN profile of author

Miscellaneous Indian Databases

- **The National Institute of Science Communication and Information Resources**, N. Delhi https://www.niscair.res.in/
- Indian Acdemy of Science, Bengalore (https://www.ias.ac.in/)

Take Away Message

- We have a plethora of databases whether paid or free. Search your institute's subscriptions or go for free databases or ask for help from authors or from the subscribers of these databases.

- Scopus and Web of Science are huge multidisciplinary subscription-based database

- Among the free database PubMed is vital for medical biomedical and chemical sciences search.

- The free national repository of Indian database includes shodhgangotri, shodhganga, eshodhsindhu, INDCAT, ICSSR data service etc.

- Do the best use of these.

Further Reading

- INFLIBNET Centre, Gandhinagar, Gujarat, https://youtu.be/asdLiBbVoXI
- Using Search Techniques in the EBSCOhost Databases, https://youtu.be/3S289Oka_Jw
- https://www.niscair.res.in/
- https://www.ias.ac.in

References

- https://en.wikipedia.org/wiki/List_of_academic_databases_and_search_engines
- Scopus - The Dream Research Database!,https://youtu.be/nkuHSASo-1U
- Getting Started with Web of Science, https://youtu.be/gpvSaq1jpVU
- https://www.inflibnet.ac.in/
- http://www.scopus.com
- http://www.elsevier.com
- http://shodhganga.inflibnet.ac.in

CHAPTER 17

Literature Review-IV
(Literature Management Tools)

Dr Ajay Semalty
H.N.B Garhwal University (A Central University)
Srinagar Garhwal-246174

In this chapter we will discuss literature management tools. We have discussed the various literature databases for literature Review. Now we will discuss literature management tools in this chapter.

Learning Outcome

After completing this chapter, you will be able

- to understand what the Literature management tools and

- to use the tools for managing your literature effectively.

Lesson Plan

- What are management tools?

- Why the management tools are required?

- Two major literature management tools with practical examples.

Literature Management Tools

These are the means of qualitative and quantitative organization and compilation of references and citations.

Need of the Literature Management Tools

These tools are a set of means which provide qualitative and quantitative organization and management of literature. Why these are needed. Let us discuss these things… very firstly.

Why they are needed?

- Easy referencing: organization and compilation

- Saves time over manual management of references

- Avoids rejection:

- Save time, effort and rejection

We have discussed a lot of literature review databases. The subscription based and free databases. A huge plethora of databases is there. What happens when we deal with the huge literature? A huge literature always needs organization and management. The organising and managing the literature is the key. A huge literature needs organization and management of all literature aspects.

The tools organize them, manage them. What happens! When we are using manual one. We use these as compared to the manual management. A few decades ago, we use to cite the literature manually. What we were doing that we were putting the citations and again manually managing them. And it becomes very typical task. It may be Ok for the research papers when the references cited are just 10, 20or may be 30. But, when we move to the review paper the references in quantitative number of 200 to 300 references, you have to manage.

And what if, you speak of thesis, a huge number of citations you have to deal with. Which reference will go where, and the things become very typical when you need correction in your thesis. Your guide/supervisor said, "Ok! remove this paragraph along with all the references. They are not relevant." Then you become puzzled.

In another case, you have communicated a paper to a journal which gets rejected somehow. Now you need to plan it for another journal which recommends different set of writing style or different set of citation style. If you are unaware of these tools as in case of earlier days, you will have to do it manually again doing the things. So, it becomes a very typical task and time-consuming task. So, the literature management tools save our time. Literature management tools make it easy for re-planning your article for a new journal. Literature management tools make it easy for you to re-plan your manuscript after rejection to another new journal. Literature management tools provide error free bibliographic styles with respect to the accuracy, style and citation location. These are very important aspects. As you come to know if you reshuffle a thing, if you are not aware of these tools, you will be lost there in the literature. It Just like sinking yourself in this ocean especially, if the body of your article is huge, body of your thesis is huge.

Section Summary

So, ultimately if I summarize that Literature management tools are certain means which provide quantitative and qualitative managements of your references and citations in your text. They save your time, efforts and avoid the rejections.

In the next section, we will be explaining how the literature management is done by two major tools.

- Mendeley
- Endnote.

If you have other sources, you can try these things from those management tools also.

So, just enjoy the practical demonstration of using Mendeley and Endnote as Literature management tools.

Mendeley as Literature Management Tool

So dear learners, let us discuss and see the very first literature management tool. Mendeley.

This is the website of Mendeley (Fig. 17.1).

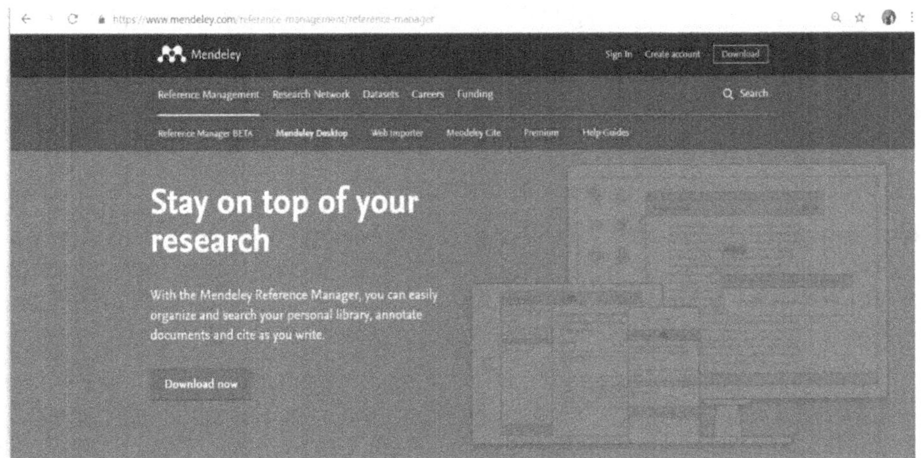

Fig. 17.1 Mendeley's Home Page

As the website states that it helps you to stay on the top of your research.

Cite as you write

With the Mendeley the reference manager you can easily organize and search your personal library, documents and cite as you write. When you move, scroll down the website down, the first thing the Mendeley does for you as you write it generates citations and bibliographies in a whole range of journal styles with just a few clicks. For your convenience the Mendeley have made Mendeley citation plugin and it is compatible with word including word format and library office. Mendeley also supports Bibtex export for use with latex.

Annotate as you read

Mendeley annotate as you read. It easily adds your thoughts on documents in your own library, even from mobile devices. For ease of collaboration you can also share document with group of colleagues and annotate them together.

Import references easily

Now the next function of Mendeley is imports references easily. Mendeley import papers and other documents from your desktop your existing libraries or websites quickly and easily. Mendeley automatically captures author, title, and publisher information.

Access your library anywhere

The next function is that it provides access your library anywhere. You can securely access mendeley on any computer through the desktop cline, web browser or using your mobile app. Your library is backed up in the cloud when you sync and is always available to you.

Now scrolling down you can see this website. Let's move up again... and go to the Mendeley desktop or you can see the reference manager beta is available. Mendeley reference manager beta, you can download it freely.

Searching publications/authors by Mendeley

Now moving ahead, we can sign in to the Mendeley. With your username and password. So this is the users' dashboard of Mendeley. You can go to the search and type your search item. As we typed Cyclodextrin complex and clicked on search. It yield 8324 items, scholarly articles for 20 pages.

We can scroll down and see all the results, all the articles which have we got after the results and we can see that the first digit is the number of readers for this particular article say for example these are the Mendeley users who have this article in their library. There is 474 readers have this article in their library and the second column indicates the citation of this article that is in this case is 1.7K citation this article is having.

Now again moving up in this search, you can add a particular article in library into your own library. You can select an article, click it, it will directly be transferred to your library. It will be stored in your library, you can scroll down, see all the articles, go to the further pages, if required.

Now let us move upward again and then see people. If you want to search a particular author, a particular researcher, type his or her name in the search bar. You will find the names. Say for example: for this surname: Loftsson, we have two author profiles. These are the similar kinds of author profiles which we have studied in case of google scholar, and Scopus, or Web of Science.

So, for this particular author, we can see all the publications by the author, his/or her H index and total number of papers, each paper's readers and citations are available. You can scroll down again; you can follow this particular author for getting the timely alerts on the publication of his/ or her researcher review papers or any kind of publication which are tracked by Mendeley. You can again go to the group of topics. What type of topics you are interested in say for example I entered Cyclodextrin, the cyclodextrin groups, so these are three kind of groups. Three groups are there, you can select, you can see the number of members there. Say for example we entered economics for a particular subject. We saw in the name of behavioural economics, there are 533 members for this group...so what this group provides...if you go to the list of groups..you can select the particular groups which is related to your research or your field of work. You can select and join a particular group if you are interested. Say for example we entered in this particular group behavioural economics. You can overview the members and kind of discussion forum among the members of the particular group in which the discussion is based on.

And you can share the articles with the help of these platforms and ask for the help from the peers about the field of your research. You can see the various databases are there. The journals and you can add the related articles on to your library from this group also.

Now we enter into the library. This is the Mendeley library on your dashboard in your user dashboard. Remember the enrolment or the registration in Mendeley is free. When you enter into the library you can add and create a folder. You can upload the document to your library. You can also create new references manually and import other libraries like EndNote, xml, BibTex, and RefWorks.

Exporting Literature from Mendeley Library

Now we can see all the folders. Say for example for our research work we have made a folder of spectroscopy

For example, we have prepared a folder of SANS 2016 indicating the research articles small angle neutron scattering related. The green color indicates that we have read it and the PDF is there for this particular article. So we have the three radicals in this folder and on the small angle scattering we can click them. And add, remove or export to MS Word function is there. We can click the article and export to MS Word by clicking it. We just saved it into a particular location for example we saved it in the desktop, this xml file which is generated, and this is gone to the desktop, this is saved to the desktop.

Importing literature from Mendeley

Now, say for example we open the word file for using this reference in the article, we clicked to references and then insert citations, manage sources. what we will do in the manage sources. We will browse locate that particular file which we saved in the desktop again and you can say we imported these references and you can add newer/ more references also. You can select them, copy them and then they can be imported to your file.

Now we copied all the references one by one to the current list from xml and then this is ready for importing into your file. Now we start inserting the citations one by one you can insert the citations into the word files by selecting it from the list and we can group them into a single group within the parentheses from the list. So individual or group way, both the way we can select. We can select the bibliography and then the references can be prepared depending on the style. Now we can update the citations also depending upon our requirements.

Section Summary

So, it was all about using the Mendeley, importing. It was all about Mendeley exporting the references from the Mendeley to the library and importing the references from library to your document or manuscript and managing it.

In the next section we will be discussing the Endnote as the literature management tool.

Endnote as literature management tool

Dear learners, now let us see how we can manage the literature with the help of Endnote. This is the website of Endnote (Fig. 17.2).

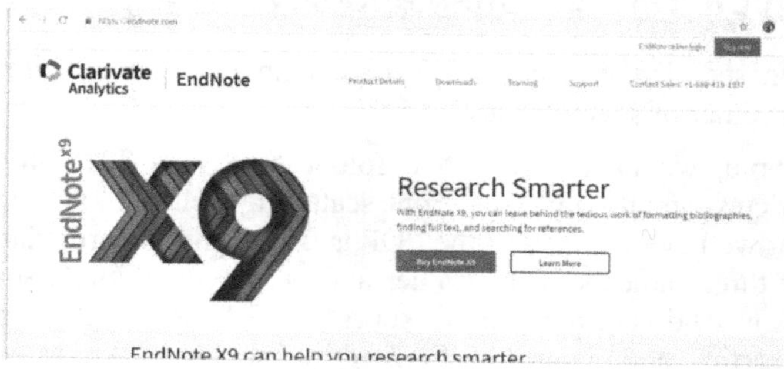

Fig. 17.2 Home Page of EndNote

Utility of End Note

With the help of Endnote, we can do couple of things.

- First is to find, collect references by searching online databases or importing your existing collections. You can search online database with the help of Endnote, create a reference manually import references and find your best potential journal with the help of endnote.

- Secondly you can store and share, you can organise group references in any way that works for yo then share your groups with colleagues. You can create a new group, share a new group and find duplicate references with the help of Endnote.

- You can create the plugins of Endnote. As the website says to form a bibliography and cite references while you write. The plugin is available. You can create a formatted bibliography with the help of Endnote and format a paper. So, Endnote does these three major things for you.

You just need to explore each aspect by going through their website.

Searching online database

Now let us enter into the search online database. In the step one, select the database or library catalogue connection. For example, it may be British Library, US National library of medicine and PubMed. So, we selected anyone i.e. PubMed then we connected it, we entered the keyword in PubMed searchbar. We can change the database or library catalogue connection by clicking here, what we will do? We will write the keywords in the bars. Say for example we entered solubility and Diclofenac in the two different bars and in the field, we selected titles for both the keywords and click the search. This search yielded six results. We can see that the latest reference the latest article of 2018 is there which is on top. which is having solubility and Diclofenac both in its titles. As we scroll down we can see in the second one also, both the terms there, in the third article also both the terms there in the title and so on in all the six articles.

Exporting literature from EndNote

We can select the number of article per page by this section, by clicking either 10 per page, 25 per page or 50 per page; or you can also refine your search with respect to the title in both the terms or with respect to the keywords. You can select one two or all articles or all the pages or you can create a group. Say for example, we created a group, Diclofenac solubility and click OK. So, what will you have?? You will have a group in which all the results have been stored in this particular library. They have been copied to your folder and they are in your group now on your dashboard.

Now, what we can do, that we can see each of this article, view the figure attachment. What is the unique feature, that you can add a figure there into that particular article, if this article is not available infull text or you can add a particular figure to this article in your library here and see it. The group members can view this record and can see the figure attachment or attach any new figure. You can also attach any kind of file, any further or more information or more dataset to that particular article for your ready reference as a literature review part.

Now, what you will do, that you can select these references, check them properly then you will see that all references you have you want to export or not…if you want to delete a particular reference, you can delete from here only, which is not of your use. You can scroll down and refine your result with respect to the authors. You can sort the list on the basis of authors. You can go to collect for adding the new references, the references which are not there online, you can add them manually in this section by entering all the details. As you can see on the screen all the results need to be filled. After filling this you can import these references into your libraries. You can also import the references which are not there in the endnote.

Importing References from EndNote

Now, let us see how we can import the references into your library from any platform. The import options are available with the help of Endnote import, google books, Library in congress, PubMed. So, by these you can import and to which directory you want to import. You can select the file from here and then import from that particular platform. Let us see, that for example the particular platform is Google Scholar. Managing functions are not there in these platforms. So the Google Scholar has got the very wide coverage. So, what we will do? We will enter the keywords into the Google Scholar search bars. Say for example, solubility of drugs we entered. We got this much results. We selected. Click the star because it is saving, keep on selecting the desired articles and when we finished with that what we will do? We will go to My Library and then when all these articles are have come there, I will click all and then click to export in the export section you will have the options of BibTex, EndNote, RefMan and RefWorks. In any of the form you can export these. So, you can select any of the form and to which you want to export

So, you go the My Library. Here your solubility of drugs. This is a kind of folder in which all your research results or search results are there and if you want to import one or all of these articles you have to click, select them and export this folder or all the articles into any of the platform if you want to export them.

For this time, we want to export them to the Endnote, so we will select Endnote. Now, we will save them into a certain location. You can also export them in the BiBtex format as shown in the screen now. Now we can save these formats. You can also select RefMan and RefWorks for exporting these. Now let us see that how we can import them into the platform of Endnote. Now we have exported from the Google Scholar. We have a file there which is ready to be imported. What we will do? We will select the particular file from this platform which we have exported from the Google Scholar. Select that particular file, that citation file, we will import, we will put the import option, Endnote import and that particular file.

File format is important

What we want to tell you that a file format is very important. If you select a bibtex format for importing into the endnote, the platform won't take it. Say for example, we selected the citation that bibtex format but it won't be imported.

So, what we have done so far, that we have exported the file in the two different file format. The file format is very important. Now we have exported the things. We want to import into the Endnote. What will happen that, if you want to import it into Endnote format, if you want to import a bibtex file you can't import it into the Endnote.You have to select any other option other than the Endnote import for selecting a bibtex file. Say for example that zero references were imported to this particular group, why? Because we didn't select the right platform in the tool section. What we have to do that we have to choose the suitable file format. Say for example this is the that Endnote format we will have to select and then the Endnote import will take if we will import only in this case as we can see now here the 8 references were imported to this group Diclofenac solubility group. Now it was successful. So, you will have to assure that your source and destination format is similar and compatible. This compatibility is very important.

Now we will move to the Diclofenac solubility group, this particular library/ that particular folder in this particular group we have imported the references. We were having the references from the endnote and we imported some more references from the Google Scholar. So, this is the entire list. Now, again check this list and now this much references have been ready for using in your manuscript. We can select them or deselect them depending upon the requirement. We can format them depending upon the bibliography. We can select the references of that particular folder, we can select the bibliographic style, whatever the bibliographic style is needed. Say for example, in this case we selected the Vancouver style, the file format we selected the text and then saved. We can see the preview of these references. Now they are in Vancouver style. You can see. You can print this page or select this page COPY and PASTE it to your required location in your manuscript.

Let us take another example, if you select the IEEE bibliographic style, you can see the preview, print it, copy it, or paste it to your required location in your manuscript. In the same way any kind of file format can be selected. Say for example APA 6 style we selected here and the file format plain text, again the file format will change automatically in just clicks. You can select and change the references styles. In the Endnote website you can use these plugins. This plugin uses the endnote plugin to insert references and format citations and bibliographies automatically while you write your papers in word. These plugins also allow you to save online references to a library in internet explorer for windows. So, this plugin is very important. Now, what we can do, we can move to the format paper section here. We can format our paper or export the references whenever required. We can select the group or the folder, say from your dashboard: say for example diclofenac solubility, the same folder we selected.

And we selected the export style what style we need say for example we selected that the style a particular style and then we can also go the match section. What is going to be matched here and find the best fit journal for your manuscript when you put your title abstract or the references by entering these data, this will tell you the suitable journal for your manuscript. So, it helps also in journal targeting for your manuscript. Other features involve the collection of references.

Summary

So, as we can see that endnote provide a good blend of tools or sub tools, which helps you not only in importing the references and bibliography into their platform but also exporting this to your work and overall managing all your references. It helps you in saving your time and converting the details of all the references in a particular folder or in their particular database. It allows you to select a journal if you enter the information like title, an abstract. So, it helps in journal targeting also.

So, we can use with the help of other databases also we can enrich the endnote literature management tool, export the data, export these references from Google scholar like platform to the Endnote and use them there. So, it is a comprehensive literature management tool.

Hope you have enjoyed, and you will try to utilize this literature management tool for your academic writing.

Thanks for your attention.

Further Reading

- Comparison of reference management software, https://en.wikipedia.org/wiki/Comparison_of_reference_management_softw are
- Importing Documents into your Library (Mendeley Minute), https://www.youtube.com/watch?v=qRiAIaqdAOg
- Organizing Your Library (Mendeley Minute),https://www.youtube.com/watch?v=VD1z0boSpQY
- Generating Citations with the MS Word and OpenOffice Plug-ins,https://www.youtube.com/watch?v=zkrVbBSrK_w
- https://clarivate.libguides.com/endnote_training
- About Endnote, http://clarivate.libguides.com/endnote_training/home

References

- https://endnote.com/
- https://www.mendeley.com/reference-management/reference-manager

CHAPTER 18

Review Paper Writing- I

Dr Ajay Semalty
H.N.B Garhwal University (A Central University)
Srinagar Garhwal-246174

Dear learners, we have discussed Literature Review as the foundation of a quality research and process of literature review in the previous chapters. Let us learn writing a review paper in this chapter.

Learning Outcome

After learning this chapter, you will be able to

- Understand the importance of review paper
- Plan a review paper
- Write a review paper

Lesson Plan

- What are review papers?
- Importance of review papers: features, timing, type
- Planning a review paper
- Commencing a review paper writing

Publications in Academic and Research

Who does not want the publications in CV in academic and research career?

We have discussed the importance of publication in very first week. Publications are important for all the stakeholders of academic environment i.e. students, researchers, teachers, and institutes.

We have discussed the individual importance of AW for all the stakeholders.

In general, the publications are the benchmark of expertise or level of academic proficiency for all the stakeholders.

Common Types of Publications

Review paper and research paper are the most common type of publications. Between the two major types, research papers can only be written after systematic research but the review can be written just by going through the body of knowledge of the field only. It seems easy but actually it is more challenging than writing the research articles.

What are review papers?

Reviews are the mirror of current status and the path finder for the future trends of a particular area of study. Student can understand Review paper as advanced form of their lecture notes.

"Systematically organized compilation of a specific body of knowledge which aims to critically study, summarize and synthesize the author's opinion or highlight the development of field with time."

Think, how do you plan for preparing your lecture notes. You employ the various kind of inputs: Class notes, books (2 or 3 atleast), pictorial, graphical presentation/ infographics, to the point coverage of topic, defined outline for each topic or subtopic. Review paper, in principle require the similar things but the quantity and quality of inputs are advanced. If you know how to prepare the quality notes then you will find it easier to learn writing a review paper.

Importance of writing reviews

* Ready reference for students
* Ready reference for new researchers
* Develop understanding of the knowledge domain
* Integration of rapidly advancing knowledge
* Elucidate trends in research and point to research gaps
* Help in origin and incubation of idea of research
* Help in improving h index of authors
* Help in policy making

Ready reference for students

Review provide a ready reference of knowledge for beginners in research or to students. Students get the Condensed, power packed compilation of the body of knowledge in one document only about the related field or topic. And this is the utmost need for them. For PG students, reviews provide a quality insight of the topic. It is better, if they study from a text first and then study from a quality review paper. Please, remember to choose the quality review paper from quality journals only. Otherwise, it would be a disaster. Because a lot of TDH journals are filled up with so called review papers.

Ready reference for new researchers

Research scholars who are planning their research topic and even the experienced researchers working on any field or subfield of research for first time need to refer reviews without fail. Review help them to plan their research. Reviews are also very helpful for multidisciplinary researchers. E.g. I am a pharmaceutical researcher and iam doing research on some drug delivery composites through small angle neutron scattering in DHRUVA nuclear reactor at BARC Mumbai. So for me to understand neutron scattering, reviews and the experience of my collaborator is the ultimate source of quality knowledge about the field. I can study the text of solid state physics for that but it would not give me the precise information and in condensed form.

Develop understanding of the knowledge domain

Reviews play an essential role in developing understanding of the knowledge domain. Well-written, critical reviews provide a necessary overview of the huge body of knowledge. It is like getting the crude oil (collected from the depth of ocean) in the refined form. We develop understanding by reading the review papers. The review provide the collective knowledge in easy to follow manner.

Integration of rapidly advancing knowledge

In the rapidly advancing knowledge field you cannot rely on the books only. You have to refer the current research and review papers. Though current research paper reading is the answer to the problem. But to save time and to integrate the rapidly advancing knowledge effectively in a specialty or subspecialty, reading review is the best choice.

Elucidate trends in research and point to research gaps

Review papers can elucidate trends in research. Existing research gaps and unanswered questions can be identified through reviews. And hence the reviews show the opportunities for future study.

Help in origin and incubation of idea of research

Review papers lay down the foundation of research. After identifying the research gaps, reviews provide vital inputs for the origin and incubation of idea. Review papers help in establishing the rationality: what & why?

Help in improving h index of authors

We have studied journal and author metrics earlier. We know how the number publications and number of citation contributes to the h index of an author. You can see in any author profile in google scholar or any other database that most of the times the reviews get the maximum citations. Reviews are cited more as compared to research papers due to obvious reasons of wider coverage and readership. The authors who write quality review on the hot topics of the research field are cited more and hence their h index increase sharply.

Help in policy making

Reviews also give science policymakers as well as researchers a clearer insight into the potential importance of emerging knowledge. They are like the editorials of a news paper. They give the central idea or the key idea and the key message in the effective manner.

Section Summary

Publications are vital source of information for academic stakeholders. Reviews are the mirror of current status and the path finder for the future trends of a particular area of study.

"Systematically organized compilation of a specific body of knowledge which aims to critically study, summarize and synthesize the author's opinion or highlight the development of field with time is called as a review paper."

Reviews are ready reference to students and researchers; develop understating, integrate the knowledge, lay down the foundation of research, identify the research gaps, increase the authors reputation and help in policy making.

After the introduction of review papers and its importance for major stakeholders, let us discuss the major features of review papers.

Major features of review paper

- **Uniquely demanding and intensive task.**

  It is not an easy task. It requires efforts and time. We mentioned that you need not to got to laboratories for writing the review papers. But you require lot more than that in review papers. Most of the researchers find it easier to write research papers and not the review papers. Most of the experienced authors runaway from writing review papers because it requires immense efforts.

  [Two decades ago, tracking down relevant references to a subject in libraries was a major task. But now the availability of online databases has changed the scenario significantly. Now searching the topics has become both feasible and affordable. But still writing reviews is a uniquely demanding and intensive task.]

- **Synthesizes ideas**

  A review paper does not describe the author's own work generally, but rather synthesizes ideas and results from other research papers that have been published in a certain subject area. However, analytical or critical reviews may have some of the author's work. A review is not just the compilation of studies it is the analytical or critical view of the study area i.e. strength, weakness, opportunity, and threats (SWOT analysis).

- **Written by experts:** Generally a person who has the experience of research on the particular area of study should only be expected to write a review. Supervisor or mentor role is important, if students or research scholars are trying to write review.

- **Focused approach**: The article must address or focus a particular specific research topic. It can not have very wide scope. More focused the approach, more it would be effective.

- **Recognition to authors (paradox):** Though it helps in increasing impact of an author or the h index, a scientist avoids writing reviews due to the lack of professional recognition. Promotion and tenure committees still give more weightage to original research articles over the review papers. Funding agencies also adopt the similar attitude in decision making.

- **Completeness:** It should be complete in nature for the specific domain being covered. It cannot not completely cover the subject but it should cover the particular topic completely.

- **Triggering the future research:** It should have the potential to trigger the new research in future and should pave the way or originate the new idea for the new research.

Time to start

We will discuss the aspect in a separate chapter also.

"DO NOT WAIT TILL THE END OF YOUR RESEARCH."

Writing review paper should be started in later stage of planning of your research topic itself. At the stage of LR where you identify the problem and outline the tentative aim and objective review writing can be started (Fig. 18.1). And the writing process can be done along with the continuing LR.

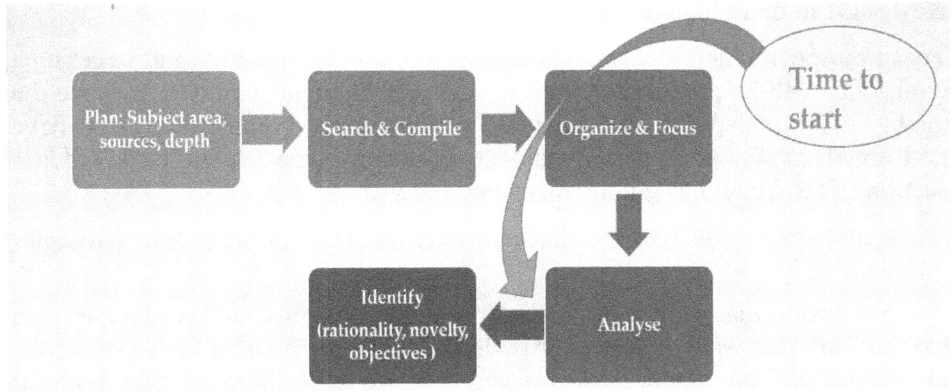

Fig. 18.1 Time to start writing a review paper

If you are not doing research, no time constraint is there, just start after doing proper LR on the desired topic. Faculty members can also plan reviews from their academic experience. It may serve as advanced compilation or overview: advanced lecture note style reviews: desired by some journals/ thematic issues

Process of Review Paper Writing

After proper LR you start writing review paper. The steps of an effective review, we have already discussed (Fig. 18.2).

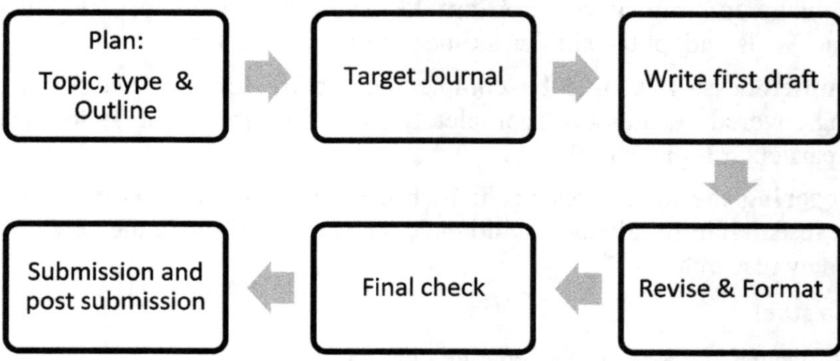

Fig. 18.2 Process of review paper writing

Plan (Topic, type and outline)

First identify the field/ topic and draw a rough outline of your paper. Discuss with your supervisor/ team members/ coauthors. Also discus and plan what type of review you want to write?

Key point in drawing outline

Try to be specific and focused while pinpointing your objective and targeting a topic. This will help in organizing LR, and your thoughts and will improve the quality of your paper. The review is not always the complete LR you have performed. Donot copy paste your LR in review paper. It is just a specific section of the LR. It has got the different approach.

Do not try to cover everything and be specific in addressing a particular topic.

Also ensure that it must be interesting to gain attention and readership; and have an optimum size of scope for reviewing. The topic should have the scope of reviewing. You cannot review a very narrow topic. So, sufficient body of knowledge must be there.

Section Summary

We focused of the major features of the review papers in this section. It is a uniquely demanding and intensive task, written by expert; It has got a focused approach. A paradox is there that reviews are not given the due importance over the research papers. Completeness and triggering to future research is required. We studied when to start the review papers. At the end of compiling organizing analyzing and identifying the aim and objectives we can start writing review papers. After the LR we can plan. Then we discussed planning. We plan a tight outline of your topic and subtopic with supervisor and your team members.

At the time of planning we mentioned to plan the type of review paper you want to write. Let us discuss the various types of review papers.

Type of Review Papers:

You must define the type of review on the basis of requirement of your subject/ discipline and level. The reviews may of following

- Argumentative Review
- Integrative Review
- Historical Review
- Methodological Review
- Systematic Review
- Theoretical Review

Let us discuss these types one by one.

- Argumentative Review: This reviews are used in social sciences research to develop a body of literature that establishes a contrary viewpoint. It is like challenging the established or deeply embedded assumption in the literature.

- Integrative Review: It is about generating a new frameworks and perspectives on the topic in an integrative way by developing the identical hypothesis or studies. It is the most common type of review in Social Sciences or lecture note style in sciences.

- Historical Review: These reviews focus on examining research throughout a period of time, its evolution, current development and prediction of future development.

- Methodological Review: When the focus of analyzing the studies are not just based on its conclusions but also cover the methodology adopted in the referred literature/ study, it is called methodological review. It means what methods adopted, how they were used to draw the result and discussed etc. These are critical in tone and review the studies thoroughly.

- Systematic Review: It is the overview of existing evidence in systematic manner so as to study the cause and effect of the main research problem. In sciences and life sciences, often this kind of review are written.

- Theoretical Review: These reviews are written to examine the corpus of theory that has accumulated in regard to an issue, concept, theory, phenomena. Which theory has developed in which way and what is the current status of the theory? Or whether any new theory emerged from the older one. These kinds of approaches are used in these studies.

Step II

Target Journal

- **Your objectives and expectations:** See what is your object and expectation. See what you want? Whether you need the impact or readership or both. You have written a very novel and the best article in your field. And you want to publish it in the best journal of the field. Can you wait for that span which that journal is taking ? or you just want the good readership? See indexing or Impact what is important for you? So identify the journal as per your need.

- **Match your topic with scope of journal:** Make a tentative list of journals on the basis of first factor. Check the scope of each and every journal of your list. If it matches with the scope of your paper then only you go for that journal. Otherwise go for searching the next journal.

- **Journal metrics:** Impact, speed and reach, we have already discussed. Your decision may vary depending upon your need.

- **Article Processing Charges (APC):** See if the journal charges for publication or processing or not. If yes, can you afford it? This cost factor is very important.

- **Level of Review Articles in the Target Journal:** See the level of reviews in the journal you are targeting and assess your level review. If it is in good tune then only you should target. You have prepared a review article on nanotechnology in drug delivery and planning it for publishing it in "Nature". First see the level of review articles in Nature. If there are the reviews on the topic they all might be by invitation only. Check whether anyone can submit a review or it is by invitation only in the journal.

- **Author's Guidelines:** Search some related article in the journal. See the guidelines for authors and general submission guidelines. If it suits you then only target the journal.

These are the major factors of targeting a journal. Enlist your expectation and match them with journal's aura. If it matches then only target the journal.

Step III

Writing First draft

Within the outline write a story, develop it properly. See some recent articles on the closely related topic/(s). See, what more you can add or how differently and more effectively you can present the body of knowledge? Think from your own brain. Do not see the topic from other's view point. You must put your own viewpoint at the first place.

Develop your own style and voice. Write the first rough draft. In each and every case, it might not be the case that you are the first person to write on the topic. See how you can be unique. See other similar articles. What extra knowledge you can add in the domain of the existing knowledge? What new the reader will gain; think, note them down and include them in your first draft.

Step IV

Revise &Format

After writing the first draft leave it for some days (but not weeks). Keep thinking. Come back to it and revise it as much as you can. Give it to someone to read. And seek advice and inputs. Revise the paper again. Identify what and where is the gap in the story you are trying to tell. Keep on noting down references for citation (Literature management). We have discussed the literature management previously in a separate chapter. Consult with your team/ supervisor, brainstorm. I cannot say precisely how many times it needs revision. It depends on your LR, subject and level.

Keep on formatting the article as per the author's instructions. Follow the writing style recommended by the journal. Stick to the guidelines by the journal. You can avoid rejection by following the guidelines also.

Step V

Final Check

After revising the final draft many times, it is the time of final check. Get the final draft approved by all the coauthors. Discuss with all the coauthors. Try to make it error free. Read the article aloud. See for factual correctness and all types of formatting correctness. When all the authors are satisfied go for the next step: submission.

Step VI

Submission and post submission

Submit the article to the target journal with all required files (images, tables, copyright statement) and formats in proper way (offline or online). Then wait for the peer review process to be completed. Post submission deals with the addressing the review comments and proof checking etc. The submission and post submission part shall be dealt in detail in separate chapter.

Take Away Message

Review papers are the systematically organized compilation of a specific body of knowledge which aims to critically study, summarize and synthesize the author's opinion or highlight the development of field with time.

Reviews are ready reference to students, faculty members and researchers; developing understating, integrating the knowledge, laying down the foundation of research, identify the research gaps, increase the authors reputation and help in policy making.

We discussed the major features of review papers like uniquely demanding and intensive task, written by expert; focused approach, help in policy making etc. We discussed the timing of writing review that is immediately after completing literature review. We studied the six step process of writing review papers. Planning (outline/type), targeting journal, writing first draft, revising and formatting, final check and finally submission and post submission.

Further Reading

- Guidance of a review paper,
  https://www.journals.elsevier.com/international-journal-of-machine-tools-and-manufacture/news/guidance-for-review-papers
- Literature review paper,
  http://websites.uwlax.edu/biology/ReviewPapers.html
- https://www.springer.com/gp/authors-editors/journal-author/journal-author-academy/15186
- Fundamentals of manuscript preparations,
  https://researcheracademy.elsevier.com/writing-research/fundamentals-manuscript-preparation

References

- Anson, Chris M. and Robert A. Schwegler. The Longman Handbook for Writers and Readers. 6th edition. New York: Longman, 2010.
- Types of reviews,
  http://library.atmiya.net/research%20commons/researchguide/literature_review.php

CHAPTER 19

Review Paper Writing-II

Dr Ajay Semalty
H.N.B Garhwal University (A Central University)
Srinagar Garhwal-246174

Welcome dear learners, we are here with the next chapter on review paper writing. We learned about the basic concept of review paper writing, its importance to all the stakeholders, major features, timing and process of review paper writing in our previous chapter. Now we will discuss the practical aspects of review paper writing with each and every domain of review paper in this chapter.

Learning Outcome

After learning this chapter, you will be able to

- Understand the practical approach of review paper writing
- Write a review paper
- Communicate to a good journal

Lesson Plan

- Identifying/ targeting journals for review papers: tips
- Type of review papers
- Writing effective title
- Author's details
- Key words
- Abstracts writing
- Writing introduction effectively
- Body of paper including Infographics
- Conclusion
- References
- Some tips with practical aspect of review paper writing

Let us open up the chapter.

It is not a lecture rather it is going to be a discussion. There is no thumb rule. It is an art as well a science to write a review paper. We have been telling this thing again and again since the beginning of this course. When you have written a review paper: First draft, revising and come up with the final check. Identifying journal now (not repeating, just summarizing it). When you are targeting journals in initial stage.

Tip 1. Donot Target Top Quality Journals

When you are targeting a journal, It is better for beginners to not to target high end journals or top quality journals of the field. Because you won't get the chance to publish your work and that to a review paper IN A HIGH-END JOURNAL. You might be a good writer. Iam not discouraging you. Review papers are expected to be written by the experts only. In many journals reviews are by invitations only. They are considered or they are submitted only from the authors by whom they are invited and they invite the experts of the particular field.

Go for Multidisciplinary or Interdisciplinary Journal

So, it is better to be active regarding the identifying journal, go for little bit multidisciplinary or interdisciplinary journal which are not exactly focusing your things, but okay, they cover your article and your article fit in the scope of the journal. It will be the best choice and obviously, if you are expert, you don't need my advice you can submit to any journal. But, if you are writing, if you are good in writing research papers, still you need special tips and techniques for writing effective review paper.

So, this was about identifying the journal and I will focus my effort for beginners only. For faculty members they can try high end journals. You will get the rejection no problem. You will get rejection in 48 hrs, may be. Good journal do not with held up the things for longer period of time and within 48 hrs you will get 4 to 8 pages of comments by the editors. I will go into the domain that rejection is also a good thing. It tells you what not to do, which improves your chances in getting published.

With respect to the identifying the journals, if you want to learn and it is not your aim to get your work published as early as possible, then and only then you go for high end journals. They may respond you after 3 months, might be possible. But they will assess your work very sharply and you will be given with a couple of pages of comments, on the basis of which you can improve the quality of your manuscript. And perhaps in next journal (obviously not to that standard but little bit of average value) you will be able to publish your work. This is the key.

We have identified where we want to target our manuscript with the help of some tips. These are the common experiences of various authors which I am compiling.

Title and Keywords Selection

Titles, some peoples say it doesn't matter, some people say it is utmost important. But I think, it has got the vital value. Many times, a gross title if you give to very important and vital article, it is just like giving a wrong address to a nicely written letter. YES, it is. It is like that only. What you have to plan? You have to plan a crisp, short and the representative title to your review article. Because, you have planned it already what you are going to address? So you can plan it, no problem. Sometimes, it is very important to have a good title, well in advance, even before completing your review paper with respect to those journals which require a proposal of a review paper before getting the manuscript submitted by author. Author has to apply that I'm going to write this review paper, this is title and this is abstract, so shall I submit it? Because this is the process in journals in which review papers are by invitation only and the editor. And what the editor does, he/she evaluates the potential of the prospective article on the basis of the title and the abstract. And then agrees/ give consult or denies depending upon the merit or limitations.

So, that's why in that case advanced title is necessary. **But what I feel that if it is not utmost essential, don't finalize the title well in advance.** You will say that I'm saying the different things altogether. No, it depends on the conditions. If it is not utmost required, plan it at the last because your review paper outline will deviate little bit when you are finalizing the thing. So, your title must be representative of your paper from beginning till the end. And you must also see the title of related review articles and try to make it unique as well as little bit similar so that the people who are citing/ referring those articles will be able to track down your article also.

And this holds true to selection of keywords. You have to give keywords. **Keywords** selection must be like that (the key to this)**don't use the same words which are there in your title** because they are already in your title. You must give the keywords which are not in your title and which are very important, and your article is liable to be searched on the basis of those words. Then, those keywords should be selected only.

Authors

Then moving towards authors. The authors we have addressed the things only genuine authors should be included along with the main authors. And all the affiliation detail should be clearly spelled out along with the contact number or email id. The full information should be in the title page.

Abstract

The next thing is abstract, abstract is about the first impression and first impression is the last impression. To submit an article, very firstly the technical person or editor will see it. But, if your abstract is not okay, they will not entertain it and they will reject it at the moment only and your article will not enter into the review phase.

So, your abstract should be a representative context compilation of your entire work. It may be structured or unstructured depending upon the requirement and the guidelines given by the journal. An abstract must not be too lengthy and in most of the cases, it ranges from 200 words to 300 words. You should comply with your word limit of abstract.

Many a times, now a day's journals are demanding **graphical abstracts**, in a graphical form a figure form which is representing your work. You should plan it very nicely so that the people can get attracted to your paper. This will draw attention to your article. Apart from that various tactics are being planned.

Summary

Identify the suitable journal on basis of various factors. Think for exploring the multidisciplinary or interdisciplinary. Evaluate and assess your work, plan accordingly. Plan a good title. Use effective key words. Genuine authors should be included only. A concise and condensed effective abstract must be planned.

Dear Learners, in the previous section we discussed how a review paper is planned, how a journal is targeted, how article title is planned, how the key word are used and how we can write an effective abstract.

In this section we will be dealing with the sections which come after the abstract. We will begin with the introduction. How we introduce our topic?

Introduction

It is about WHAT and WHY: As the name suggests, it deals with the WHAT and WHY? What is the topic and why we are addressing it. These two prime questions must be answered in this section. And the answering strategy depends on the type of review. What and Why and then moving to type of review paper. **Type of review papers:** Type of review must be identified in the sense, which kind of review you are writing. As we have mentioned in the type of review paper earlier argumentative, systematic, methodological etc. You see, what kind of review you are writing and what is expected in that kind of review. Because you are writing for your reader. Always remember that thing.

Length: The next important thing in type of review is to check whether you are going to write full review or mini review. The length (the next feature of review) of the paper depends on the choice of review paper: full or mini review.

Short and concise: The introduction must be short and concise. Even in the full review it is not more than one and half pages when you are writing in A4 pages written in double spacing (12 font). Because it is just introduction. Donot try to put each and every thing in introduction part.

No separate headings: Introduction must not have any kind of separate headings.

Tell a story: Give the importance of the topic, seek the attention of the reader, create the interest, tell a story. As we can see in this example: It is one of my papers published on Expert Opinion on Drug Discovery journal devoted to review papers only. In this particular article entitled "Techniques for the discovery and evaluation of drugs against alopecia" we started the introduction with these lines

"Hair has been recognized as a sign of beauty and a socio-psychologically important component of the human personality throughout the ages."

The first line is a simple line. Then we put a specific term "socio-psychologically important component". By this way we are trying to seek the attention, in the very first line. First, we have to seek the attention. When we grab the attention from the readers then we develop the interest and thereafter we show the importance that what we are actually going to communicate in this paper. So, it is about telling a story systematically and in the organized way.

Cohesion and coherence: You have to comply with the basic principle of cohesion and coherence. You cannot jump from one topic to another topic immediately. We described the cohesion and coherence in a previous chapter in English in academic writing. Hope you remember it.

Body of the Paper

Stick to your outline. First and foremost criteria is to stick to your outline. Define the heading and paragraph and then write sentences for these. And these sentences/paragraphs should be systematically organized. I will again emphasize on cohesion and coherence here. Throughout the manuscript it is very important. Your first line should hint the second line. The last line of your para should hint the first line of the next paragraph. This is called cohesion. When it is about grammatical level it is cohesion and at idea level it is called coherence. You will have to not only have the good coordination or combination of words and sentences/ rhythm of the sentences, not only with respect to grammatical aspect but also with respect to aspect of the idea development.

You will slowly move to your focus point. Again the convergent strategy will be here also. You will start focusing the things to your particular topic and then develop your topic. Remember not to give much of the background here. Start from a particular point from which you can target your focal point easily so that you can develop your main idea in the body itself.

Critical and coherent presentation

Coherent with respect to the idea level. You will have to be critical. You cannot move in a same tone thought out the review paper. Develop your own voice. Do not just summarize the things. Synthesize your idea, what you actually want to communicate. That in this scenario this much result is there. You will have to identify the gap which you are going to address or develop. So you must have your own voice and develop it properly.

So, critical nature, coherent nature and (using your) own voice are the three key aspects of developing the body of the review paper.

Infographics

Easy visualization: These are set of figures, tables and blend of things other than the text. The Infographics must be used very effectively. The quality of info graphics should be able to provide the information in a concise way so that the reader can understand it quite effectively and easily. It should be able to provide the easy visualization of entire things.

Take care of figure plagiarism: The figure plagiarismis also a vital issue. You might remember the example of stem cell researcher who copied the figure, got caught and he committed suicide. Donot ever copy the figure exactly until you have taken the permission to reproduce it or it is available as open content or free license, creative common license, free images, figure in free domains, Wikipedia/wikimedia etc. The best strategy is to draw you figure yourself. Draw your own Infographics using the inputs/software available in your PC/laptop.

Total number of figures: You cannot have the bombardment of figures in the review paper. Generally, in the author guidelines the maximum number of figure/ table and figures are given. Sometime they relax but excessive is not allowed. A thing which has already been discussed in text should not be there in the table and vice versa.

Placement of Infographics: You must ensure where your Infographics is going to be located in the paper. Give hint about the location within parenthesis (Fig.1) or Fig. one near here/ this point. In many manuscript, figures are located at the end of text in one figure in one single page, when you are submitting for review. So, you have to mention in the text where would be location of the figure along with all the other informations like legends and footnotes.

Citation: Proper citation of figure in text must be there for easy identification of the location of Infographics.

Section Summary

We studied about the introduction and the body of the review paper in this section. Introduction deals with what and why of your review paper. Make it short and concise, without any **heading.** Seek the attention, create the interest and give your idea, why you are going to address the particular problem and maintain cohesion and coherence. Maintaining cohesion and coherence holds true for the body also. Body should be in good tune with the defined outline. Develop the idea as per the outline. It should have the quality infographics in optimum number.

Let us discuss conclusion, acknowledgement, references and some tips.

Writing Conclusion

It provides the result of your entire review paper in very concise form. It should be answering your objectives. You restate your objectives and then you say that by this way we met our objectives. Avoid any vague statement here. If possible put the line indicating expected outcome and future trend of the study in one or two lines. Always remember, conclusion should never contain any reference.

Acknowledgements

Acknowledge your resource, acknowledge who have contributed full text of any article, funding agency and others who have made considerable impact on your review paper. Donot use this section for greasing or obliging.

References

It must be current and updated. Atleast last ten years references must be covered in a balanced way.

Let me clear you. I was going through a paper as a peer reviewer on nanotechnology. For just sake of putting a few current reference what authors did, there was one reference of 2018, 1 of 2017, 02 of 2016, 01 of 2015, 01 of 2014 and rest of the references were of 1990s. This indicates that the author has not covered the references properly. He just took an old paper of around 2013 or 2014 and just copied the cross references in his paper.

Being a reviewer, I feel that it is very important and keep in your mind that you cannot deceive a reviewer. Have a balance of coverage of references in your review paper. In an optimum way, you should put the references. Do not avoid pioneer/ historical references. We have mentioned earlier also. But do not put the excessive use of very old references.

Sometimes some journals like Expert opinion ask authors to highlight the major references which are of utmost importance. So you must have in your mind what are those references.

References must be written as per the particular guidelines of the journal. A simple punctuation error in references, may lead to rejection sometimes.

After references writing, see for any punctuation error, omission of any comma/semi colon, pagination, doi. Ensure it. Then finish your manuscript. Go for final check and then submit your manuscript. We will be dealing with the submission and post submission in a separate chapter. So not dealing with these issue here.

Major Tips

- Have balanced/current references.

- See the number of references in general, a review paper contain in any journal. If you are submitting a review paper with 30-40 references to a journal which contain 100-250 references in a review paper in a review paper in general, obviously the journal will not entertain your paper at all,

- Having a tight outline; very focused/critical approach; compilation and organization should be unique; the Infographics quality is very important.

- Your abstract should be like the face of your paper. Many readers just decide to read only after reading the abstract. So give maximum time in writing, rewriting and improving the quality of your abstract. It should be a power packed presentation of your review paper.

- Title: Should be attractive. Search the related articles, see the trend, take two three title, think upon them and then finalize one.

Summary

In this chapter, we covered planning and targeting a review paper to the journal. Making an effective title, Writing effective abstract, using effective key words, writing introduction effectively, answering what and why of the related field of research. Body of review paper answers and develops the idea of your review paper systematically with cohesion and coherence with quality Infographics **and inclusion of latest references** in optimum/required number which is in good tune with the article level and type. Then conclusion summarizes your work and tells whether it has satisfied the objectives or not. While the references must comply with the journal's guidelines.

I will be happy if you plan a review paper after going through this chapter.

Further Reading

- Guidance of a review paper, https://www.journals.elsevier.com/international-journal-of-machine-tools-and-manufacture/news/guidance-for-review-papers

- Literature review paper, http://websites.uwlax.edu/biology/ReviewPapers.html

- https://www.springer.com/gp/authors-editors/journal-author/journal-author-academy/15186

- Fundamentals of manuscript preparations, https://researcheracademy.elsevier.com/writing-research/fundamentals-manuscript-preparation

References

- Anson, Chris M. and Robert A. Schwegler. *The Longman Handbook for Writers and Readers.* 6th edition. New York: Longman, 2010.

- Types of reviews, http://library.atmiya.net/research%20commons/researchguide/literature_review.php

CHAPTER 20

Research Paper Writing-I
IMRaD Style- I

Dr Ajay Semalty
H.N.B Garhwal University (A Central University)
Srinagar Garhwal-246174

Dear learners, we have discussed Review paper writing in the previous chapters. Let us learn writing a research paper in this chapter.

Learning Outcome

After learning this chapter, you will be able to

- Understand the importance of research paper
- Plan a research paper

Lesson Plan

- What are research papers?
- Importance of research papers
- Steps in research article writing
- Planning a research paper

A research article is a written document which contains the critical presentations of the results of an original research and is published in a peer reviewed scholarly journal.

It is a primary source: It reports the methods and results of an original study performed by the authors. Review papers are theoretical in approach while research articles are practical in approach. Review writing is an intensive task. But research is not that much intensive task. But the publication of research work depends on the novelty, quality and presentation of research work. It is about presenting / justifying your objectives with your own work.

Importance of research papers

In academic and research career it is important for all the stake holders i.e. students, researchers, teachers and institutions

- For submission of Ph D a published paper is mandatory.
- For direct recruitment and for CAS publications have due weightage.
- For researchers it is the source of information for planning novel research and exploring collaborations.
- For Institutions it builds rapport in academic world.

Steps of writing research papers

We must follow the sequence as mentioned in Fig. 20.1 for research paper writing.

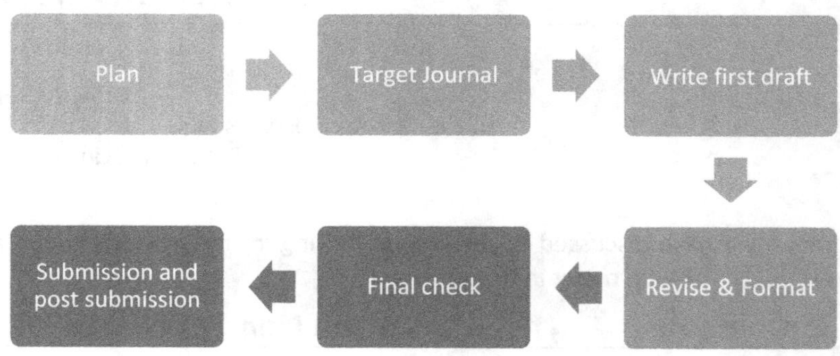

Fig. 20.1 Steps of writing research papers

I Plan

We should effectively plan paper writing in terms of

- Time
- Type
- The studies to be included
- Language
- Author's Name
- Outline

(Discuss with your supervisor/ team members/ coauthors)

Plan

Time: Start early, do not wait up to the end of your entire research work, report preliminary study even at early stage of research work.

Studies to be included

- Plan the preliminary promising studies which yielded good and novel results.
- Chalk out initially how many papers can be derived from your work.
- Do not cut the work intentionally in smaller pieces of work.

Language: English is the most widespread language with respect to readership in international scientific communication.

- English for obvious reasons of impact speed and reach is the first choice.
- Some journals which are published in other languages (e.g. Swiss journals) also allow submission and publication in English along with the other official language of the journal.

Author's Name

Authorship denotes an "intellectual contribution" to the work, and that an author should be able to explain and defend the work. At this stage itself, it is better to decide on the authorship. The definition of who should be an author (and in what order the list should be provided) varies with the field, the culture, and even the research group. Aspects like order of authors should be clearly agreed upon at the outset of a research work.

Outline

Plan the outline of your article as per the scope and type of your article. See the inputs you have and plan the outline:

- Body of knowledge (LR),
- Rational,
- Aim/objectives,
- Methods, and
- Results& discussion

II. Target Journal

For targeting your manuscript to a suitable journal you must take care of the following points.

- Match your topic with scope of journal
- Refer Databases for targeting
- Reputation (IF or other journal metrics: Impact, speed and reach)
- Time
- Availability (online/ offline)
- Indexing
- Format (See the guidelines for authors and general submission guidelines)
- APC

Match your topic with scope of journal: Failing to match your topic with the scope of journal leads to rejection at the editorial level only, even before sending it to peer review.

Observearticles in latest issue of journal, Theme of issue or conference, Additional information required so as to decide.

Refer Databases for targeting

As we discussed in LR, you can refer the LR databases like Endnote, Web of Science for searching a suitable journal. Enter you title, abstract and / or keywords, the database will give the list of suitable journal for your manuscript. Springer Author Academy also provide the opportunity for the authors to target the most suitable journal on the basis of title and abstract. Apart from Springer's Journal suggester, Elsevier also provide Journal finder for the authors (Fig. 20.2 & 20.3). These are AI based journal finder which provide many information like impact factor/ cite score, journal's speed, acceptance rate etc. You can decide as per your need.

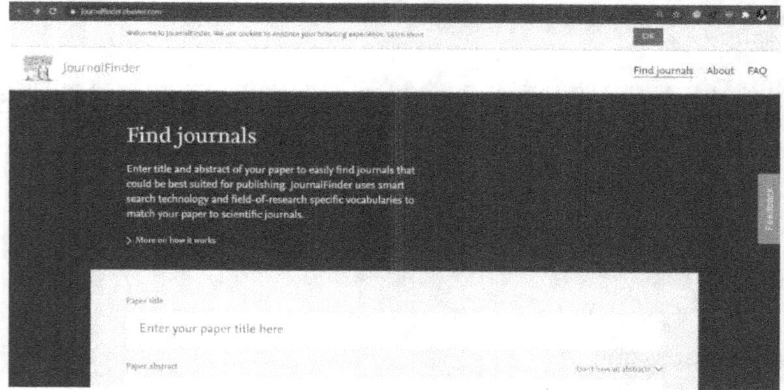

Fig. 20.2 Journal Finder by Elsevier

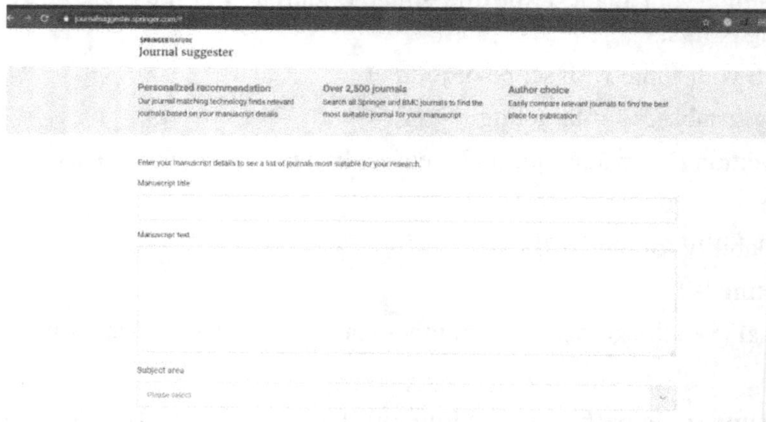

Fig. 20.3 Journal Suggester by Springer

Reputation (IF or other journal metrics: Impact, speed and reach): The reputation of a journal can be assessed by several ways. Your own perception along with the opinion of your seniors (by their experience), type/quality of recent articles published in the journal, their importance, the presence of field experts in members of the Editorial Board, the acceptance rate of the journal and the journal's *impact factor* and other journal metrics may be the criteria of judging the reputation of a journal.

An author must overview these things before deciding to submit a paper to particular journal.

Time: Check how much time the journal takes for the peer review process

This is very important to check the time taken by a journal in review process. Target a journal which takes less time for the same.

You can not wait for years if you are planning article with your Ph D thesis submission.

In general, the minimum time is 2 to 3 months taken by a journal's review process.

Fast track processing of article: If you can afford, avail it.

Availability (online/ offline): The availability of the journal is also very important. Some journals are broadly available both online and in print version. On the other hand some journals are online only or in print only. So, the article submission process may also be easier or the complex depending upon the availability of the journal.

Indexing: The journal indexing in the major electronic databases such as Medline, Biological Abstracts, Chemical Abstracts, or Current Contents must be checked before deciding to submit a paper to a journal.

Format (See the guidelines for authors and general submission guidelines):

Does the appearance of published articles – the format, typeface, and style used in citing references suite you as per your study or the research work? If relevant, does the journal publish short and/or rapid communications?

APC: Apart from the various open access journals of leading publishers, various print journals also bill the author for *page charges*, a cost per final printed page. Most journals have a separate *charge for color plates*. This may be as much as $1000 per color plate. Many journals will waive page charges if this presents a financial hardship for the author; color plate charges are less readily waived and would at least require evidence that the color is essential to the presentation of the data (e.g., to show a double-labeled cell).

Further Reading

- Selecting a Journal: https://youtu.be/XILmZp84Qf8
- https://utah.instructure.com/courses/306223/pages/picking-a-journal-for-the-manuscript
- https://www.springer.com/gp/authors-editors/journal-author/journal-author-academy
- How to Maximize Your Study's Visibility by Choosing the Right Journal, https://www.wiley.com/network/researchers/discover/how-to-maximize-your-studys-visibility-by-choosing-the-right-journal
- Find the right journal to publish your research, https://authorservices.wiley.com/author-resources/Journal-Authors/find-a-journal/index.html

References

- Fowler, J. (2011). Writing for professional publication. part 8: Targeting the right journal. British Journal of Nursing (BJN), 20(4), 254.
- Henly, S. (2014). Finding the right journal to disseminate your research. Nursing Research, 63(6), 387.

CHAPTER 21

Research Paper Writing-II
IMRaD Style-II

Dr Ajay Semalty
H.N.B Garhwal University (A Central University)
Srinagar Garhwal-246174

Welcome dear learners, welcome again in the next chapter on research article writing IMRaD style 2. In the last chapter, we discussed what are research articles, what are the importance of research articles, how to plan research article and how to target a journal for your research article. Now, let us discuss drafting or writing the article.

Learning Outcome

After completing this chapter, you will be able to

- Learn writing the draft of the research article

- Learn to focus on every section of the research article.

Lesson Plan

So, we will be covering this chapter under the heads

- Writing first draft,

- Writing effective title,

- Author's affiliation planning,

- Keyword planning,

- Abstract writing.

- Introduction writing.

Writing First Draft

Now, let us start our discussion with the writing first draft. You will plan your article with an outline planned in the planning stage that defined the outline. Now, what do you have to do that you will have to focus your efforts in the realm of that outline only and we'll start the first draft of research.

Now decide style accordingly. The style may be IMRaD, the most commonly used research article writing style and IMRaD stand for introduction material methods results and discussion. Empirical studies will be dealing in a particular topic in which we will be discussing the writing the empirical studies. However, most of the empirical studies are also written in the IMRaD format.

In writing the first draft, your planned outline or the outline you planned in the planning stage helps a lot to start writing as per that outline.

General points to consider

What are the general points to consider while you are writing your first draft?

1. Develop your idea systematically. Maintain a systemic flow in writing and presenting information.
2. Follow guidelines stick to authors guidelines of the journal while drafting the manuscript. It will help you out in the latest date.
3. Uniform language use either British or American English. We have discussed this aspect of English in academic writing. It should be error free.
4. Avoid any type of grammatical error in this stage only.
5. The next aspect is the reader friendliness and clarity.
6. Use simple and reader friendly language with good clarity. Attribution don't forget to give due attribution.
7. Avoid plagiarism. Do not copy. We have discussed the issue of plagiarism and this becomes utmost essential to be vigilant in avoiding the plagiarism.

So, what have you have to do in writing the first draft? You must focus on every section of your article including title of your article, author's details, corresponding author's details, abstract, keywords, introduction, materials and methods, results, discussion, conclusion, acknowledgement, references and infographics tables and figures.

Title

Now, let us discuss the very first thing-**Title**. We discussed earlier also, in my opinion, plan it at last. Because you might have many ideas while you're writing the manuscript or finalizing the manuscript. Keep in your mind that it should be simple, catchy, representative of your work, reader friendly and suitable for the journal. A good reliable and simple technique is to briefly write the question that was asked or the answers that was arrived at the end of the study for your title.

Tips

Let me give you some tips.

- Search for similar titles or highly cited related articles title in the Target journal or related journals or in the relevant databases like scopus, ICSSR, pubmed or Google Scholar.
- Get the idea to focus on hot, trendy and relevant words.
- Plan two to three titles.
- Finalize one, after consultation with your team.

- Plan brief running titles also (in 50 spaces or few words) as the journal recommends. It is needed by many journals as a header of the article as you can see below.

We have given the example. This is the title.

"A cyclodextrin inclusion complex of a second order effect of drug beta cyclodextrin ratio and method of complexation"

And what is the running title? You can see at the top left corner "cyclodextrin inclusion complex of Racecadotril" (Fig. 21.1) So, this is the running title which is representative and used as the header in the article's PDF or you can say in the print version or online version.

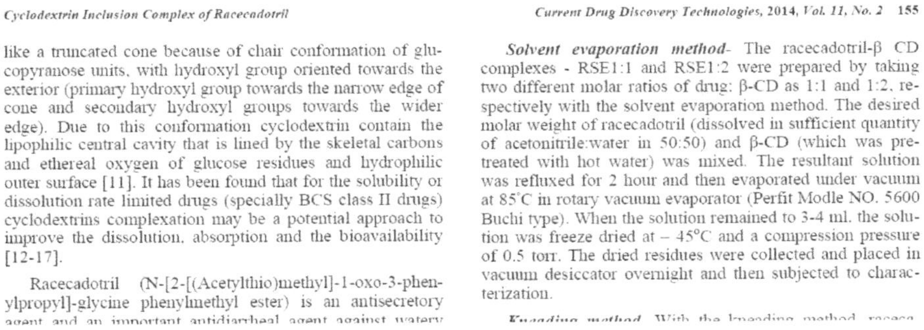

Cyclodextrin Inclusion Complex of Racecadotril   *Current Drug Discovery Technologies, 2014, Vol. 11, No. 2*   155

like a truncated cone because of chair conformation of glucopyranose units, with hydroxyl group oriented towards the exterior (primary hydroxyl group towards the narrow edge of cone and secondary hydroxyl groups towards the wider edge). Due to this conformation cyclodextrin contain the lipophilic central cavity that is lined by the skeletal carbons and ethereal oxygen of glucose residues and hydrophilic outer surface [11]. It has been found that for the solubility or dissolution rate limited drugs (specially BCS class II drugs) cyclodextrins complexation may be a potential approach to improve the dissolution, absorption and the bioavailability [12-17].

Racecadotril (N-[2-[(Acetylthio)methyl]-1-oxo-3-phenylpropyl]-glycine phenylmethyl ester) is an antisecretory agent and an important antidiarrheal agent against watery

Solvent evaporation method- The racecadotril-β CD complexes - RSE1:1 and RSE1:2 were prepared by taking two different molar ratios of drug: β-CD as 1:1 and 1:2, respectively with the solvent evaporation method. The desired molar weight of racecadotril (dissolved in sufficient quantity of acetonitrile:water in 50:50) and β-CD (which was pretreated with hot water) was mixed. The resultant solution was refluxed for 2 hour and then evaporated under vacuum at 85°C in rotary vacuum evaporator (Perfit Modle NO. 5600 Buchi type). When the solution remained to 3-4 ml. the solution was freeze dried at – 45°C and a compression pressure of 0.5 torr. The dried residues were collected and placed in vacuum desiccator overnight and then subjected to characterization.

Kneading method- With the kneading method raceca

Fig. 21.1 Running title

Section Summary

So, in this section, we learned that in writing the first draft you must be there within the defined outline. You should decide on a particular style and focus every section. You should choose a good and effective title. We will be moving ahead with the other aspects in the next lecture.

You have identified the title, or you have planned the tentative titles from which you are going to finalize one title at the end of your manuscript writing. Now there come author details.

Authorship

Authorship must be defined at this stage only, not only the names but also the order of the author's in the manuscript. And authorship must be assigned to the people only, who have given their intellectual contribution in the manuscript preparation and the research work. It should never be used as a favor or as a gift.

See the Journals guidelines for knowing the presentation of author detail like name and affiliation.

As you can see in this example of one of our paper of Fitoterapia, the full names are there. The first & last name are separated by the comma and the affiliations are marked with a and b and given after their names alternatively (Fig. 21.2).

Fitoterapia 81 (2010) 306–314

Contents lists available at ScienceDirect

Fitoterapia

journal homepage: www.elsevier.com/locate/fitote

Review

Supramolecular phospholipids–polyphenolics interactions: The PHYTOSOME® strategy to improve the bioavailability of phytochemicals

Ajay Semalty [a,*], Mona Semalty [a], Mohan Singh Maniyari Rawat [b], Federico Franceschi [c,*]

[a] Department of Pharmaceutical Sciences, H.N.B. Garhwal University Srinagar (Garhwal), India
[b] Department of Chemistry, H.N.B. Garhwal University Srinagar (Garhwal), India
[c] Research and Development Laboratories, Indena S.p.A. (Settala), Italy

Fig. 21.2 Author's names presentation from different affiliations.

REVIEW ARTICLE

HERBAL HAIR GROWTH PROMOTION STRATEGIES FOR ALOPECIA

Semalty M.*, Semalty A., Joshi G.P. and Rawat M.S.M.

(Received 21 May 2008) (Accepted 30 July 2008)

ABSTRACT

Hairs are considered to be a major component of an individual's general appearance. Hair loss creates the psychosocial impact and results in a measurably detrimental change in self-esteem. Extensive researches are going on to explore the effective and safe drug for hair growth. Angiogenesis (through endogenous substances), androgen antagonism, potassium channel opening and 5-alpha reductase inhibition are the major non-surgical therapeutic strategies of hair growth promotion. Only two drugs minoxidil and finasteride have been approved by US FDA for hair growth promotion. Herbal drugs are also being investigated for potential and safer alternatives. The article focuses on causes and factors affecting hair loss. The developments in hair rejuvenation strategies are discussed along with the potential of herbal drugs for hair growth activity.

Fig. 21.3 Author's names presentation:
Indian Drugs (SCOPUS indexed journal)

In another title, it is a paper of our team from Indian drugs journal. In this, last name is followed by the initials of the first name, separated by comma. So, it depends on the journal's guidelines on how the name should be mentioned in author details.

Affiliations

As far as the affiliations are concerned, the authors may have single or multiple affiliations. Corresponding authors details like name, affiliation contact details like phone and email ID must be there. Apart from institutional address, give email which is used more frequently rather than the one which you do not use normally. We can see in this example (Fig. 21.4). We have given with asterisk mark on the corresponding author. The details of the corresponding author are given as "For correspondence".

<u>**REVIEW ARTICLE**</u>

HERBAL HAIR GROWTH PROMOTION STRATEGIES FOR ALOPECIA

Semalty M.*, Semalty A., Joshi G.P. and Rawat M.S.M.

(Received 21 May 2008) (Accepted 30 July 2008)

ABSTRACT

Hairs are considered to be a major component of an individual's general appearance. Hair loss creates the psychosocial impact and results in a measurably detrimental change in self-esteem. Extensive researches are going on to explore the effective and safe drug for hair growth. Angiogenesis (through endogenous substances), androgen antagonism, potassium channel opening and 5-alpha reductase inhibition are the major non-surgical therapeutic strategies of hair growth promotion. Only two drugs minoxidil and finasteride have been approved by US FDA for hair growth promotion. Herbal drugs are also being investigated for potential and safer alternatives. The article focuses on causes and factors affecting hair loss. The developments in hair rejuvenation strategies are discussed along with the potential of herbal drugs for hair growth activity.

Keywords: alopecia, hair growth, Hair loss, minoxidil.

INTRODUCTION

To maintain good hairs is an inevitable need in rapidly changing life styles. Hairs are important sociologically and make the vital part of a human personality. Hair disorder, especially when severe, often profoundly affects the lives of those afflicted. Alopecia, a dermatological disorder that has been recognized for more than 2000 years is a common and distressing problem in cosmetics as well as primary health practice. It is common through out the world and has been estimated to affect about 2 % of the world population [1-3].

Minoxidil was introduced in the early 1970s as a treatment for hypertension. Hypertrichosis was a common side-effect in those taking minoxidil tablets and included the regrowth of hair in male balding.

*For correspondence
Department of Pharmaceutical Sciences,
HNB Garhwal University Srinagar
(Garhwal)-246174; Uttarakhand.
E-mail: monasemalty@gmail.com

This led to the development of a topical formulation of minoxidil for the treatment of androgenetic alopecia in men and subsequently in women[4,5]. Currently, minoxidil (useful in both male and female pattern baldness) and finasteride (useful in male pattern baldness) are two U.S. FDA approved synthetic drugs finding concomitant use for treatment of androgenic alopecia, but their side effects have reduced their usage [6,7]. Therefore it becomes vital to search for the more precise therapies with the newer synthetic drugs or the drugs of plant origin. Herbal drugs have been widely used for hair growth promotion since ancient times in Ayurveda and Unani system of medicine. In Ayurveda, Indrayan (*Citrullus colocynthis*), Bhringraj (*Eclipta alba*), Brahmi/mandukparni (*Bacopa monnieri*) and Nagarmotha (*Cyperus rotundus*) have been reported to be effective in the treatment of "Indralupta" that is baldness. Natural products are very popular and well accepted in the cosmetic and hair care industries and about 1000 plant extracts have been examined for hair care usage. There are many products available in the market, which are prepared by combination of one or more herbal drugs and find acceptability as hair tonic, hair growth promoter, hair conditioner, hair cleansing agent, antidandruff agents, and for the treatment of alopecia and lice infection [8,9].

INDIAN DRUGS 45(9) SEPTEMBER 2008   689

Fig. 21.4 Corresponding authors' details

Title page

So, this was all about the title page. You can say a title page is a single page in which the title of the article, authors names, their affiliation, a running title and corresponding authors details are given. Sometimes, it also contains a type of article on the top left corner like original research article or short communication, as per the journal's guidelines now;

Now, we will move to planning the abstract.

Planning the abstract

Abstracts are just like the trailer of a film. They provide a summary or central idea of your entire work. For writing the abstract, follow word limit200 to 300 words. In general, the format must be either unstructured or structured. Unstructured means, in just a running paragraph, the entire abstract is there.

And in the structured abstract: it is mentioned under some heads like background name and objective method result and Conclusion.

Never use abbreviations in the abstract. And you should also not give all the quantitative results in the abstract itself. Instead, one should give only the main quantitative results which are representative of the study and cover the complete study. It is very important. Do not use even a single unnecessary word. Make short sentences with clarity. Write in Draft stage. An abstract must be written in the draft and finalized at last in manuscript writing before the final check, just like the title we considered.

So, let me give you some tips. Remember it is the face of the article. Editors peer-reviewers make a preliminary opinion about the article just by reading the abstract. The reader decides to read your article only after being convinced by the author in the abstract. Take this very seriously. Check it word by word for hundred percent accuracy.

Keywords

After the abstract, there come keywords. Keywords are the words which are used by the researchers or the readers to search an article. So, what you have to do, you will have to choose the word which is not there in the title. Use relevant and hot or most popular Words which are often used to search the work as yours. Think from the viewpoint of a reader and choose the best 5 to 6 keywords; generally, just 5 to 6 keywords are allowed maximum. In the paper, they are placed just after abstract or in the online system some special sections are there for inserting the key words.

Section Summary

So, we have covered upto the keywords, now, IMRaDstarts. IMRaD means introduction, methods, results and discussion. So, we will be dealing this in the next section. After planning up to the abstract and keywords, we move to the IMRaD style.

We know the abstract, we know the central idea.

We have covered so far up to the abstract and keywords.

IMRaD

After the title page, the abstract and keyword come and then the IMRaD starts. The abstract is the central idea of your research work while the introduction deals with what you did and How you did. It deals with defining problem and objectives, the rationality of work while the materials and methods are about the way to solve the problem. Results about the findings of the experiments discussion is about an analytical study of obtained results conclusion are about concluding remark on the outcome by the acknowledgement is for the support received in references you mention sources of studying and infographics tables and figures are presented.

Introduction

So, dear learners in the introduction, we try to answer what is the problem? What is the hypothesis included in the research? Give background and the origin of the idea from where it came.

When writing an introduction, remember that you are not narrating a story. And neither is the introduction is a review of the subject. It should be mentioned clearly. What do you want to do and why you did it?

Start with the brief background and brief literature review, introduce the topic, identify research gap, define problem, present rationality and novelty and then present your way out what you did in the study.

Problem definition: The logically, sequentially, systematically you have to introduce your problem and give your hypothesis; spell out the question or hypothesis that your research aims to settle.

With respect to **novelty and rationality of your work**, explain why the research was undertaken; Were there gaps in the existing knowledge or work; any existing data conflicting in its conclusions. And therefore, the present study came into existence.

Proper citations: Focus on pinpointing, to give the proper citation to latest and important previous studies. Don't use very old references, until they are indispensable, or they are must to discuss. Do not miss the current and most important related study, especially, of the most cited researcher of the field.

End statement: The last para of the introduction must address your hypothesis leading to aim & objective and methods of the present form.

Length: What should be the length it should be about two pages in length one and a half pages in most cases.

Tips

So, let me give you some tips.

Be concise, develop the cohesion and coherence in the entire introduction, follow the logical transition. Remember the reader is an expert, write for him not for a layman.

In sciences and life sciences and in Engineering, the writing introduction is well versed, but in arts Humanities and other subjects and social sciences, writing introduction depends on completely on the field and specific subject being covered.

What you have to do specifically, if you are targeting introduction of any type, for any type of study, at least refer 10 to 20 research articles of related field and then try writing the introduction. Remember, there might be some figures which may need to be introduced into the introduction. So, you can also add a figure which is representing, and which is unavoidable to put into the introduction.

Summary

So, in this chapter, we learned planning the title page, writing the abstract, keywords, and effective introduction with the help of cohesion and coherence and presenting aim and objectives of the work. Never forget this aspect that your idea must be translated into the brain of the reader. We will be addressing other aspects of material & methods on work in the next chapter.

Further Readings

- McMillan, V.E. 2001. Writing papers in the biological sciences. 3rd Edition. Bedford/St. Martin's Press, Boston and New York. 207 p.
- Faber J, Writing scientific manuscripts: most common mistakes, Dental Press J Orthod. 2017; 22(5): 113–117. doi: 10.1590/2177-6709.22.5.113-117.sar

- Borja A, Writing the first draft of your science paper — some dos and don'ts, https://www.elsevier.com/connect/writing-a-science-paper-some-dos-and-donts
- https://www.springer.com/gp/authors-editors/journal-author/journal-author-academy
- https://www.apa.org/pubs/authors/new-author-guide.pdf
- https://projects.ncsu.edu/labwrite/index.html
- Maxim S. Pshenichnikov,Academic skills, file:///C:/Users/DELL/Desktop/Writing-paper-SEPOMO3.pdf

References

- Ohwovoriole AE, Writing biomedical manuscripts part I: fundamentals and general rules. WestAfr J Med. 2011 May-Jun;30(3):151-7.
- Ohwovoriole AE, West Afr J Med. Writing biomedical manuscripts part II: standard elements and common errors. 2011 Nov-Dec;30(6):389-99.
- https://writingcenter.unc.edu/tips-and-tools/scientific-reports/

CHAPTER 22

Research Paper Writing IMRaD Style-III

Dr Ajay Semalty
H.N.B Garhwal University (A Central University)
Srinagar Garhwal-246174

Welcome, dear learners. Welcome in our last chapter on research article writing. We covered how a research article is planned. How targeting a journal is done. How the title page is planned, how abstract is written, how the introduction is written. Now, we will move to the other aspects, other sections in this chapter.

Learning Outcome	Lesson Plan
After completing this chapter, you will be able to	Dear learners we will be covering this chapter under the heads

Learning Outcome

After completing this chapter, you will be able to

- Write the first draft of your article
- Revise format and final check it

Lesson Plan

Dear learners we will be covering this chapter under the heads

- materials and methods- writing effective material and methods section,
- result, discussion,
- conclusion,
- infographics,
- acknowledgement,
- revising& formatting
- final check of your manuscript.

Materials and Methods

Let us begin. What do you do in the materials and methods section in IMRaD style? It is the second most crucial aspect. Try to answer.

How was the problem studied following a hypothesis?

In the materials and methods section, sometimes, it is written just methods or sometimes in sciences it is written as an experimental section. The nomenclature maybe a little bit different but discuss the methodology here. Discuss the methodology logically and in a concise way. In the materials give the materials used and other sources, subject chosen, Area of research etc in case of field Study. We will be dealing with the field study in a separate week as we have earlier mentioned also. Describe methods. If anew method is used, it is needed to be discussed in detail. Established method maybe just mentioned by name with reference of method or just you can briefly discuss them. Don't miss discussing any modifications made by you in the standard method. This section contains very few citations, only for previously reported methods.

Have clarity in reporting methods. It is important because it is the point which makes your work reproducible and reproducibility is the key feature of any research article. A reader should be able to reproduce your research by following your methods. So, it must be clear.

Instrument software name, model, version, company etc must be mentioned wherever used. Statistical method and level of significance/acceptance criteria must be mentioned clearly. Necessary permissions should be included like ethical committee approvals, herbarium numbers. You must include or if you are following a particular protocol from an official guideline, you must mention that.

In material and methods, you can also include the figures like a flowchart of a process for easy understanding, the image of subject study area, plant map etc. You can add as an image but remember the image must be plagiarism free.

As far as the tense is concerned you must write this section in the past tense or past perfect tense. And in most cases, it is in the past tense.

Results

Now, let us move to the result section. Here try to answer what was the outcome of problem treatment from a hypothesis. It must be evidence-based. This section provides evidence that leads to the answers of the study to the question you post at the start. The presentation must be professional. State the results without any bias. Do not exaggerate the results. Do not be afraid of reporting negative results. So, even if some paradox is there, some contradicting results are there, report them also. Statistical support must be mentioned. Use good statistics to show the result. In the modern era, there is no research article which is accepted, and which is not having statistical tools or use of suitable statistical tools to show the significance of the results. Equation and special characters must be given due attention. The presentation of the equation must be given due attention. It is better to write the equations in Microsoft equation. Here again, you will write it in the past tense or past perfect tense, and the last feature is completeness. Cover all the parameters. You must cover all the parameters, and the results must be complete.

Section Summary

Dear learners, in this section, we learned how we can plan the material and methods. Clarity and the reproducibility are two key features. As far as the result is concerned result is a very crucial section you provide your outcomes here in a professional way. We will be discussing other aspects in the section.

Dear learners after covering the materials and methods and results section in the earlier section, we will be dealing with the infographics. However, we mentioned earlier in the sequence that infographics come at last but they are the part of results. So, we'll be discussing here because we justify our results with the judicious use of infographics in the form of tables and figures.

Infographics

Infographics are used in the section, but maybe or may not be inserted in the results section. In the manuscript, many journals demand one table or one figure in one single page after references. So, the table and figure must be self-explanatory with the suitable Legends or footnotes. A reader should not need to refer the text to understand a particular table and figure. So, it must be self-explanatory.

All related information should be there table title should be on top of the table. You should comply with the guidelines refer guidelines for table format many journals provide that particular format like no vertical lines only major horizontal lines, etc. For the presentation of tables. You can also refer to the recent articles and the table there. And follow the style Don't repeat the text in figure or table or vice versa. Redundancy must not be there. Cite the table or figure in text. It must be indicated in the text that where the table or figure should be inserted in the text. You can put "Table 1"or "figure 1" within parentheses or some journal requires "insert table one near here".

Make the table and figure self-explanatory as we earlier mentioned and use the suitable legends or footnotes. In the case of figures, legends may be required with the title of the figure. And many journals don't require them inserted into the figure.

So, always remember and refer the guidelines for this thing. For the images proper type like jpg,png,tif file format that size legends and font of the image should be selected properly.

Discussion

Dear learners, let's move to the discussion. Many journals combine results and discussion sections together. And some journals require separate sections for result and discussion.

In the discussion, try to answer how was the outcome of problem treatment correlated and or contradicted with the previous studies?

Now in discussion take the result one by one and discuss them. Discuss the results, in light of existing knowledge.

Be critical: Critically analyze the result, don't be biased. Don't make your opinion, so as to suite your hypothesis only.

Discuss the results with clarity and reader friendliness: These two attributes are the most important and vital characteristic of a research article.

Use the references freely to support discussion: And this is a section which must have good number of relevant references supporting and contradicting. Both kind of studies should be discussed discuss in tune of aim and objectives and try to converge to the final outcome and it's explanation.

Conclusion

Now let us move to the conclusion section. Here you conclude your study outcome in concise manner conclude in tune of aim and objectives laid down in the introduction. The conclusion should not just be copied from abstract and you may give future directions also.

Acknowledgement

In the acknowledgement section try to answer who helped you and supported you but not eligible to the author or may acknowledge only to product grants, gift samples, analyical or other service providers, logistic support.

You can also acknowledge this course in your future publication if you feel that it triggered research article writing in you. Do not use this section for grazing do not acknowledge senior authors and direct stakeholders like department principal, head etc.

References section: Correct, and quality references should be cited and be listed at the end of the article as for the recommended style referencing will be discussed exclusively in the next chapter.

Section Summary

So dear learners, judicious use of infographics, discussing your results critically and then concluding your outcome is the prime focus of good academic writing. We discussed and then acknowledging those who helped you out like project grants and other logistics supports. So, we'll be covering the next aspect of revising, formatting and final check in the next section.

Dear learner you have written your first draft now it is the time to revise edit and format your manuscript.

Revision

In revision, make major alterations to make your manuscript attractive and easy to read. Fill in gaps. Correct flaws in logic restructure the document to present the material in most logical order.

Polish the style, refine the text. Focus on correct grammar and spelling polish the style, refine the text and focus on correct grammar and spelling. Ensure cohesion and coherence we have talked a lot about this aspect. Format the document and page setting like one inch margin from all the corners of a4 page, 12 font, times new roman font, double spaced or as recommended by the journal. Refer table and figure guidelines and follow them.

It is important to do the tasks in the stated order. Otherwise, you may find yourself spending a lot of time revising material that you later delete.

Revision is an ongoing process. The most practiced strategy is to write out an initial draft and leave it for a day. Then again come back to it and revise and always be ready for multiple revisions.

Getting feedback

Getting feedback is one of the most important thing. For improving your article be sure your co-authors have had the chance to read and comment to the draw. When ready, give the manuscript to a third person or a colleague, ask for his/her comments. And what level of information you like to ask for it. After you get the comments revise your manuscript to address your concern, but do not submit your manuscript until you feel it is ready for publication.

Final check

Friends now, we will move to the final check. Final check must be done by all co-authors. Co-authors should check the final draft to make sure it satisfies their expectations. Check the overall appearance identify any break in inconsistency page breaks etc. Check compliant to formatting guidelines.

Check table and figure number and their text citation. Sometimes it happens that table and figure are there but they are not cited in-text, you forget it. Do not do this thing. Finalize abstract and check the accuracy of the abstract. Lastly finalize the title and see the accuracy alphabet by alphabet.

Dear learners, you know the best strategy once you finalize your manuscript, read it aloud. Reading it aloud points out the mistake. And read it aloud from the hard copy.

Summary

So, let me summarize this model for you. We learned that in drafting stage, what and how we should plan materials and methods. In methods, clarity and reproducibility should be ensured. In results, the professional representation must be there. The presentation of your result must be critical. The description should be in good tune with the laid down aims and objectives. And again, it should be critical, and the free use of references must be there. Conclude your outcome. Acknowledge the supporters and the help of the grants. References, we will be dealing with this aspect in the separate chapter. And then you revise, format, and edit it. Remember that revision is an ongoing process. Multiple revisions may be required. Finally, check for the overall consistency and quality of the article. Read it aloud and you are ready for submission. We will be dealing with the other aspects of article writing in the next week. So, try it out try to write a research article. Happy learning.

Further Readings

- https://libguides.usask.ca/writing-help/disciplines/humanities-social-sciences
- Writing for Scholarly Journals, Publishing in the Arts, Humanities and Social Sciences, https://www.gla.ac.uk/media/media_41223_en.pdf

- Borja A, Writing the first draft of your science paper — some dos and don'ts, https://www.elsevier.com/connect/writing-a-science-paper-some-dos-and-donts

- https://www.springer.com/gp/authors-editors/journal-author/journal-author-academy

- https://www.apa.org/pubs/authors/new-author-guide.pdf

- https://projects.ncsu.edu/labwrite/index.html

- Maxim S. Pshenichnikov, Academic skills, file:///C:/Users/DELL/Desktop/Writing-paper-SEPOMO3.pdf

References

- Faber J, Writing scientific manuscripts: most common mistakes, Dental Press J Orthod. 2017; 22(5): 113–117. doi: 10.1590/2177-6709.22.5.113-117.sar

- Ohwovoriole AE, Writing biomedical manuscripts part I: fundamentals and general rules.WestAfr J Med. 2011 May-Jun;30(3):151-7.

- Ohwovoriole AE, West Afr J Med. Writing biomedical manuscripts part II: standard elements and common errors. 2011 Nov-Dec;30(6):389-99.

- https://writingcenter.unc.edu/tips-and-tools/scientific-reports/

CHAPTER 23

Referencing and Citation

Dr Ajay Semalty
H.N.B Garhwal University (A Central University)
Srinagar Garhwal-246174

Dear learners, we have discussed Review, research paper writing in the previous chapters. Let us learn referencing in this chapter.

Learning Outcome	Lesson Plan
After learning this chapter, you will be able to • Understand the basics of referencing of research/review papers	• What are references? • Types of references? • Basics of accurate and effective referencing.

What are references?

- **Source of information or studies:** References are the group of information source from where we gathered the information. We refer other articles to develop our idea. These are references. We are going to discuss that quality of reference give the quality to the article itself.

- **Citations are very important in any article:** The quality, quantity and coverage are the three decisive aspects (of relevant studies in citations) for acceptance of a manuscript. So, don't over look this aspect. I agree that you are tired after doing the exercise of revising editing and going into your details of your technical aspect but don't ever skip giving due attention and time in referencing section.

Sources/Types of references

Let's discuss what are the types of references. The references sources may be journals articles, it may be conference paper, it may be research article or review paper, URL or a website, it might be a book or book chapter, or it may be a patent. Refer the various style of writing which we have discussed very early in our course: The APA, CMS, MLA and their kind of styles. The style which suits you best, which is recommended by journals; See in that particular style, how journal article is written in reference section. You can't write at randomly so you must know the way to write each of type of references in the reference section. For example, how you are dealing with sur name and first name, whether sur name is first or first name is first, in writing comma, colon journal names.

Remember one the one most importance thing is that in each and every subject, a particular journal has got the standard abbreviation. Use the standard abbreviation. For example in Pharmaceutical field journal for International Journal of Pharmaceutics, we write Int. J. Pharm. So, abbreviations are standard. You need not to give full name of journal. Remember this and practice.

[Refer List of Title Word Abbreviations (LTWA) that contains all the standardised abbreviations used for words in scientific citations. It is based on ISO 4 which is an international standard which defines a uniform system for the abbreviation of serial titles, i.e., titles of publications such as scientific journals. The ISSN International Centre, which the International Organization for Standardization (ISO) has appointed as the registration authority for ISO 4, maintains the "List of Title Word Abbreviations", which are standard abbreviations for words commonly found in serial titles. For LTWA, refer https://www.issn.org/services/online-services/access-to-the-ltwa/.]

Right Referencing

Major factors to be considered in right referencing are

- Quality
- Quantity
- Uniform Styles/ guidelines
- Coverage:
    - type
    - updated

Right Referencing

Major factors to be considered in right referencing are

Quality: Refer articles from quality journals. Why to focus the quality journals? We have discussed earlier that wrong or substandard article and its content cannot lead to good quality research. If you are referring to a substandard journal's article it cannot be of good quality. So, reviewers make their perception on the basis of quality of article which you are citing; that what kind of work you have done and the perception is like that only. So, refer only quality journal articles. And quality, we have discussed with respect to journal metrics decide which journal is good for referring one.

Many journals in peer reviewing validate the references. Validation is done in terms of indexing. Less the number of validated references lesser will be the chances of acceptance. And this validated list of references is provided to the peer- reviewers. So, reviewers just go through the list and see how many validate references are there.

So, focus on quality journal for referring.

Quantity

Many journals have some **trend**, it may be **written guideline**, or it may be a trend that require **set number or range of references in research or review article**. Many journals, for research they require lesser number from 20 to 30 or maximum 50 (sometimes, they are very precise 25or 30).So, limit your references as per the trend and within range. Otherwise, reviewers' comments will come again and you have to reduce number of references.

So, see the trend. In most of the cases of review, limitation is not there. Because reviews are the exhaustive body of knowledge. So, you can add as much as references, in most of cases, but still the journals are there which limits the number of references for the review articles.

So, what you have to do? You must see the articles of journals for general number of references in research and review articles and limit the number of references as per the trend.

Section wise number of references cited must be justified. Why we are saying it because we cannot add 75-80 % of references in the introduction and just 10to 20 % in the result and discussion. It means we are just doing formalities. Reviewers see that thing, that you must have a balance of distribution of references/ citation in sections. So, have a well balance.

Most references be there in "Result and discussion" followed by "Introduction" and the least in "Methods" section.

Section Summary

Dear learners, References are the compilation of source of information which you have adopted or which you have referred in your research work. The quality and quantity of references must be ensured for your research article.

Uniform Styles/ guidelines

Dear learners, we discussed that we must refer good journals, must refer good quality references and should also have limit on references and distribution of references. Now we will move to uniform guidelines and styles to be practiced in writing references.

APA, MLA, CMS etc. anyone of the styles recommended: We have discussed the writing styles in a particular chapter, when we are discussing in English in academic writing. We discussed many styles and we focused major three styles: the big three: APA, MLA and CMS. We must focus how particular type of reference is written in a particular style.

Comply with the style and formatting guidelines of references including the text citation: Nowadays, it is trend to be unique in a formatting. Many Journals recommend a unique style which is a mix of 2-3 type of big style houses. So, you have to focus from each and every aspect that referencing, the formatting, is complying with the guideline of the journal.

Remember, to use right abbreviation of the journal (Refer LTWA: https://www.issn.org/services/online-services/access-to-the-ltwa/)

Do not skip comma, colon, semi colon when you were writing as per the style.

Sample APA style: Let's have a quick look on APA style.

- The reference list should be alphabetically ordered that listed by last name of first author like

- Smith,G. (2008)

  or in case of multiple author,

- Smith,G. Canady,J. , and, Roberts, J. (2008)

  for more than 6 author list first six than use a etal.

- References should be double spaced,

- Use hanging index

- Second and subsequent lines indented one half inch.

So, by this way, every style has somewhat little bit difference and you must go through them.

Web Resources: Now, let us move to web resources. URL and date of access must be noted because page may be changed after some time. Do not overuse the web resources. It looks awkward.

Among all the type of resources, the research article from a journal should be largely cited. You should not go entirely for review papers. It looks awkward and it is not justified also. So, you must have the good proportion of journals article into your reference list.

Coverage

Type: First is type of reference you mention. Whether you are referring review paper, research paper, patents, web resources, it should be good mix of all kind of resources. When you are discussing your research paper it is not supposed that you are skipping patent, or some very recent and relevant research article or very recent state of the art review article also. So, you must have **a good blend of resources but focus on journal article** specifically.

Updated: Use updated/current references in references. The distribution of references must be uniform i.e. according to time period the references used. You cannot have just one or two current references and next latest references of 5 years back. No, because reviewer give due attention to the chronologically distribution of references also. It shows that how seriously you have gone through the literature. So, this is key of avoiding rejection.

Section Summary

Dear learners, in this section we discussed the implementation of the guideline/ writing style recommended by the journal. You must practice writing in particular style, observe it and practice it; refer good quality journals with respect to quality, quantity, distribution, updation, coverage and then it will have a good mix of references. Literature management tools will be discussed in next section.

Literature Management Tools

Dear learners, after covering the major aspects of right referencing, we will move using the literature management tools. We discussed literature management tools aspects in a entirely separate chapter. Hope you remember that.

You are having a lot of references, you should manage the references with help of literature management tools. Take the help of one or another kind of literature management tools like EndNote, Mendeley, We discussed those, hope you remember that.

Export and import of references

You have to take help of these tools for exporting reference from those databases into that particular style which is required by the journal and then you will import from the library to your text.

So, by this importing or exporting of references with the help of literature management tools, the accuracy of particular style is ensured. It makes the process of referencing easy for you. Can you do the manual referencing when you are writing a review article or a research article which is having 50 references or 60 references? For one journal, it may be ok. But, when you are again redesigning or reformating the references for targeting another journal/ next journal, what will you do: Will you do it again manually? No.

If you will do with help of these management tools it will save your time, efforts and will ensure accuracy of the references.

Dear Learners, just recall how we exported the reference to library of endnote and then imported to your manuscript's word file.

If you will not manage the references through the tools you will just waste your time in managing and will also increase the chances of mistakes. Chances of mistakes means you are inviting rejection of your article. And I am sure you wont like it.

Tips

- Stick to type of style referred by journal.
- Don't increase unnecessary number of journals references.
- Follow trend of journal.
- Spell author name properly he or she may be one of reviewers and he/she would not like if their name is written wrong.
- Don't forget to cite latest and relevant references and;
- Do not use self-references very much. You can use one two depending upon the need, but don't overuse self-references.

Summary

What we learnt?

References are source of information which you are giving at the end of your manuscript.

The quality must be there from quality journals. The quantity should be as per the need of journal and as per need of topic. It should stick/comply to recommended style.

With respect to coverage, it should cover good mix of different types of references and at the last, the references must be updated.

So, with respect to coverage, the type and updation is required and you must use literature management tools for accurate referencing. It will save our time, efforts and avoid rejections for our manuscript. We will discuss submission and post submission in the next chapter.

Further Reading

- Cross JO, Impact factors – The Basics, the e resource management Handbook, https://www.uksg.org/sites/uksg.org/files/19-Cross-H76M46 3XL884HK78.pdf. https://researcheracademy.elsevier.com/writing-research/fundamentals-manuscript-preparation/guide-reference-managers-effectively-manage

References

- https://www.issn.org/services/online-services/access-to-the-ltwa/
- http://www.elsevier.com

CHAPTER 24

Submission

Dr Ajay Semalty
H.N.B Garhwal University (A Central University)
Srinagar Garhwal-246174

Welcome, dear learners. You have written your first draft, revised it and formatted it, gone through the references with the accuracy and now it is the time to submit your manuscript. In this chapter, we are discussing the submission part.

Learning Outcome	Lesson Plan
After learning this chapter, you will be able to know • Submission steps of your manuscript to a journal.	• Filling your copyright agreement • Drafting the cover letter • Going through the check list. • Different types of submission systems (offline, semi-online and online) • The basic of all the type of submission.

Let's begin this chapter. You have written your manuscript. It is ready. You have checked it, completely revised it, and the draft is final now. Now you are going to submit it. What will you do?

Copyright agreement

First you will go to the journal's website.

Download the copyright agreement.

Fill it. Before filling it, see it, read it. Read the conditions, if you are satisfied, get the signature of all the authors, co-authors and if you are going to submit online scan it otherwise get it ready.

In parallel, you must also see the article processing charges. Many times, it is mentioned in the copyright agreement and in author guidelines. Sometimes, it not there in author's guidelines and it is in there in copyright agreement. So, go through it. It should not happen with you that you communicated the manuscript and at 11th hour you come to know that you have to pay this much amount of money, if you want to get it published. So, you will be waiting for the entire period and then you will have to withdraw your manuscript. So, see carefully at this moment only.

Cover letter

Plan a cover letter: Next comes planning cover letter. For each submission a cover letter is required. In cover letter what you do? You will plan a good cover letter. You will **address cover letter to editor or chief or the editor**. And I suggest that address the editor with his/her name Like: Rather than writing dear editor write dear Professor Murthy (If Prof. Murthy is Editor in Chief). Everyone likes his or her name in print.

Give the subject: Submission of manuscript. Then give the **details of manuscript** that you are submitting the manuscript with manuscript details like **type of manuscript, title of manuscript.** And then in a paragraph most of the journals require that you justify **how your manuscript is suitable for journal.**

Rationality and novelty are also required sometimes. You mention the rationality and novelty of the work, in just one paragraph (in 3-4 lines).

At the last, you canopt for processing charges: If you are opting for the **fast processing charges or APCs.** You can mention it here or you are **opting for "color on print only"**. It means you don't want to pay for the color figures in the print. So, you give that particular option that color online only. So, the figure will be given only in the online version, in color. In print, it will be in grayscale.

So, you must mention here your details in your cover letter and sign it. And if you need online submission some journals require scanned copy or some journal just required the text.

Final checklist

Now, we come to the final checklist. You will have to be ready with these things. What are these things?

- **Manuscript in the desired file format**. i.e. in which version it is required as whether in word or pdf file, in word in which version it is required. Note it down and prepare your file in that version only and that to with author or without author. Many journals require manuscript without any indication even of authors. So, author profile/ author's name must not be there in manuscript for blind peer preview. So, you must be ready with the without author manuscript file. Sometime, journal require both with and without.

- **Separate title page:** Many journals require a separate title page to be uploaded in online system.

- **Graphical abstract:** The graphical abstract is a new thing that is coming in the online journals. What they do, apart from text abstract they require graphical abstract, as the name suggest a graph, a figure, an image, a diagram, a ray diagram, picture that can represent your study; can be put in the form of graphical abstract. So, plan a good graphical abstract which can grab the attention of the eyes, in the online platform.

- **Image Files:** Prepare the images files in desired file format jpg, jpeg, png, whatever it required and in the defined size.

- **Total number of tables/figures etc.:** Total number of tables/figures, total number of words you must have ready with you

- **Suggested reviewers**: Some of the journals require a list of suggested reviewers. You can't give name of person who supervised you. And it is mandatory that they must not be of from your institute itself.

- **Author details**: You must have all the author detail of each author, their affiliation, their contact details, email id for filling details in online system.

Section Summary

Dear learners, we discuss that for preparing yourself for submission of your manuscript first go through the copy right agreement. See APC (rather check it earlier) if not checked earlier, this is the point to check it out. Plan a good cover letter. Go thorough check list which we have mentioned. You will be dealing other aspects in the next section.

Now you are ready with agreements, cover letter and all related files in desired formats and information and this is time to submit it.

Submission type

Submission type may be-

(a) Offline

(b) Semi online

(c) Online

(a) **Offline:** However, gone are those days when we had to send the print copies in hard copies by the surface mails to journal office. But still there are journals which require hard copy of manuscript. So, if it is there you are having all the files, print them and send them with the cover letter to journal.

(b) **Semi- Online:** Haven't heard about it? Do not worry. It is just sending file in attachment through online to editor's email id. However, this practice needs the perfection because it doesn't indicate where you are going wrong.

(c) **Online:** While in the online system the system itself pokes you that you have reached the word limit, you have not filled the mandatory fields, that's why the online system are very popular system now a days.

Manuscript Handling System

Most of the journals are having their manuscript handling system specially with respect to the prestigious publishers like Elsevier, Springer, Bentham Sciences, Willey, Taylor and Francis all the publishers have their own manuscript handling system. Manuscript Handling system? why it is called so? Because author as well as reviewers use that platform. Author use it for submitting manuscript and reviewers use it for accessing the manuscript for reviewing and submitting the review report.

So, register yourself in that manuscript handling system by going through the **"submit manuscript" Link.** You will be directed to register into that particular handling system with all your detail as a lead author. Enter your details, contact details then your page will be created. So, you will be directed to the dashboard of journals' manuscript handling system.

Then first thing is **type of article.** You will choose whether it is original research article/ short communication/ major review article or mini review article.

Then **title of your article**. What will you do? You will just paste. Please don't try to write in manuscript handling system. However, it allows you, no problem, but writing directly on manuscript handling system may have some chances of mistakes.

So, what you do? Keep your files open in your laptop or computer and just copy paste desired things on the manuscript handling system.

So, once you have pasted your title. Go to the **keywords**. Again, paste the key words in desired number.

Then comes **abstract**. So, abstract is pasted on the section. If you are reaching or crossing the word limit the system will indicating in the red that you are going above the limit. So, it is better to see in advance to see the word limit. So, that you do not need to correct the abstract here. Avoid modifying the abstract at this particular moment because then it will be cut in some wrong places, which may lead to disaster.

I would like to remind you **keep on saving** work when you are filling it or pasting it into manuscript handling system. If system allows and most of system allow so in different setting you can process the submission.

Now once you have filled the details the **author details** are asked earlier. But you have all the detail in your check list. So, go to one by one in the order you have defined already. You will define who is first author what will be sequence. The corresponding author should be marked. Even you can submit and you may also be the corresponding author and others as co-author. Some journals also allow more than one corresponding author.

After finishing author detail, you may be asked for **suggested reviewers.** Sometimes it is optional and some time it is mandatory. The number of suggested reviewers is generally 3 to 5. If they ask suggested reviewers, it is to mention that they must not be associated with you and must not be from your institute. You have to enter their details, their email id.

After that **uploading files** section comes. You have gone through the check list, then start uploading files. So, you will upload (sequence may differ from one platform to another)

(a)   Copyright agreement

(b)   Cover page

(c)   Title page

(d)   Manuscript with or without author as described by journal.

(e)   Image files or infographics.

Some journals require each of the image as a single file and all the files are to be uploaded. Some journals just ask all the figures in the manuscript word fill only.

Upload: Click upload for uploading it.

Final preview and submission: For many journals there are final steps but for some journals combine it in one PDF and then allow you to have preview on your submission. Once you gone through preview, you have downloaded it in pdf form and go through it. And I suggest you take print out. See carefully in hard copy and then if you feel some problem is there, then revise it, edit it, and again upload it. Then preview again and give a final submission to it.

So, dear friends you have submitted it and quite relax now.

We will discuss some the remaining aspects in the next section.

We learned the basics of submission of manuscript. After submitting your manuscript, you will get **automatic acknowledgement** mail in your email id and remember to **use only that email id which you often / regularly check** and see mail should not go to spam folder.

Checklist: Meanwhile to **check the editorial policy of journal.** You must have idea that today if I will submit it, So at which particular time or day I can expect communication from journal. So, let me clarify you that each journal has standard time of communication after the submission. Some journals communicate you regarding the article processing charges, fast processing charges, and sometimes you immediately get the rejection. It will happen only when you have not targeted journal properly and something went wrong out rightly. It might be improper targeting.

So, be alert after submission. You should be vigilant from the communication being received at your end and you have to properly give the answer to the journal in a time frame.

So, you should be alert even after submitting for communication received from journal and meanwhile see your manuscript for accuracy or search if there is mistake. Prepare yourself to answer any query/questions/comment regarding it.

Summary

Prepare the final checklist.

Check out: Be ready with the following files and follow these steps.

- Copyright agreement.
- Cover page.
- Title page
- Manuscript file
- Author details suggested reviewers.
- Image files or graphics.
- Before submitting your file.
- Choose your submission type/system some time it is defined also (just copy from your manuscript)
- Preview it.
- Click upload for uploading it (if satisfied with preview)
- And submit it finally.

Further Readings

- https://researcheracademy.elsevier.com/writing-research/writing-skills/writing-persuasive-cover-letter-manuscript
- https://researcheracademy.elsevier.com/publication-process/finding-right-journal/identify-right-journal-publish

References

- http://www.scopus.com
- http://www.elsevier.com

CHAPTER 25

Post-submission

Dr Ajay Semalty
H.N.B Garhwal University (A Central University)
Srinagar Garhwal-246174

Welcome, dear learners. Welcome in the last chapter of article writing. In this chapter we will be discussing post-submission.

We have covered in earlier chapters, how to plan a research article, how to target a journal for your manuscript, how to write, how to focus all the aspects of research article, all the sections from the title page, to the abstract, introduction, methods, results & discussion, conclusion acknowledgement and references.

We learned how to submit your manuscript and now it is the time for addressing post submission steps.

Learning Outcome

After learning this chapter, you will be able to understand

- What should you do after submitting your manuscript?

- How can you avoid rejection?

Lesson Plan

We will be covering the chapters under the heads

- Knowing the editorial process.

- Responding to the reviewers' comments.

- Proof reading, and

- Discussing tips for avoid rejections of articles.

Let us begin our discussion.

You have submitted your manuscript. Now, what will you do? Little bit things we have discussed in previous chapter also, that what will you do after the submission. After submission, you will be vigilant that what communication is being sent by the journal. You will check your emails.

Editorial Process

Meanwhile, you must see editorial process, you must learn the editorial process of the particular journal and the trend.

What happens when you have submitted your manuscript, the editor or the technical editor just go through your manuscript with respect to the formatting aspects, with respect to the scope of your manuscript, matching with journal or not. If formatting is okay, that's fine. If scope is okay, if manuscript is matching with scope of journal, fantastic! Then the editor will just forward the manuscript to the peer reviewers. Peer reviewers are selected by the editor in chief, not mandatory, as per your suggestions.

So, it may be double or single blind peer review. Double blind means, neither author nor peer reviewers know about the names of each other.

Single blind means, peer reviewers know, who are the authors. Mostly, single blind peer reviewing is done.

So, in some journals also, suggested reviewers and non-suggested/preferred reviewers are required. You can exclude some researchers who may have the problem with your work and with you. It happens, everywhere. You may give a list of authors who are not desired by you. But very few journals have this kind of practice.

What does the peer reviewer do? When the peer reviewers are assigned the particular task. He just goes through your manuscript and checks

- The quality of your content
- Your presentation
- References
- Novelty and rationality of work and
- Overall how you have presented your hypothesis, is evaluated.

Depending on their expertise, they may accept in one read only, they may recommend "accepted with minor revisions with some comments" or "accepted with major revisions" or if you are not fortunate enough it may be rejected, at all. Sometimes, it is rejected in this format and can be resubmitted. Sometimes, the decision may be like this.

Peer reviewer send the comment to the editor and the editors ends the comments to the author.

Now, authors have got the comments from reviewers. Most of the times, there are two or three reviewers.

We will discuss responding the comment sent by reviewers in the next section.

Friends, you have gone through the editorial policy, the processing. Now, you have got the comments on your articles.

How to respond to comment sent by reviewers?

- Don't be nervous. It happens. Relax, sit back.

- Observe the comments.

- Discuss with your supervisor, colleagues, research team. Each and every point you must discuss.

- Formatting compliance: Modify your manuscript as per the comments. First thing, if you are agreeing on each and everything, its OK, No problem. What you will do? Use the Track changes in MSWord and start correcting the things one by one.

  o Correct all the formatting objections.

  o Typo errors and grammatical error should be corrected

  o Even if you got anything else, which was not pointed out by the reviewers you can also correct it here. You have the opportunity here.

- Run spell checker but don't rely upon it too much.

- Again, see, sit back and check whether cohesion and coherence is there or not. Read it aloud.

- The transition must be smooth from one point to another point, from one idea to another idea. The idea development must be smooth here. However, you are not supposed to change much. You should stick to the response to the comments addressing only. Smaller things you can do.

- Now work on the idea level things. The objection/comments may be small mistakes as for abbreviation, full forms etc. These are the minor things.

- Sometimes, you are asked to elaborate/ condense any method. You can do it. No problem, at all.

Problem occurs when there is the comment that "You should have done this kind of study also". Then, at this stage you can not do the study. So, you will have to humbly accept that thing. And be humble in responding that comment that "Yes, we could have done it. And we will do it in our next exercise. Or the scope of this particular study was limited to this aspect only. So, we didn't practice or did that particulars experiment/test."

But discuss with your supervisor, mentor.

There might be some condition that the reviewer might be objecting on your rational or some mistakes. If the blunder is there, you cannot defend it. Obviously. What will you have to do? You will have to skip that part. Not skipping the responding! You will have to delete it and admit it humbly, if you have done some blunder.

If you have some differences in justifying the rationality/ some technical or idea level issue, then you should put your idea humbly to the reviewer so that it is not awkward to the reviewer, also. ***And never ever out rightly reject the comment by the reviewer***. And admit it humbly and then present your idea, humbly. This is the key.

Plan the rebuttal letter

What you have to plan.

You will plan the rebuttal letter/ response to comment letter, addressing the editor. Mentioning your article number, you will write that you are giving the response to comments. However, you will be inserting those comments/ modifying your manuscript in the word file in track changes so that each change is visible to the reviewer or editor.

But a file must be maintained, in a word file you will mention the response to comments like this.

As you can see here that we are addressing to editor, subject and then after the few lines with reference to the revision, response to comments are as followed.

Comment from Reviewer 1

Comment 1:

(That particular comment you will just paste from the mail from which you have got the comments)

Comment 1: ....

Response 1:.............

Comment 2:...........

Response 2:...........

By this way every comment by every reviewer, you will address. It might go upto 2 pages, 3 pages. It does not matter. And remember! your tone must be humble and assertive.

Resubmission: After you finish the response to comments letter or rebuttal letter, go to the manuscript handling system, again. And then submit your revised manuscript, along with rebuttal letter and upload it.

If you are practicing or following offline mode, then you will have to prepare this letter, get the printout and send it by surface mail.

Response from the editor: You have responded to the comments. Again, wait. This time you will have to wait less. Might be in a week maximum, you will get the answer from the editor. If he/she is satisfied they will give decision. Sometimes, it is again sent to same reviewers, then it may take time. It depends on the policy of the journal.

And on the basis of the reviewer's expertise, the reviewer give their decision, which is passed on by the editor to you.

Now, you have got the acceptance letter. Enjoy the moment.

I remember my moment when I got the very first acceptance, that to in the form of an inland letter. I have kept it today itself along with my testimonials. You feel like you are in seventh heaven, specially, when it is your first publication.

So, enjoy the moment. And be ready for the next steps. We will discuss the next steps in next section.

Dear learners, you are happy with your acceptance.

Now, what to do next. Now, the editor will take next step. The editorial team finalize your manuscript, type set them in the format in the paginated or non-paginated form in the journal's style and send it to you for proof reading.

Proofreading

In this proof reading step/ the final step, remember, in this step you can't do major change here. Focus on-

- Number of table and figures
- Text citation.
- Any break or skipping in any para (during typesetting it may happen)
- See the author name and affiliation properly.
- See the title.
- Observe for any typo error. (some small typo error can be corrected)
- Observe the legends/ figures/ their placing; if wrong report them.

You may submit word file or pinpoint comments with required/desired modifications. And remember, this is time bound step. Once you get the galley proof for proofreading, in most of the cases, you will have to return the corrected galley proof in 48 or 72 hours max. If you skip the mail, your manuscript may get delayed or sometime may get rejected also. So, follow the timeline and immediately/ as early as possible get back to the editor with the corrected galley proof.

You have returned the galley proof. Now you don't need to do anything. Now, when the article will be online, you will get the DOI of the article (ahead of print section without pagination). You can share that link with your friends, colleagues and other researchers. Here, you got published your paper online and in a short period of time it will come in print.

So, you might have enjoyed the process. The hard work pays.

What are the major tips to avoid the rejection? Let us pinpoint them one by one.

Tips to avoid rejection

    (a)   Present the article in simple language with clarity.

    (b)   Rational of the study must clear and be focused entire study.

    (c)   Aims and methodology should be clearly explained.

    (d)   Provide the results systematically and in totality.

    (e)   Discuss the result critically and with logical reasoning without any general vague statement.

    (f)   Conclude study as per the objective of study.

    (g)   Provide the scope of further study required, if any.

    (h)   Provides supplementary data or spectra also in support of your study. It will improve the quality of your article and will also help you in getting the higher citation for your that particular paper.

    (i)   Always stick to manuscript guidelines prescribed by the journal with respect to formatting and all other aspects of the research article (to prevent failure or rejection of the manuscript).

Summary

Let me summarize the chapter for you.

Once you have submitted your manuscript, after the peer review you get the comment from the reviewer through the editor.

You prepare point to point response to comments, without out rightly rejecting even a single comment by the reviewer. Address each and every comment with humbleness and logical presentation. Your logic must be there when you are answering the idea level comments.

Thereafter, when you get the acceptance, you will be given the final form of your article for proofreading. Get back with the corrected galley proof to the editor. And last, enjoy your published article online and then on print.

We learned how we can avoid the rejections. Simplicity, clarity, logical sequence, cohesion, coherence are the keys. Stick to the formatting guidelines. It will help you to avoid the rejections.

Best of luck for your future endeavors for writing the research articles. And please do not skip acknowledging our course, if you get a publication after completing this course. I wish best of luck for your manuscript writing. Happy learning.

References

- http://www.scopus.com
- http://www.elsevier.com

CHAPTER 26

Thesis Writing-I

Dr Mona Semalty
H.N.B Garhwal University (A Central University)
Srinagar Garhwal-246174

Learning Outcome

After reading this chapter, you will be able to know

- About synopsis
- What are different parts of synopsis?
- How to write synopsis?

Lesson Plan

- What is synopsis?
- How to write synopsis and what are important things to be kept in mind?

There are different segments of thesis writing

- Introduction
- Literature review
- Aims and objectives
- Methodology
- Time frame
- References

Synopsis

This is presented in the form of a brief research proposal for approval of the topic from the authorities (like research degree committee). For this we should provide very brief detail of all the parts in specific and concise manner. The length of each part of the synopsis may vary depending on the requirement of individual institutional guidelines, trend and as per the instructions of supervisor.

Title

Choosing a research topic or title (i.e., getting started) is perhaps the most difficult part of writing a synopsis. Title tells us the full story and it reflects the ascent of the complete research. Title of a synopsis must be same (word to word) the title of your thesis.

Title should be -

- Catchy
- Should not be too lengthy
- Broadly cover the topic
- Should easily create interest
- Clear
- Precise
- Justify the time limit

Introduction

Introduction of synopsis should cover the little background information of the topic. In brief you must explain that what work already exists in this area. And What are its strength and deficiencies? Writing introduction of a synopsis for sciences is significantly different from other disciplines. For example, sciences may justify addition of figure or graph for better explanations where in case of arts and humanity you may be need some data to prove relevance of the topic. And by the end of introduction in both the cases you should state how your area of research is different from others.

Thing to remember

- Should be started reflecting background
- Justifying the need of research
- Statement of problem
- How your research is different from existing one
- References should be cited

The introduction should not have a strict word limit, unlike the abstract, but it should be as concise as possible. Sometimes, scientists and researchers prefer to write it last, to make sure they haven't missed anything important.

Things which you should never do

- Should not write about yourself and your supervisor
- Don't expose the whole idea and about your expertise
- Don't forget to relate your work with basic literature

Literature review

Literature review (LR) can be defined in many ways.

"Systematic and organized compilation and critical study of related body of knowledge is literature survey."

"A literature review surveys books, scholarly articles, and any other sources relevant to a particular issue, area of research, or theory, and by so doing, provides a description, summary, and critical evaluation of these works in relation to the research problem being investigated."

What are the criteria for selecting literature...?

- Should be restricted to 10- 20 research paper (Important)
- References should be latest
- Research paper should be preferred over review
- Should be 100 percent relevant to the topic

For synopsis there can be two ways of writing literature review

- Paragraph wise little description of paper
- Running discussion of all the selected papers

Aims and Objectives

Aim of the study should be in the form of a statement which somewhere reflects the hope what you want to do and where you want to finish. Aim is about what you want to do and objective is about how you are going to do that.

Aim should be -

- Brief and concise
- Realistic (Means weather you can achieve them in the given time period)

How you can start for sciences-

- To develop...
- To design...
- To prepare...
- To evaluate ...
- To synthesize.... likewise

For other discipline

- To track
- To generate
- To build ...
- To compare
- To critically analyze

Objectives

These objectives should be SMART (Specific, Measurable, Attainable, Relevant and time bound). In totality the set of objects should meet the aim of the study. Objectives are list of things (Point wise) you are going to do to achieve your aim and it visualizes the sequence of each step of your research.

It should start like…you will… collect, construct, produce, test, trial, measure, and document, pilot, deconstruct, analyze… and in case of sciences it should go like…Preparation, formulation evaluation statistical application and comparison.

Objectives should not be

✓ Be too vague, ambitious

✓ Repetitive

✓ Lengthy

✓ Messy

Methodology

Methodology involves the process of carrying out the actual research. In case of sciences state the method which you used to prepare. synthesize, to analyze and mention how did you perform it and describe the materials and equipment used in the research.

For humanities you should state which method you used to collect data /information/cases etc…and explain if any randomization techniques/statistical methodology you are going to apply. Methodology can usually be divided into a few sections depending upon the area of research.

Methodology should be-

- Clear

- Concise

- Conclusive

Things to be kept in mind

- If you are mentioning one single method, you have to be specific in description and if you are

- This in case of if you are mentioning multiple strategies (It may be coz you didn't yet decided the actual method or you want to compare different methods).Let it be little messy so that you have enough space to confine the final strategy.

- Don't discuss the established technique. Rather discuss how did you apply it in your research for example

✓ Don't discuss single blind and double-blind method of sampling. Everyone knows it. Discuss how you applied in your case.

✓ Don't discuss any equipment or machine. There is no use of doing... that state how did you used them

Time Schedule of Work

This is just like a simple time table of your work .This part should tell when you are going to finish a task in the entire duration and also tells how the completion of a phase of the work is related with the commencement of the next phase. Once your problem statement / formulation and objectives are approved by your supervisor, all details should be added to your time plan.

This is either given by CPM (Critical path method), PERT chart (Performance evaluation review technique) or by simple tabular presentation. It covers -

- Literature search,
- Potential fieldwork (e.g. interviews and/or questionnaire administration),
- Procurement of raw materials
- Compilation of data/preparation and synthesis
- Data analysis /Evaluation
- Result writing /Discussion

References

A synopsis should have limited references depending upon the prescribed guidelines. In general, a synopsis should have about 10-15 latest references. References should be properly and uniformly cited in the text. The way of references writing should be uniform as there are number of reference writing styles.

All references should be

- Relevant to the topic
- Latest one
- Research articles
- Authentic one (should be of reputed journals and books)

And tips for synopsis

- The length of synopsis may vary depending on the set guidelines of university/instructions of supervisor

- Show raw draft of synopsis to your supervisor in hard copy
- Discuss aims and objectives and rational in detail with your supervisor
- Refer latest references and cite them properly
- Plan coverage properly (…Title, University logo…sequence of names…(institute /supervisor/co-supervisor etc.)
- Take final approval from supervisor before taking final printout of the same.

Further Readings

- http://betterthesis.dk/literature-search/test-and-summary
- www.nitttrchd.ac.in › sitenew1 › comp_sc › pdf › PhD-Synopsis-template
- https://www.youtube.com/watch?v=Gv6_HuCYExM. Semalty A., Thesis Writing, Lectures on ARPIT/ Refresher Course: http://tiny.cc/ARPIT

References

- https://youtu.be/3S289Oka_Jw
- http://archive.mu.ac.in/syllabus/4.71%20Guidelines%20thesis%20dissertaio n%20report-%20Ph.D,M.E,B.E%20%20(AC%204-3-14).pdf
- http://intra.tesaf.unipd.it/pettenella/Corsi/ReaserchMethodology/ResearchSy nopsisWriting.pdf
- https://documents.uow.edu.au/content/groups/public/@web/@stsv/@ld/doc uments/doc/uow195641.pdf

CHAPTER 27

Thesis Writing-II

Dr Mona Semalty
H.N.B Garhwal University (A Central University)
Srinagar Garhwal-246174

Learning Outcome

After reading this chapter, you will be able to know-

- What are important things to be covered in introduction?

- What are different ways of writing literature review?

- What all things you need to focus while writing material and methods?

Lesson Plan

- How to write introduction of thesis?

- How to write material and methods for thesis?

Introduction

Introduction of thesis should cover the background Information of the topic. So, you must explain that what work already exists in this area. And What are its strengths and deficiencies? The introduction should not be restricted to word limit, unlike the abstract, but it should be as concise as possible. Sometimes scientists and researchers prefer to write it last, to make sure they haven't missed anything important.

And writing introduction in sciences and life sciences is significantly different from other disciplines. For example, Sciences easily justify addition of a figure or graph for better explanations, if the topic is related to the formulation which is used for hair growth. You can discuss anatomy of hairs to understand it more clearly.

In case of arts and humanities you may need some data to prove relevance of the topic. You can also relate it to any established theories for example; if we are analyzing juvenile delinquency then we can relate it to the conflict perspective or anomie by Robert Merton, a well-known sociologist.

So, there is little difference in that, but in both the cases by the end of introduction you should state how your area of research is different from others and connecting it to the research problem by pointing out the relevance of the subjects.

Thing which should be kept in mind

- Always Start with 'eye-catching' opening sentence that will keep the reader's attention focused
- Keep it broad, but not too broad.
- Provide only helpful, relevant information
- Relevant background, but don't write anything that reflects argument.
- Not to say everything you have to say in the introduction – save some of your good material for later.

Now have a look on few things which you should never do-

- Should not write about yourself and your supervisor
- Don't expose the whole idea and about your expertise
- Don't forget to relate your work with basic literature

Literature Review

In the previous chapter we have discussed about detailed literature review and different methods of doing literature review but. Here I will focus on how to write literature review …you may find several books and write-ups on how to write literature review…but it's difficult to conclude anything from such waste text books. That what you must do

This also we have discussed in brief in how to write literature for synopsis…for thesis we must cover it in more and more deft and you should do proper referencing of this part along with writing ….

There are two different ways of writing literature review

- Paragraph wise little description of papers

This paragraph wise literature writing is generally used in synopsis writing/ or for dissertation. It gives complete details of the paper whether it is research or review in one small paragraph. But it does not look nice and is not effective method of presentation.

Literature review must have -

- Introduction-Little introduction about topic as to form background following name of author and year of publication
- A body- consist of brief about actual work done
- Conclusion-final reporting (summary and significance of the study) of paper

Example

"Adhirajan et al. (2001) studied the hair growth activity of mixture of petroleum ether extract of Eclipta alba Hask. (compositiae), Citrullus colocynthis Schrad. (cucerbitaceae) and Tridax procumbens Linn. (compositeae) in various concentration in the form of herbal cream and herbal oil82. The ratio of E. alba, C. colocynthis and T. procumbens in 3:1:2 showed excellent hair growth activity with 35% more anagen hair follicles as compared to 20% with standard drug (2% ethanolic solution of minoxidil)."

- Running discussion of all the selected papers

It means in running text form discussing one paper after another in continuation relating one to other. In this case you can categorize thembased on method, material and target area etc. It May be done like a review paper.... Or if you have already planned a review paper, just put it here.... Or you can write this in review paper style.... You may use figure or table judiciously here. ...can also use flow diagram, pictorial presentations using smart art or paint for better understanding. Cite references properly, updated (latest) and quality review should be written. Patent survey should be done if relevant or applicable.

Example

"Rho et al. (2005) examined the effects of 45 plant extracts that have been traditionally used for treating hair loss in oriental medicine in order to identify potential stimulants of hair growth. Asiasari radix extract showed hair growth-promoting potential87. In another similar study, Rho et al. studied the hair growth promoting effect of Sophora flavescens and showed that it can be used as potential hair growth promoter."

Materials and Methods

Materials

In sciences material involves the raw material required for the preparation, synthesis and development of any formulation or any compound.

You should clearly mention what all material (Only important or critical material u need to write ...not normal organic salt or routine ingredients) used for research and from where it is procured (source) so that someone who is going to refer your thesis should come to know about details.

In case of any specific material for example if you are using any pharmaceutically active drug. Provide full monograph (complete detail) of the drug.

Regardless of the outcome, the raw material you relied on to come to your conclusion will decide your work credibility. Credibility of the sources and proper documentation of materials and sources will strengthen your dissertation. Do not skip minute details like name of software, version, equipment brand model, chemical source etc.

Methodology

Methodology involves the process of carrying out the actual research. In case of sciences state which method, you used to prepare...Synthesize, to analyses and how did u perform it. Giving full detail will help you in future and will be a quick and complete reminder of your work to you.

For humanities you should state which method you used to collect data /information/cases, sample size etc...and explain if any randomization techniques/statistical methodology you are going to apply and weather it is qualitative or quantitative. Methodology can usually be divided into different sections depending upon the area of research.

Methodology should be-

- Clear

- Concise

- Conclusive

Things to be kept in mind-

- If you are following any other authors method, you have to cite that reference or if you are modifying it you have to mention this is modification of so and so....

- You are using animal in your study discuss protocol and mention about institutional ethical committee approval

For animal ethical committee CPCSA approval

- If you are mentioning one single method, you must be specific in description

- In case ...if you are mentioning multiple strategies. Then you must prioritize what to write first and what later.

- Don't discuss the established technique. Rather discuss how did you apply it in your research for Example

    ✓ Don't discuss single blind and double-blind method of sampling. Everyone knows it. Discuss how you applied in your case.

    ✓ Don't discuss any equipment or machine. There is no use of doing... for example if you have used diffusion cell for permeation analysis don't discuss about diffusion cell discuss about how you usedit.

For effective presentation of methodology for any discipline use figures, Plan flow diagram and pictures to make it easy to understand. you can also use ray diagram to explain methodology of survey /sampling and/or for questionnaire setup in other disciplines other than Life sciences.

Flow chart -A flowchart is a visual representation of the sequence of steps and decisions needed to perform a process (Fig. 27.1). Each step in the sequence is noted within a diagram shape.

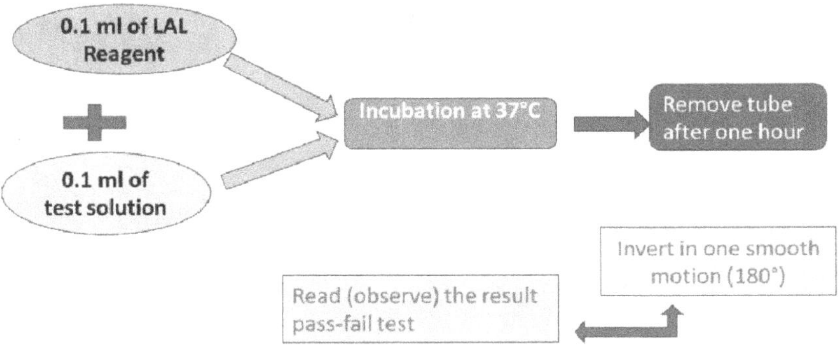

Fig. 27.1 Flow chart presentation of method

Table and figure-You can also represent in the form of table (questionnaire setup /deciding different sample size for different conditions). You can also draw original figure or diagram by your own or can modify the existing one, by providing reference.

Further Readings

- http://betterthesis.dk/literature-search/test-and-summary
- www.nitttrchd.ac.in › sitenew1 › comp_sc › pdf › PhD-Synopsis-template
- https://www.youtube.com/watch?v=Gv6_HuCYExM
- Semalty A., Thesis Writing, Lectures on ARPIT/ Refresher Course: http://tiny.cc/ARPIT

References

- https://youtu.be/3S289Oka_Jw
- http://archive.mu.ac.in/syllabus/4.71%20Guidelines%20thesis%20dissertaion%20report-%20Ph.D,M.E,B.E%20%20(AC%204-3-14).pdf

- http://intra.tesaf.unipd.it/pettenella/Corsi/ReaserchMethodology/ResearchSynopsisWriting.pdf
- https://documents.uow.edu.au/content/groups/public/@web/@stsv/@ld/documents/doc/uow195641.pdf

Thesis Writing-III

Dr Mona Semalty
H.N.B Garhwal University (A Central University)
Srinagar Garhwal-246174

<table>
<tr><td>

Learning Outcome

After reading this chapter, you will be able to know-

- How to frame results and discussion

- What are important things to be focused?

- How are you going to present it?

</td><td>

Lesson Plan

- How to write results and discussions separately?

- How to write it simultaneously?

- Other parts of thesis writing

</td></tr>
</table>

Results

The result part is very easy to write, as it is just compilation of facts and observations. And discussion is discussing the observed facts and data with the help of supporting papers (literature). In general practice you can write results and discussions in two ways

- Write results and discussions separately

- You can write the results and discussion simultaneously

Or there may be specific guidelines of writing the same (it depends) ...or...you may have to follow the instructions of your supervisor. But in any way, it is up to you that how effectively you can present it so that it is very easy to understand. You should guide the findings of your study in a logical sequence by using non-textual elements (mostly if possible), such as, figures, charts, photos, maps, tables, etc.

You can start the section stating in brief that with what you have done in the study. Then take the results of experiments one by one. And if you are little confused about certain results just recall your research question and then decide what and in which sequences it must be written. The sequences of the results should follow the listed methodology, or you can write in order from most relevant to least important one.

Points to remember while writing the results-

- Always write results section in past/ past perfect tense
- Make sure that each table and figure has a number and a title
- There should be one single way of writing table
- Focus on writing just observations and data (interpretation relevance and justification of observed data u can do in discussion part)

Tabulation of Results

The results should be tabulated whenever it is feasible. Even when the result of a parameter is not of more than one set, the results of this parameter should be grouped and tabulated with the result of the same parameter for other subjects. This may also be grouped and tabulated to observe any correlation or relevancy with other different or relevant parameter. A table should possess the following parts essentially.

1. **Type:** Tables should be of uniform type in the thesis. (Various types of tables can be found in MS Office tool bar of table).

2. **Layout:** Preferably should be in portrait (not in landscape)

3. **Title:** A brief, clear and easy to understand title

4. **Column and Rows:** Number of divisions and subdivisions of column or rows should not be confusing.

5. **Heading:** Heading of each column or row should be very brief and complete in all respect (unit of measurement etc.)

6. **Footnote:** Any extra information (p value, abbreviation etc.) for a set of data should be given in footnote, by providing asterisk or any other mark to the relevant data in the table. This is not compulsory to provide foot note to each table.

7. **Grid lines:** The grid lines should be clear enough to show the partitioning between column and rows. Sometimes hiding the gridlines make the table confusing.

In tables, unnecessary colors in column and rows should be avoided. You can directly paste any table from MS excel worksheet to MS word file also.

Plotting Graphs

Graphs should be plotted using MS excel worksheet. The Graphs may be of various types: -

Pie Chart: When data can be shown in percent or parts the pie chart can be used. The pie chart is presentation of data in a circle (2D or 3D) to show the data or readings in percent or parts (Fig. 28.1).

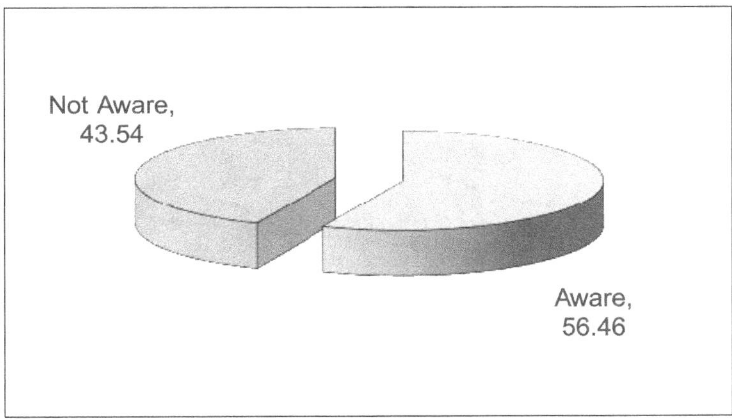

Fig. 28.1 Gender distribution of immunized population (Pie chart)

Bar diagram: The column form of data presentation is done in between X and Y axis (Fig. 28.2).

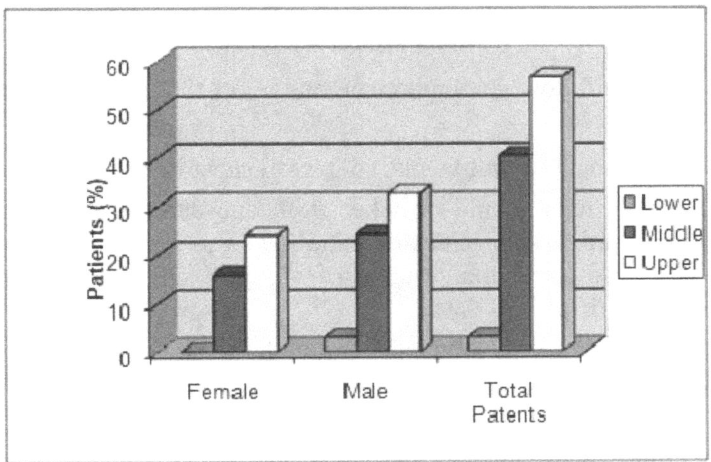

Fig. 28.2 Occurrence of Diabetes mellitus in different socioeconomic groups (Bar Diagram)

This is useful in giving the direct comparison between two set of observations with same set of variables. For example: Population living in different district of a state (Population in Y axis and District in X axis), Initiation time and completion time of hair growth by test standard and control (Time in Y axis, set of initiation time and completion time for test standard and control group of animals)

Line curve: This is the presentation of progression or change of a variable at different set of another variable (Fig. 28.3). These are the most common chart used to show the growth of microbes with time, release of drug with time etc.

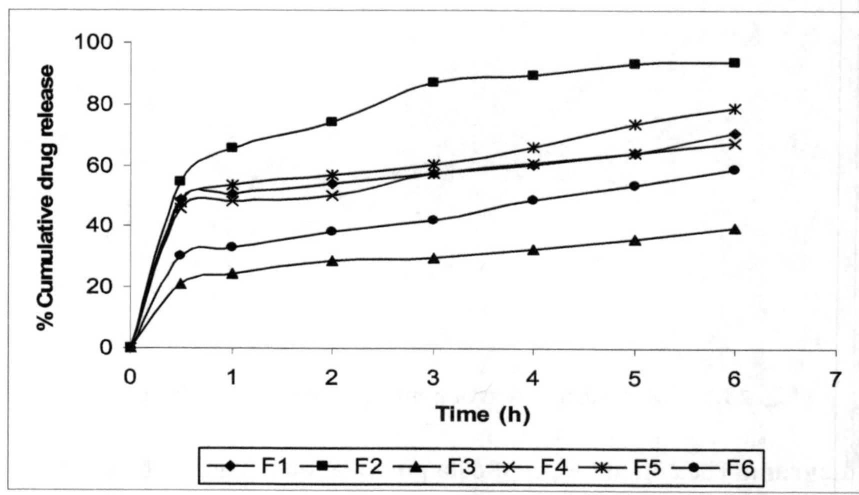

Fig. 28.3 *In vitro* drug release study of floating microspheres of Ofloxacin. (Line Curve)

A graph has the following component, and all of these should be given due attention. -

1. **Data:** Accurate data should be placed in worksheet in the right way.
2. **Axis:** X and Y axis should be taken with due care with the appropriate maximum and minimum values of scale (so as to give a right view to the graph). The major and minor units of scale should be according to the desired and full display of data.
3. **Labeling:** Axis should be labelled along with the unit of measurement.
4. **Gridlines:** X and or Y axis' gridlines may be shown if desired for certain purpose.
5. **Legends:** Legends should be clear, brief and easy to identify. The colors or the sign of legend of a set of data should be unique in a single graph. The

layout and positioning legends should be uniform for all graphs (i.e. Vertical layout and Bottom right positioning).

6. **Error bars:** Many high impact journals demand inclusion of error bars in each line curve and bar diagram. This can be done with the help of MS excel.

While pasting the graphs from excel to word sheet, you can either paste the graph as entire worksheet, or with link to work sheet or only as a curve with no link. If you paste the curve with entire worksheet, you can edit the graph from the word file itself. You do not need to go and search the desired worksheet in excel. However, if you opt for graph only paste option, then no one can access the data and edit the same in any case (from the word file).

Explore more software like Origin for effective presentation of data. For every new software you must learn importing and exporting data. The data may be asked by any reviewer at any point of time. Moreover, you may need the same at any point of your career for planning the extension of the research. Therefore, the data must be preserved after the report writing or paper writing. Now a days the supporting data/ dataset are also required by many journals along with the manuscript. Any fabrication or falsification of data may lead to the rejection or retraction of paper.

The Table and graphs must be shown with all the statistical test parameters (p value etc.). Without relevant statistics no paper is published. Required statistics software must also be practiced (like R, SPSS etc.)

Last but not the least, the image plagiarism must be avoided. Do not just copy paste the figures. It may lead to rejection and/ or retraction of your paper/ thesis. We have discussed the methods of avoiding image plagiarism in Plagiarism (Chapter 9).

Things you should never do while writing-

- Do not discuss or interpretate your results
- Do not manipulate the finding (An expert reviewer can easily trace the manipulations, and this may result in bad consequences. Even a small manipulation can make your whole work questionable and unreliable.
- Do not ignore negative results
- Do not over actualize the findings
- Do not repeat the same information
- Do not confuse table with figure

Discussions

Discussion of a thesis is the centric part that everyone wants to read. It should be strong enough to deal with the all probable queries of readers and evaluator. And if you write it strong enough you need not to do this exercise again for writing your research paper of the same thesis. So doesn't just finish it fast take enough time to write this part. Mark the most supportive research papers read them exhaustively frame the raw background by connecting and correlating it with your data.

Discussion involves the interpretation of your research findings and its significance in the related area of research and finally stating the additional new findings along with existing one.

The discussion part grooms your findings and enforces the reader to think critically and analyze the research problem and of course it brings out the space for further extension of the study. Overall it reflects the importance of your study in the existing area.

- You can start by writing your research problem in brief following importance of your study in the context-write it in brief…. just to bring rhythm .

- Mention similar studies (compare and contrast Your research findings with existing one /and Discuss possible reasons of similarity and dissimilarities with other related studies.)

- Follow the same sequence of writing as you did in results section

- Discuss how other researches supported your findings(other researches include references of reputed and high impact journals.

- Mention if any clinical relevance (supported by literature)

- Consider alternative findings if you can justify with the supportive literature

- State limitations and weakness of your study (mention wisely)

- Make suggestions for further research (In brief)

- Lastly concise the summary of the principal findings regardless of their significance.

Things you should not do …while writing discussions

- Do not over interpretate (interpretation should be supported by literature)

- Do not rewrite results

- Do not add new results

- Do not repeat the things

- Do not criticize others (if you find anything in contrast, just mention, professionally)

Results and Discussions

When you need to write results and discussion together what you have to do...You can start by writing your research problem in brief following importance of your study in the context-write it in brief.... just to bring rhythm. Firstly you have to follow the same sequence which you have mentioned in the objectives (experimental point of view) ....then state results...(sequentially in the form of table, graph ...or a figure) and start discussing immediately with the help of supporting literature. One thing you must take care...the location of table and figure...be appropriate so that one can easily connect them.

Conclusion

In some cases, the conclusion can be given at the end of results and discussion. But alternatively, a one-page conclusion can be givens separately. You should just mention the conclusion in three stanzas. First: dealing with aim and need of the study; Second: dealing with the experiment done and Last with result (sometimes the future or further studies needed to be done are added).

Summary

In Ph D thesis you are required to submit the summary of the work in the same number as that of the chapter. Plan chapter wise summary or abstract and compile them. Each chapter's summary should state the contents, or the matter discussed in the chapter. Never ever take the summary lightly. Many a times the thesis reviewer makes the summary as the basis of the evaluation. The length of the summary should be around 8 to 12 pages depending upon the number of chapters. Each chapter's summary should neither be too brief nor too lengthy. At least, important results must be mentioned (as you provide in preparing the abstract of a paper).

- Style and Typesetting common uses of computers
- Set ting margins
- Font
- Headings and subheadings
- Chapter Titles
- Header and Footer

- Paragraph setting
- Table and Figures
- Equations and special character

This may be written as per the guidelines of the institution/ university/ supervisor. And if there are no such guidelines you can follow the standard format (like: double line spacing/times new roman font/12..one inch margin from all corners of the page.)

Cover or Title Page

The cover or title page is generally standard for a University. So, refer any previous years' (or current) thesis, put all the required things (title, your name and enrollment number, subject, Logo of University, year of submission, Supervisor's name and the name of the institution). Give utmost attention to the title. It must be matched word to word with the title mentioned in your registration letter. Even addition or deletion of simply "a", "the" or "some" may lead to problem in submission. If it happens you have to unbind all the thesis again, get the changes done and resubmit it. This shall delay the submission only.

Auxiliary parts of reports (Acknowledgement, Index, appendixes etc.)

Final checking and proofreading

Binding and submission

Supervisor' Role

A constant supervision by supervisor in execution phase avoids any blunders. This is the duty of the student to get in close touch with his or her supervisor. As supervisor may be busy in his or her various other academic duties, the suitable time slot should be requested from him by the student. Daily, weekly, fortnightly or monthly reporting should be done.

The reporting should be in complete form. You should carry all the raw and processed data to show the supervisor. Any serious problem must be discussed with supervisor after a good homework. The routine reports which are to be sent in the University should be discussed, checked and approved by the supervisor. No change in work should be made without the consent of the supervisor.

Further Readings

- http://betterthesis.dk/literature-search/test-and-summary
- www.nitttrchd.ac.in › sitenew1 › comp_sc › pdf › PhD-Synopsis-template
- https://www.youtube.com/watch?v=Gv6_HuCYExM
- Semalty A., Thesis Writing, Lectures on ARPIT/ Refresher Course: http://tiny.cc/ARPIT

References

- https://youtu.be/3S289Oka_Jw
- http://archive.mu.ac.in/syllabus/4.71%20Guidelines%20thesis%20dissertaio n%20report-%20Ph.D,M.E,B.E%20%20(AC%204-3-14).pdf
- http://intra.tesaf.unipd.it/pettenella/Corsi/ReaserchMethodology/ResearchSy nopsisWriting.pdf
- https://documents.uow.edu.au/content/groups/public/@web/@stsv/@ld/doc uments/doc/uow195641.pdf

CHAPTER 29

Empirical Studies

Prof. Rajat Agarwal
Indian Institute of Technology
Roorkee-247664

For a good research, we must have various ethics which are the various characteristics required for publishing writing in top ranking journals. But it is very important to understand that only those academic writings which are based on empirical research are normally accepted in top ranking journals.

Empirical Evidence

Direct or indirect observation or experience or experimental observation.

Those researches which are not based on empirical findings do not find much place in leading journals. You can publish a book on the base of some intuitive thoughts, some beliefs which you have, but this is not supported have empirical researchers. If you want to publish a research paper based on some your research it must have proper empirical evidences. That is what we are going to discuss in these sessions. This empirical research is based on empirical evidences. The issue is what is empirical evidences?? Empirical evidence is the evidence which is achieved / obtained through observation or experience, through experimental set-ups. These can be the direct observations/ experiences or indirect observations/ experiences. So, those things which can be observed / experimented with those kind of experimental set-ups are known as empirical evidences. We are going to accept those things which are empirically validated that is empirical research. There may be other various sources to validate the beliefs. The other sources may be the mythology book available on my religion and in that book something is written in that way: Good things done, then you will go to heaven and if you are bad things then you are going to hell. Nobody has seen what is heaven and hell. But since it is written in text, which we cannot question, therefore we are believing. So that is not empirical research. If we have not experienced and not observed that how people are going to heaven or hell, which is beyond the scope of empirical research.

Those things which can be validated where observations/experiences and experimental set-ups are possible. These are known as empirical evidences. This is a perquisite for the publication for the top publishing journals.

EMPIRICAL EVIDENCE can be categorized :

- Exploratory Researches
- Conclusive Researches

Exploratory Researches: We have some kind of unstructured problem, when we do not know what the research real problem is. We are not able to identify a clear problem statement, objectiveness. When you have some hesitation of the problem we go with exploratory researches.

Exploratory Researches are used to:

1. *To provide insights and understanding. (Objectives)*

   There are persons like C.K. Pralad who mentioned that there is lot of value at the bottom of the pyramid. You can reach the bottom of the pyramid. In order to reach the bottom of the pyramid you need to develop that kind of products and services. Now for the tapping the bottom of pyramid. A large number of products came to India, which offer products for those rural customers. But many of them failed as they were not having clear understanding of Indian rural market. Indian rural market is not only a learned market also a complex socio-economic market dimensions where caste, religion, gender comes to picture. For a MNC marketing manager, all these things are almost unknown that how caste can claim some role in Purchase of product. But in rural market it is a reality. When you do not have clear understanding of the problem, we need to develop more insights. We have exploratory researches.

   Why every year some many forest fires happen in Uttarakhand forests??There are so many things that are discussed when summer season comes. But we are yet to get complete insight and understanding about forest fire problem. To understand that problem we need to go for exploratory research.

2. *Information needed is defined only loosely. → Methods*

- Brainstorming session, focus group decision is done for information which is analyzed later on.

3. *Research process is flexible and unstructured.*

- Research problem should be identified. What type of data is to be required, what are the variables in research process is not clear, so it is unstructured.

4. *Sample is small and non-representative.*

- It is very difficult to collect data in exploratory data. Because the researcher also knows that am unaware of complete problem. So need to do long interviews and need to do brainstorming/group discussions. Small sample means about 10-20 samples.

5. **Analysis of primary data is qualitative.**

6. **Findings of primary data is qualitative.**

7. **Findings are tentative: Findings are subject to modification and correction. These findings are subjected to further investigation by descriptive researches.**

8. **These are followed by further exploratory or conclusive research.**

From Above points we conclude that Research Methodology Related issues:

- Information
- Research Process
- Sample

Empirical means things which can be verified with observations, with experience, and with experimental set-ups. In that category, we discussed the exploratory researches.

The other type of empirical researches are conclusive researches. In case of conclusive research, we have well-defined problem.

Conclusive Research

To test hypothesis and examine relationship with the variables. (<u>OBJECTIVE</u>)

- Develop relationship between dependent and independent variable. When we write research papers, you will see 5-6 independent variables. In economics, when we apply economic metric techniques, there are multiple dependent variables also. There are situations where we are unable to identify which variable is dependent and independent. Therefore, we have multi-variate analysis like factor analysis where we do not know the relation between the dependent and independent variables. We have large number of the techniques to find relation between variables.

- Information needed is clearly defined.

- Research process is formal and structured.
    - o Because you re clear about objectives and information is required. So therefore, the research process is well structured.

- Sample is large and representative of population.

- o   Sample is properly designed as we know what the population is and how carefully you design your sample, so it is representative to population.
- Data analysis is quantitative
  - o   Statistics is the primary for data analysis.
- Results are conclusive.
  - o   Example: If your weight is more than a limit (say 90kgs), then you can have blood pressure. So, you can draw a conclusive statement. So, this the conclusive research.
- Findings are used as input into decision making.
  - o   The output of exploratory research because that is subject to verification using the conclusive research methodology. The output is firm in nature and can be as input for various decision making process.

Conclusive Research Design

1. Descriptive:
    - (a)   Cross-Sectional Design
    - (i)   Single Cross sectional Design
    - (ii)   Multiple Cross Sectional Design
    - (b)   Longitudinal Design
2. Causal Research

1. **Descriptive:** When we are testing hypothesis then it is Descriptive Research. It is basically used to describe the characteristics of the population. The population under your investigation to verify/ describe characteristics with respect to why population feels happy and which type of clothes they like, which type of cars they like, which type cars they like. All these types are various types of characteristics which will you like to define with respect to your population and for that purpose, you may do these two types of study:

(a)   *Cross-Sectional Design:* It is used when you're taking one type of data collection from the population, you do single data collection that is cross-sectional design. In that single data collection, two types of issues are possible:

(i)   *Single Cross sectional Design:* You take a sample which is representative of the population and you have done one type of data collection; that is "Single Cross-Sectional Design."

(ii) *Multiple Cross Sectional Design:* From the same population, you are making multiple samples, and collecting data from these multiple samples. So this is known as "Multiple Cross Sectional Design".

(b) Longitudinal Design: When you are going again and again to the same population and to same sample, in that population. Suppose Sample 'S' in population 'P' in the data collected in year 2010, in year 2011, in year 2014; after every two years, I am going to collect data from the population. This is the longitudinal study. This is also called as descriptive research.

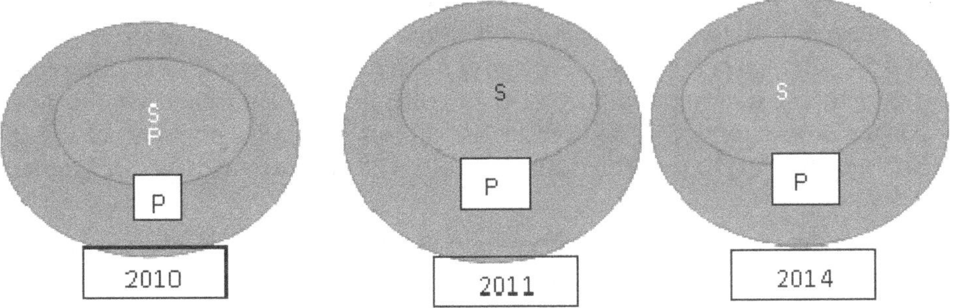

Causal Research

When we are deriving relation between variables of our interest, then it is causal research.

So far, we have discussed about basic idea of empirical evidences. We discussed two types of research under this umbrella that is exploratory research and conclusive researches. We discussed the characteristics of these research. We need to have better understanding of the issue. Those researches which come under conclusive researches are easy to publish, but those which come under exploratory research are difficult to publish. There are many issues under such as it is a big challenge to write a good paper under exploratory research. Another issue is that it is more pertinent to Indians who are good in Mathematics and computer processing. So, therefore we must do conclusive research where more statistical analysis and more mathematical analysis is required to give enough leverage in academic writing. In case of exploratory research, it is more qualitative analysis which is done and because of variety of issues we may lack qualitative analysis. So, based on my own experience, I suggest going for conclusive research activities, which will help us in publishing in good journals.

Relationships among Exploratory, Descriptive, and Causal Research

- When little is known about the problem situation, it is desirable to begin with **exploratory research**.

- **Exploratory research** is appropriate when the problem needs to be defined more precisely, alternative courses of action identified, research questions or hypothesis developed, and key variables isolated and classified as dependent or independent.

- **Exploratory research can be followed by descriptive or causal research.** *For example, hypothesis developed in* **stage 1(Exploratory research)** *can be statistically tested using descriptive or causal research in* **stage 2**. You can use a structured way of data collection to determine whether this particular assumption or hypothesis is true for particular population or not. So, **exploratory research can be followed by descriptive or causal research.**

- It is not necessary to begin every research design with exploratory research. It depends upon the precision with which problem has been defined and the researcher's degree of certainty about the approach to the problem. If as a researcher you are confident and have sufficient knowledge, understanding of the problem and what approach and what process, can be adopted for the research. In that case, there is no need to go for exploratory research.

  For example: I want to design a turbine for a new power project, I know which are the parameters which can affect the design of the turbine so I already know with available literature, documents, research; that these are parameters which are going to affect the performance of turbine. So I will design a turbine using all these available literature. I need not go for exploratory research, to identify the parameters which can affect the design of the turbine.

- Although exploratory research is generally the initial step, it need not be. Exploratory research may follow descriptive or causal research. Therefore, we need to know how to use them in a sequence. In some cases, exploratory may be followed by descriptive or causal or vice-versa. In some cases, you can require only descriptive or causal research, when you have a very good understanding and the certainty that what type of research process you need to adopt.

Further Readings

- Jeffrey Dubin, Empirical Studies in Applied Economics, 2001, Springer US, ISBN 9780792373957
- Joe Zhu, Data Envelopment Analysis: A Handbook of Empirical Studies and Applications, 2016, Springer US, ISBN 9781489976826

CHAPTER 30

Challenges in Indian Research and Writing

Prof. Rajat Agarwal
Indian Institute of Technology
Roorkee-247664

There are two types of challenges which you can see in Indian research. These challenges are with respect to:

- Quantity
- Quality

Quality means that we have already discussed in earlier sessions including the impact of the research, so the quality is basically related that how impactful is your research, how others are going to follow your research, what kind of societal value is being generated by your research. So, these are the different issues which are basically focused on quality aspect.

Second aspect is with respect to the quantity. These numbers which are not significant with respect to research. Both are the imp challenges in Indian Research & Writing. We need to improve on our numbers and need to improve on our quality of research. So, our challenges are two-fold.

In this chapter, we are going to discuss the numbers. When we see the numbers, one of the most respected databases is Clarivate Analytics. Now as per this Clarivate Analytics where you have the very popular name, Web of Science. As in particular field of writing in Engineering & Science, SCI indexed Journal, basically those journals which are a part of Web of Science. They have a product, the company which is doing the whole data basing is Thomson writer. The Thomson writer has developed this Clarivate Analytics. As per CA, see imp research area at the global area; that is the biochemistry & molecular biology, Chemistry, Electrical & Electronics Eng., Medicine, Research & Experimental, Economics, Management. These are the areas where we publish our papers.

As per the comparison available from June 2019, institutions which are affiliated to USA published in field of the biochemistry & molecular biology: 978,855 papers while in India it is less than USA: 54,684 only. Therefore, there is huge number gap. Similarly, around Chemistry in USA universities have published: 665,598 while in India In field of Chemistry: 85,117 papers. In the area of Electrical & Electronics Eng. In USA universities have published: 742,679 while in India In field of Electrical & Electronics Eng.:142,714 papers.

Similarly, so on, around Medicine, Research & Experimental in USA universities have published: 349,414 while in India In field of Medicine, Research & Experimental: 14,889 papers. Soon, in Economics in USA universities have published: 269, 406 while in India

In field of Economics: 7088 papers. Also, in Management in USA universities have published: 127,445 while in India In field of Management: 4612 papers. So, there is huge gap in writing in academic area.

Our number in top journals which are globally recognized is a matter of concern. We need to create more awareness why this number is less as we are not writing, and we are not publishing. We do not know what the benefit of publishing is. We can recognize that what benefit the publishing will give to us and therefore we are not interested in writing research papers, not interested in publishing in quality journals. The most popular institution, MIT, globally recognized institution. From MIT till JUNE 2019, the total number of publications which are available around 2 lakh 90 thousand. The complete IIT system present in India which is reputed institution with respect to engineering, technology, science and education. The cumulative output in terms of publications from the Web of Science database is less than 2 lakhs. Based on citations, MIT has somewhere around 9 lakh citations but in case of IITs have 3 lakh citations. So, this talks about the output quality which we are producing.

The major Challenges in Indian Research are:

- The lack of a scientific training in the methodology of research.
  - In graduation & post-graduation, we are not able to develop the temperament. Though we have world-class syllabus yet cannot develop those entities, therefore not able to develop a good research equipment. We are creating those which obey directions well and are not thinking as thinking is required in the research.

- There is sufficient interaction between the university research departments on one sideband business establishments, government departments and research institutions on the other side.
  - We are having limited interaction between various entities of the society. When you will not interact with the entities around you, you won't be able to identify the new research problems. When you do not have any research problem in hand then what you will research & what you will publish as well. Therefore, that system needs to be developed. If am working in government, then I know much about the government, but the academicians do not know much of the government. So, am not open to academicians. When am in academics, those others do not know much about the research publications. Therefore, we cannot develop that ecosystem where collaboration and partnerships can be developed. Good research publications and writing

can only happen when you have collaborations with various stake holders.

- Most of the business units in our county do not have the confidence that the material supplied by them to researchers.

    - In India, most of the organizations have their own trading department, own R&D set-ups. They have very limited confidence on academia. The academia can also provide the valuable input to them. Therefore, this interaction among them is very weak.

- Research studies overlapping one another are undertaken quite often for want of adequate information.

    - Overlapping is occurring because we are not making proper collaboration. We are not interacting so may be many would be doing similar kind of studies. In India, the Himalayas, the Ganga and beautiful flora and fauna also in this area, yet because of limited coordination between the agencies, may be many would be doing similar kind of studies in this area for the sake of saving the Himalayas, The Ganga and beautiful flora and fauna also in this area. If all these agencies collaborate and they do that research with proper synchronization, then impact will much higher. Overlapping studies are not letting possible for good publications. As you cannot bring novelty in your research, that will decrease your numbers and quality of publications.

- There does not exist a code of conduct for researchers and inter-university and interdepartmental rivalries are also quite common.

    - We do not have proper research architects and we do not have proper research ethics also. Therefore, a lot of issues are happening. For example, you are going to PhD examination, you got a good idea in PhD presentation. You come to the university and you can give similar kin of idea to your research scholars or team, to pursue that idea further. Now this may create a lot of interuniversity and inter departmental rivalries. So, very imp thing that how can we write white papers, working papers. So that whatever work am doing so that we can protect it right from the beginning interuniversity and inter departmental rivalries can be minimized. Therefore, we can avoid overlapping of research.

So, if we start working proper in Indian organizations, it will certainly improve the quality of the research output and will also avoid the overlapping. There is a need of more collaborative research in our country.

Academic Challenges in the India

In the second session, related to academic challenges in India, we already discussed about quality vs quantity issue that we are poor with respect the numbers and our impact of research is also a questionable issue. Now in earlier session we were discussing about some imp challenges related to research. In this we discussed about lack of co-ordination between various agencies. We have a lot of overlapping research, there are inter university and inter department rivalries. Because we do not have a code of conduct, improper system of working of papers in our country. So, we need to have a good enabling environment for the research and here I would like to Quote our Ex-President, Shri Pranab Mukherjee, recently he mentioned because of lack of proper research environment in India, we are unable to get any Noble Prize in Science so far. People residing in India getting Nobel Prize should dream this. Why we are unable to get the Noble prize in Science because their certain Challenges.

We had discussed 5 Challenges in previous session, now the sixth challenge is:

- Many researchers in our country also face the difficulty of adequate and timely secretarial assistance, including computer assistance.

  - You can be a very good researcher in chemistry or biology, but it is not necessary that you are also an expert in computer processing. So, you require adequate secretarial assistance for getting all those writing work, documentation work, communication with journal. So, all this kind of secretarial assistance is also very imp thing, unfortunately, in our country we hardly think of these issues. While in other top-ranking universities, professors get a very good secretarial assistance and that secretarial assistance take away lot of non-vale added activities from the head of the professors. In our India, a professor is all the time busy with lot of secretarial activities and all the sesecretarial activities up to an extent are non-value-added activities. You are going for a big research, which needs to purchase some equipment costing around Rs. 50 lakhs in India. So, for purchase of that equipment, you need to run from pillar to post because of lot of bureaucracy in purchase process. So how will you purchase that equipment??

- When your entire energy is drained in completing the purchase process. Then this very unfair for me to ask you to publish paper in top ranking journals.

  - If there is proper secretarial availability which can take care of that purchase process, then need not worry about it. You just raise your indent that what equipment u require then there is a complete office available like material management which will take care of the purchase of the equipment. So that you can concentrate on your output. Your output is not getting that equipment purchased but it is the academic writing that is your output. Your output is the impactful research publications. So that is almost a zero culture in our country. We hardly think of any secretarial assistance to people involve in academics. Once u become a Dean in your organization, Head of the department in your organization, Vice Chancellor in your organization then you will get a lot of secretarial assistance. But if you area professor then no secretarial assistance is there. If a proper computer assistance is there for proper browsing of internet which can help in e-mail management, in your desktop publication activities which will be a big help to academia in India.

- Library management and functioning is not satisfactory at many places and much of the time and energy of research are spent intracing out the books, journals, reports, etc., rather than intracing out relevant material from them.

  - There are real issues about our library management and its functioning. A proper classification books is required. When you to the computer in the library to the tracking of book and that system tells that book is available in this almirah. You go to that almirah and it is your good luck if u find that book in that almirah. You will not find books most of the time at right place and that talks about the functioning of our libraries and therefore lot of time & energy is spent in just identifying the books. It is like purchase of equipment. Most of the energy is spent on just getting the right book rather we should spend more time in reading those books that what is written in it and what important conclusion can be drawn. Similarly, when we are browsing our internet, most of us our browsing research papers from the google only.

- There is a need to know how to access good databases.

  - Various places also we have limitations in funds and not enough investment that u can have subscriptions of all leading databases. But there can be some ways and means through which we can develop some collaborations and with the help of those collaborations, we can access those tools and databases which are not available in my institution. But that is not our objective. Our objective is lost because of poor functioning of all these materials, the system; therefore, by the time we get the paper, that material, that book, we are already exhausted that means we feel the completion of our task.

- Thereisalsotheproblemthatmanyofourlibrariesarenotabletogetcopiesofoldand newActs/Rules, Reports and other government publications in time.

  - Lot of government publications also are imp source of information for research purpose. We are unable to get that material in most of our libraries. Therefore, we are more dependent on internet sources. Many a times, this internet is not reliable sources. We do not know how to differentiate between are liable internet source and an unreliable source. When you access data and information from these non- reliable sources, it may result into poor research output.

- There is also difficulty of timely availability of published data from various government and other agencies doing this job in our country.

  - Data is most authenticated from the government agencies because there is a whole system available behind collecting the data but if timely availability data is not there it will not much use when the time has gone. So, in various institutions as there is a lack of collaboration with the government organizations and academia, cannot access the databases timely.

- There may, at times take place the problem of conceptualization and problems relating to the process of data collection and related things.

  - These are issues related to research methodology. There is no proper training in research methodology, no training in publication approach for good research. So, we are not able to conceptualize the limitations about how to collect data for the research. This becomes a barrier in research publication activities.

There are various phases in research writing:

1. **Pre-writing Process:** Problem: No clear thesis and no idea of what to do your research on.

- You must plan that what are you going to write and that clarity in that writing process is important. To bring **clarity of objectives,** need to know and have clarity on your topic. If clarity is not there in your topic then you may not be able to publish good journals. There should be clear idea about what are we researching, and this becomes even more important in case of social science, humanities, and literature so that we can know what we are researching on and hence we have severe issues due to improper training that Occur. We are unaware of from where to get resources and how to interpret and analyze your finding. Data may say different, but you will interpret, analyze and the way you are thinking. In Indian research, many a time, u will find that the findings, interpretation, which you are interpreting may not have any relation with the data which you have collected for the research.

2. **Planning and Logistics:** Problem: You have not considered where to start your research nor have developed a research writing plan.

- You should know what type of review is required for the problem. What type of methodology is required for the problem because others are doing this,soamalsodoingthis?MostoftheIndiansfollowthislogicinplanningand logistics aspect. In America, an approach was followed so am also following the same approach in India. But this won't going to work in India. You have to be very specific why this particular methodology and steps are suitable for your research. So planning and logistics need to have a careful activity.

3. **Research and Data collection:** Problem: You have not enough resources nor do you know how to interpret and analyze your findings.

- This is a corollary of lack of research training. When you do not have proper research training, these types of problems occur. We are unaware of from where to get the resources and then do not know how to interpret and analyze the data. In India, this usually happens that the findings and interpretations which are presented may not have any relation with collected data for research. This is a poor indicator of the research activity.

4. **Drafting and Writing: Problem:** You've collected your information but don't know how to organize it in to a well written research paper.

   You assume that you have clarity and research training and you have been for some empirical study, you collected data, you used some latest statistical packages for analysis of data. Now you have enough things with you but because of your poor ability to organize the things in proper way so you are unable to organize it into a well-written research paper.

5. **Revising, Editing and Proofreading:** Problem: You have written your draft but still have several logics, organization, grammar, spelling and punctuation mistakes.

   When you are submitting paper to some good ranking journal which being published from Europe or America. You get the setypes of comments that papers need to be proof read from native English-speaking person. This is a major challenge for many Indian researchers. We need to have good skills of writing, good command over the language and when you have command over language, you will be able to express your ideas in better way. So, you are knowledgeable but putting your research with these words which can explain your research lucidly. So that others can understand and take interest in reading your research. Research writing must create interest in readers' mind so that they can read your entire writing. We need to work on how we can improve our language and need to have training from education from schools so that they can have good control over the language. Language development is very important.

6. **Weak organization: Problem:** The longer a paper, the more challenging it is to ensure that the paper is well organized and unfiled.

We should have a decent length of the research paper. If you try to write too many things in a paper, there may be problems related to proper organization of research paper.

7. **Poor Support and Development of Ideas:** Problem: Excessively short or long paragraphs are one indication that the ideas in a paper re not well developed.

   If paragraph is too short or too long, that means you do not have clarity on that idea or on what you're writing in that paragraph. So, it must be moderate length paragraph to express the ideas into limited words can be highlighted. The proper support of idea and development of ideas is also required in good research writing.

8. **Weak Use of Secondary Sources:** Problem: We use literature review as a ritualal ways.

   We do not understand the purpose the secondary sources, i.e., literature review in our research that how it is going to strengthen the logic of your research.

9. **Excessive Errors: Problem:** Already discussed errors are involved in with respect to logic, grammar, language, spelling, sentence formations with writing activities.

   So, proper command over language is required for improving the research writing and thereby, reducing errors in our research writing. These errors reflect poor part of researchers. So, there is need to improve these aspects so that we can overcome the challenges of our research.

Chapter 31

Team and Time Management

Prof Rajat Agarwal
Indian Institute of Technology
Roorkee-247664

It is always a matter of concern not only in India but also at different universities in different parts of world. What are the essential ingredients to publish your paper in extremely in top ranking journals? Journals which are having Q1 category as per the web of Science, Clarivate Analytics and there are different answers for this question. Some people say that based on your academic contribution; you will be able to publish papers in top ranking journals. There are people who say that your affiliation is also equally important for publishing in top ranking journals, that if you are coming from top Q1 ranking institutions, it will be easy for you to publish in those journals. But at the same time, a good team is an equally imp for publishing papers in top ranking journals. Good team means you need to have goodmentor,needtohavegoodcollaboration,needtohavegoodsponsorsforyourrese archandwhenyou have all these people in your strategic team, then you will have good publications. So, in this session we are going to discuss about how to develop good team management and what are the model for effective teamwork.

Now, the meaning of "team" is a group of people who work together toward a common goal.

Explanation: Here, Common goal is good academic writing which can lead to publications.

- Common goal aspect of team is critical.
    - If different members in the team having their own different objectives and goals, then the common goal I will be defeated. So, there must be a common goal for all the members. Sometimes research is related to get a patent that is also a research, here it is related to publications. So, if the goal of the team is to get a patent or good quality publication. If half of the member of the team work for patenting and rest half work for good quality publications of research, then these are 2 types of objectives or goals for the organization. Team must have a common goal then only it can work with the characteristics of the team. Otherwise, it will just as a group of people.

- Individual ambitions Vs Goal of team.
  - o As we all can have Individual ambitions, but we need to subordinate our individual ambitions in respect of objectives of the team. So the objective of the team is primary and your individual interest is the secondary that should be the goal of the team.

There are different types of teams:

Work Teams

- These team do the normal, everyday work of technology companies.
- Example: A work can be assigned to a work team consisting of a project manager and several engineers, CAD technicians, and clerical support personnel.
- We have a department like Department of Management Studies. So, all the professors in that department, they have a team. Their team has a routine to follow a particular time table to follow as allotted to them. So, these teams are doing the routine kind of work and their importance with respect to our academic writing is not very significant because their job is a routine job.

Improvement Teams

- These are cross functional teams assigned the task of improving a given process or some specific function of the organization.
- There can be people from different departments, different areas, and different organizations. They're assigned a task of improving a process or some specific function of the organization. So, when we are in the publication activity for top ranking journals. We need to develop some specific teams for those research projects and these people can be drawn from different areas so that you can develop some impactful research.
- Cross Functional means the team is composed of representatives from a variety of functional area.

Temporary V/S Permanent Teams

- Normally these teams are made on temporary basis that you have a particular project in hand to complete that project, we need to have different people from different areas. For example: We have got project from Ministry of Forest, in that project we need to find some solutions for pine needles in Himalayan Region. We started working on developing solution.

Now, in that in process of developing of solutions, I, come from Management background. So, I require a lot of people who can help me. I contacted mechanical engineers, biochemistry people and petroleum engineering background, contacted the social science and humanities point of view as well. So, these are coming from different functional areas. We started collecting pine needles so that we can study their biochemistry. We started converting those pine needles into bio-brackets. It can be an alternative source of fuel in upper Himalayan regions. We started developing machines so that the brigading can be done without many efforts. We started developing so kind of value chains for those pine needles. The point which am needed to say that we started as a temporary team but over the period we relegalized we have bonding. Those temporary teams started becoming permanent teams where we can do various subsequent projects also. All those projects resulted into good academic writing also. These teams started as temporary, worked cohesively and how each team supporting each member.

Standing Teams

- These teams are responsible for carrying out on going assignments related to specific functions or disciplines.

- They are typically advise higher management about decision relating to the functional area in which they specialize, and they keep management up-to-date on the latest information relating to this functional area.

- These teams are the parliamentary teams. These are the permanent teams that are assigned kind of work which is suitable to that particular team. These are specific teams which are created in the organization to carry out the specific work. Such as there is a discipline committee is there for specific academic year. All the activities related to discipline are brought to that committee and that committee meets and discuss about those issues.

Working together as a TEAM

- Large number of factors work against effective team work
- Some of them can be:
  - Personal identity of team members (If some member of the team is very tall, not in terms of height nutal so in terms of personality. So that tall personality will overshadow the other members of the team. So this is not going to help the entire teamwork. There is a need to see individual personalities. So no member should work against the spirit of the team.)

- o Relationships among team members (Many a times, the team members sometimes do not have good relation with each other, there maybe some personal rivalries within the team members.)
- o Identify within the organization (How team is viewed in organization. Sometimes this discipline committee can have identity that it is a very tough committee, if any such activity goes to them they can take harsh action.)

These are the Character traits that promote effective team:

- Honesty
- Selflessness
- Dependability
- Enthusiasm
- Responsibility
- Cooperativeness
- Initiative
- Patience
- Resourcefulness
- Punctuality
- Tolerance/Sensitivity

This is the second section of team management. In the first section we discussed that common goal is the most important characteristic of the team. The team members need to understand that they need to subordinate their Individual ambitions for the sake of common goal of the team. We discussed different types of team, some challenges for effective team management, and some traits for effective team management.

Need of teams: Teams are important for completing any bigger task.

Writing research papers for top ranking journals also require effective teams.

$$1 + 1 = 11, \text{not } 2.$$

- If you want to write a paper in non-review journals or willing to write in predatory journals, then probably a team is not required. You alone can write a paper in predatory journals. But if we are talking of publishing in Q1 category journals, or in a high impact journals, then a team is required. If you go to "Nature" and you will find papers which are published by as many as 50 authors which shows that entire lab is working as a team contributed in writing that paper. So, team is required for bigger tasks for tasks which are beyond capabilities of individuals.

The Ten Step Model for Effective Teamwork:

1. Establish direction and goals for the team
2. Establish clear roles and ground rules for the team
3. Establish accountability for the performance of the team.
4. Develop team leadership skills.
5. Develop communication skills for the team leaders (mentor) and team members.
6. Develop conflict-resolution skills for team leaders and team members.
7. Establish a well-defined decision-making process for the team, and empower team members to be part of the process.
8. Establish positive team behavior, ethics, and trust among team members.
9. Recognize and reward effective team performance.
10. Continually evaluate, improve, and build the team.

1. Establish direction and goals for the team

- Demonstrate how to write a team charter
- Explain how to write a team mission statement
- Demonstrate how to write team goals.
- Explain how to write team assignments
- Demonstrate how to write team schedules and deadlines.
- WHAT IS TEAM CHARTER?
    o Mission Statement
    o Goals
    o Assignments
    o Schedule/ Deadlines

2. Establish Clear Roles and ground Rules For Teams:

- Explain the role of team leader (mentor)
    - In India, when you take a new PhD scholar, the PhD scholar will come to you and will tell on which topic of research he/she wants to do. The teacher should have the topics available with him/her and out of that list one topic will be assigned to that research scholar. Because, we don't have the clarity of team leader. If that faculty or supervisor is the Team leader, you can expect that the he should be able to guide you to the integrity of the research. He is not the solution of the problem but because of

more experience, is knowledgeable and has accessibility of information, capable of guiding you.

- Explain the role of team members.
    - The clarity of the role of each team member should be present for good team management. But we don't have the role clarity.
    - A student will come and say about to submit a manuscript in XYZ journal. Suppose that you are expecting that your supervisor or your mentor will do editing and proof reading of that particular paper, so there is a mismatch of role clarity.
- Demonstrate how to develop ground rules for team members.
    - Guidelines for Developing Team Ground Rules.
- Form a cross-functional ad hoc committee to develop a standard list of issues teams should consider when developing ground rules.
- Circulate the draft list of issues among all employees companywide and ask for their input.
- Have the ad hoc committee to finalize the standard list of issues based on employee input.
- Give all team leaders the standard list of issues to use as a guide when working with their respective teams to develop ground rules.
    - You as a team leader need to develop team involvement so that there will be no resistance to accept the rules and then it's easy to implement the ground rules.

3. Establish Accountability:

- Explain how to build a foundation for accountability.
- Distinguish between formal and informal accountability.
- Explain the concept of accountability for team leaders.
- Explain the concept of accountability for team members.
- Explain the concept of peer accountability.

> When we know everybody in a team, everyone is accountable to each other in a team, then the goals can be achieved.

Elements That Make Up Foundation for Accountability:

- Team members understand what is expected of them individually and of the team as awhole.
- Team members understand how fulfillment of expectations will be determined.
- Team members have been given the training necessary to fulfill the expectations
- The team is given the resources and support necessary to fulfill the expectations.
  - As leader guides, accordingly the whole team is going to function. The role of leader is to develop good leadership skills which is a Herculean task. The leader should be able to take the team to a new heights and to higher goals that has not been achieved by others that is the true leadership skill.

4. Develop Team- Leadership Skills:

- Leadership is single most important ingredient in achieving consistent peak performance in teams.
- With good leadership, ordinary people can achieve extraordinary results.

The Ten Step Model for Effective Teamwork:

1. Establish direction and goals for the team
2. Establish clear roles and ground rules for the team
3. Establish accountability for the performance of the team.
4. Develop team leadership skills.
5. Develop communication skills for the team leaders (mentor) and team members.
6. Develop conflict-resolution skills for team leaders and team members.
7. Establish a well-defined decision making process for the team, and empower team members to be part of the process.
8. Establish positive team behavior, ethics, and trust among team members.
9. Recognize and reward effective team performance.
10. Continually evaluate, improve, and build the team.

These 5 points (In red) have been already completed in last session. Now we will start with the rest.

5. Develop conflict-resolution skills for team leaders and team members.

- Conflict occurs when a person's desires are frustrated, or needs are threatened by another person.

- A simple example can be sequence of authorships in a research paper

    ▪ Political parties discuss in Parliament about large number of issues but the common objective of the members of Parliament is to serve the country, how to develop the country, how to work for the welfare of the country. But many a times, we all have experienced that individual ambitions, aspirations become so high that takes the primary stage.

- Conflicts are based on perception. Probably this person going to take over my interest, or this person going to do something which will do harm to my interest. That is why the conflict occurs.

- This is an issue in most of the universities the first author and second author. In University A, the first author is supervisor, and second author is Researcher.

    In University B, the first author is researcher and second author is supervisor. Now some researcher moving from University A to University B system or vice-versa, this will be create a conflict. In University 'A', the supervisor will say that I will be the first author and the researcher means you be the second author as he/she has come come that culture of University. Now researcher will fee that supervisor is threatening the interest of the researcher. Because in many cases, we have different weightage for different stage of authors. If two or three authors are there, the first author will get maximum benefit comparatively the other second or third author. So, the sequence of author is a good example of conflict.

Common causes of Conflictin Teams

- Limited resources

- Incompatible goals:

    o We are not able to have a concrete or common goal. Therefore, there is compatible goals.

- Role ambiguity

    o If the ground goals are not achieved then role ambiguity is there.

- Different values

    o You are giving more value is given to technology transfer, but, am giving more value to publications. So these can lead to conflict.

- Different perspectives:
  - We all have different perceptions and way of thinking, which may create again conflict.

- Communication problems:
  - When researcher not being able to understand what the supervisor is thinking or vice- versa that results in conflicts.

How people react to conflict?

- Competing
  - When we start working individually within the team, the output will be not be achieved, when competition starts within the team starts.
- Collaborating
  - There may be one smaller team and another sub-team, so within the team also collaborations should be there.
- Avoiding
  - When in a team such as a team member 'A' is not in terms of another member 'B', then the conflict occurs.

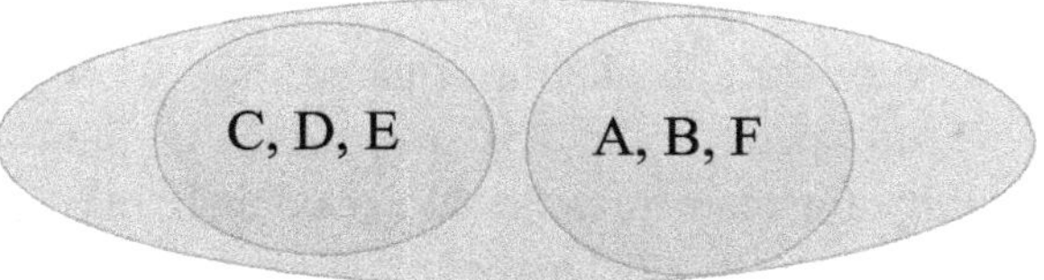

- Accommodating
  - When you start accepting others logic. If you start accommodating, then its philosophy for the team members. If we understand the perception of other person and we start respecting those perspectives, then this will help for effective team management.

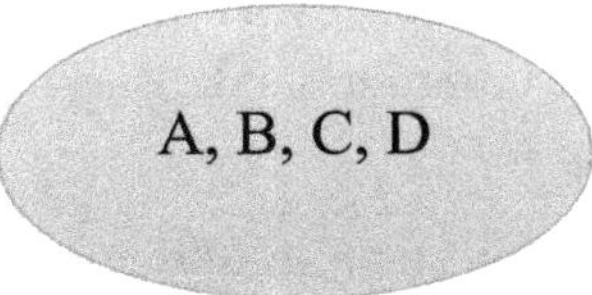

5. Establish a well-defined decision-making process for the team and empower team members to be part of the process.

- The best decisions come from involving the people who must carry them out.

Decision making Models for Teams

1. Objective approach:

(a) **Logical and orderly:** In this approach, more accountability in teamwork, more logical, orderly, transparency and it is based on facts and figures. So, this is preferable.

2. Subjective approach:

(a) **Intuitive and based on experience:** In this approach, it is based on individual experience. Sometimes there is no data or information available to you, then you may go with subjective approach.

Techniques of group involvement:

- Brainstorming
- Nominal Group Techniques
- Quality circles

7. Establish positive team behavior, ethics, and trust among team members.

- The most important measure of how good a game I played was how much better I had made my team mates play.

  o You are very skill in playing a game, but, when you're playing in a team. Your performance is measured in how you creating an opportunity for team members to play their maximum efficiency. So need to have positive behavior in team where team members can play with the best of their performance.

- Selflessness
- Give to other team members
- Refuse to gossip

  o The team should be more involved in value-added activities and motivated towards the common goal.

- Avoid Territorial Behavior

  o This is particularly important in Indian context. We have different types of Cultural issues. This territorial behavior is not effective for development of effective team.

We need to create an environment where team members can contribute effectively to the team. You can use a team for value added activities.

Further Readings

- Charles J. Margerison, Team Management: Practical New Approaches, Dick McCann Management Books 2000, 1995

- Charles Baden Fuller, Strategic Innovation: An International Casebook on Strategic Management, Martyn Pitt, Routledge, 1996

- Patrick Forsyth, Successful Time Management, Kogan Page Publishers, 2007. Semalty A, "Developing need based Microenvironment in Higher Education and Research", https://youtu.be/BMXVBxxaPQY

CHAPTER 32

Authorship and Competing Interest

Prof. Rajat Agarwal
Indian Institute of Technology
Roorkee-247664

Authorship

We are starting the discussion on promoting integrity in scholarly research and its publication, and because of improving our integrity into the publication activities, UGC has also started a course on ethics in research. Therefore, we can understand there is a paramount significance of knowing the ethical dimension of the research we know tools techniques we have discussed during the last many weeks of this particular course but at the same time the ethical issues involved in the research be known to us. In this particular section, we are going to discuss a crucial issue which is related to authorship. Now you all may think that what is so important what is so great for authorship but there are various potential issues, there are various potential conflicts which may happen because of our unknowing related to authorship.

You are a student, you have a supervisor, now, there may be a small conflict about the order of authorship: who is the first author, who is the second author. There may be a conflict about the corresponding author. Sometimes benefits are given to those who are the corresponding author for a particular paper. All these are the type of conflicts, just two I mentioned, but there may be many more so, we would like to know the meaning of authorship and how can we resolve these conflicts, how can we follow some of the standard guidelines to avoid such conflicts.

Now, authorship is basically the naming of authors on a scientific paper so that all individuals associated with that research who are associated with the writing of that particular paper get due credit for that publication and then also held accountable for that research. So, it is both things that we want to make accountable also, and at the same time whatever benefits are possible from that particular publication, so, you all must get, all the authors must get due credit of that particular paper. So, that is the first important thing about the authorship that we have to name all people associated with the development of that scholarly writing. Now, the starting point of this discussion is the ethical dimension of the authorship. Sometimes, we deliberately misrepresent the scientist's relation to their work and it is considered a form of misconduct and undermines the confidence in the reporting of the work itself.

Deliberate Misrepresentation, Representation and Author

What is this deliberate misinterpretation or deliberate representation of the relationship maybe you have not appropriately contributed, but I am still quoting your name as one of the authors or, have significantly contributed to the work, but I am not willing to write your name. I wrote that paper without you giving the due credit so all these kinds of misrepresentations are possible and we need to see that we must avoid all these conflicts. There are different types of guidelines that are being available to you one is COPE that is Committee of Publication Ethics and then different publishing house; the most important out of that is Science Direct. They are also giving some guidelines concerning authorship.

According to them an author is generally considered to be an individual who has made a significant intellectual contribution to the study now it very important that intellectual contribution is essential. If you dictate your manuscript and some typist is doing the desktop work, he is not contributing intellectually to that paper writing. You can understand he/she cannot be the author of that particular paper. Only those who have intellectually contributed in that paper they will be given the authorship. So the intellectual contribution you have to identify what are the different forms of intellectual contribution.

According to one very reputable source: International Committee of Medical Journal Editors (ICMJE), they say that "all persons designated as authors should qualify for authorship and all those who qualify should be listed." So those papers where you see that more than one authors are there these issues are related to that particular thing that all those who can be designated as an author should qualify for the authorship means whatever type of intellectual contribution you have done. Somebody may have done help in designing experimentation; somebody has helped in the data analysis, somebody has done the conceptual development of the idea. All these are the different types of people who are making different kinds of intellectual contribution in the development of that research so they qualify for the authorship and all those who qualify for the authorship must be listed in that academic writing. So that is the guideline given by the International Committee of Medical Journal Editors.

Criteria for Authorship

The criteria that are generally used and are like a guideline for giving the authorship that substantial contribution to the study conception, study design, data acquisition, data analysis, and interpretation of the results are the first important thing. The second is drafting or revising the article for intellectual content, helping you or those who are writing the manuscript. Then your supervisor, your mentor may help you give the final approval of the manuscript that yes now this is in a form that can go for the publication. The fourth essential criteria is an agreement to be accountable for all aspects of the work related to the accuracy or integrity of any part of the work, that they all those who are involved in this work they all are taking the responsibility about the accuracy, about the integrity concerning the entire work.

Often, it happens that we submit a manuscript it get published, and then some question comes on the manuscript; some question comes of the content of that particular paper. Then I can say that I am a supervisor of the manuscript submitted by my student, so I am not very much about the content. When it was the authorship matter, I was very much willing to take authorship, but when it comes to becoming accountable for that particular finding, I say that I do not know it is my scholars' work. So it is not the right thing I need to be appropriately accountable; I need to have the full confidence, I should be adequately aware of the content of that particular work. So these are the critical criteria for the authorship.

General Guidelines

Now the general guidelines in this particular case are that some of the essential issues are concerning the order of authorship that what should be the order which is the first, second, third, fourth author of the manuscript. It usually is advisable that all the author take a joint decision about the list of authors on how the order will be decided. There are no set criteria that the supervisor should be the first author, or the student should be the first author. In many places, the supervisor is the first author, and in many places, the student is the first author, but if the supervisor-student relationship is not there, two independent researchers are working for particular academic writing so they can mutually decide that who should be the first and who should be the second author.

The second important issue is about individuals who are involved in a study but don't satisfy the journal's criteria for authorship should be listed as a contributor. So when we are talking of authorship, another word comes into the pictures that are contributors. Some of the criteria we just discussed like, making some intellectual input. If somebody cannot fulfil that criteria of intellectual contribution to the paper, you can list them as contributors. You can acknowledge them in the paper that they were also helping in this research, like assisting the research by providing advice, providing research space, departmental oversight, and obtaining financial support. Sometimes in my department, we support our students who are not my direct students but are the students of some colleagues, they come for some advice they take our inputs. So just for some input, we can not be said to be the author of their manuscript. They can acknowledge our help that these professors or these students have helped me in my research to be known as contributors or acknowledge their support in our manuscript.

Then further you can also have some researches which are conducted over multiple centres, it is not done at a particular location. Multicenter trials are possible in pharmacy, biotechnology, medical science, agriculture, etc. So the list of clinicians and centres is typically published, along with a statement of the individual contributions made. Some groups list authors alphabetically, sometimes with a note to explain that all authors made equal contributions to the study and the publication. So sometimes you have to recognize the contributions of a large number of people in big research we follow a practice of alphabetically listing the author so that no conflict is there and then you can give a note that all authors are given equal weightage they have made the equal contribution in this study and the publication. This is how you can avoid any possible conflict about the order of authorship when it is not possible to have a joint meeting or a joint kind of decision, so you can follow this alphabetical system.

Authorship 2

In the last section, we discussed a few essential aspects related to authorship. We discussed that authorship is an essential aspect of publication, and some potential conflicts may arise because of not understanding or not knowing the issues related to authorship. So we tried to address the criteria for getting the authorship we must contribute intellectually for getting the authorship of a paper. So that may be related to the conceptualization, that may be related to setting the experiment, that may be related to data acquisition, that may be related to data analysis, that may be related to the interpretation, that may be related to drafting the manuscript, that may be related to finally approving the manuscript. So these are the criteria we discussed that are going to help in deciding about the authorship. In some extensive research where some trials are happening at multiple locations, multiple experiments are happening, and many peoples are involved in collecting the data. So if that kind of situation is there, then it is good to have authorship alphabetically, and then you can note that all authors have contributed equally to this study and publication. That avoids unnecessary conflicts. So these are the issues which we discussed in the last section. In this section also we are going to discuss some of the essential aspects of authorship. Some of you may not know that these kinds of issues are possible, or these kinds of terms are associated with authorship issues. So let us see what some of the unexpected types of forms of authorship are. If I write a paper, my name will come as an author, but there are certain unacceptable forms of authorship.

Unacceptable Forms of Authorship

1. **Ghost authors:** What are these unacceptable forms of authorship? One is ghost authors; now it is exciting that what is a ghost author. Now, these are the authors who contribute substantially, but they are not acknowledged. So they are kind of a professional author those who write for the sponsors. They are professional authors, but you will not be able to know who is the author. Sometimes if you see some books, I should not name nut in some books, mainly guide books. If you see the guide books, so some authors write these, but many a time you will not know who the author of this particular guide book is, there will be only the publisher's name, but there will not be any name of the author. So they are the ghost authors, they are professional authors they write books on behalf of somebody who is sponsoring that work, but their name is not acknowledged, so that is unacceptable form because nobody is taking the responsibility of that academic work. That may happen sometime in your research publications also, that some of the research publication is done by professional authors and we will see in our subsequent sections because of some interest we sponsor research, and we hire professional authors those who work in such a manner.

2. **Guest authors:** Then we have guest authors, now guest authors who make not a significant contribution but are listed to help increase the chances of publication. You invite some celebrity author to become your co-author because you feel that adding that name may increase the chance of publication of my paper. It is also very unethical that because of improving the chances of publication, you are giving the authorship to somebody who has not contributed in your manuscript. So that is the second form of unacceptable authorship.

3. **Gift authors:** The third form of unacceptable authorship is gift authors. Who was the gift authors? Whose contribution is based solely because of some affiliation with a study that you are a friend of mine and I am writing a paper and now because of my friendship with you I include your name in my manuscript. So that is a kind of gift authorship that is also unethical it is also unacceptable.

So we must avoid these kinds of ghost, guest and gift authorship because these come under the unethical dimension. Though it is difficult to say that because once the manuscript is published, somebody says that it is a guest author you can always say that he helped me finally shape the manuscript and therefore his name is there. So it is complicated that as an outsider how can I substantiate my this kind of claim that he is the guest author or gifted author because I cannot prove that he is a guest or gift author you can always say that they have done some kind of intellectual contribution in the development of the manuscript. Therefore, it is your own ethical behaviour towards your's integrity, towards the publication world that you do not get into this kind of unethical behaviour of guest and gift authors.

Now some critical issues for these things involved in a study but not listed as an author or contributor. Someone is taking your idea of publishing a paper claiming full authorship; these are some of the critical issues we need to understand when discussing ethics in the authorship. Then other important issues are you find your name on a publication without your permission. Therefore, most of the useful journals when you submit your manuscript will send a confirmatory email to all the listed co-authors because you may be trying to involve somebody's name as a guest author in your manuscript to improve the chance of authorship. However, that person has not given the consent, and without his consent, without her permission, you have included the name of that personality. So it is important, and if somebody denies that I am not a co-author, I am not given the consent, and I am not contributed significantly in this manuscript then it is an embarrassment for the submitting author and corresponding author. These are important issues that you must take all your co-authors into confidence when submitting a manuscript that you have taken proper permission from them.

Guide to Authorship Dispute and How to Prevent them?

Now some essential authorship disputes and how can we prevent those authorship disputes let us discuss these aspects.

Misrepresenting a scientist's relationship to their work: Now first is misrepresenting a scientist's relationship to their work that is scientific name is listed who took little or no part in the research and you are listing that name, omitting names of people who did take part that is also possible or the "ordering of a byline that indicates a greater level of participation in the research than is warranted". So these are three types of misrepresentations which are possible. You have included the name of somebody who has not done any kind of work or very little work. You have omitted the name of a person who has taken part in the research or you are trying to show the contribution of somebody then it is required so here the fundamental issue is you are not taking the proper permissions from all your co-authors and all the authors who have listed in that particular paper, therefore, this type of conflict arises. According to ICMJE, "all persons designated as authors should qualify for authorship, and all those who qualify should be listed." We have already discussed this kind of guidelines previously. Misrepresentation also includes all these ghosts, guest, and gift author, so these three types of categories discussed at the beginning of this chapter are related to ghost, guest, and gift authorship.

So what can we do in this kind of situation? Review the journal's instruction for authors before submitting a paper and be forthright about all contributors. So you need to be very clear about the journal's instruction because it may vary though the general principle is clear, but there may be some variation that how can you acknowledge the contributor's name, how can you write the author's name, so a minor difference maybe there from journal to journal, but otherwise overall policies remain same. This includes substantial contributions, paid writers, any other who contributed to the study, and therefore you can acknowledge their name in the form of acknowledgement. To avoid disputes, set clear expectations from the outset about who is doing what and how authorship will be handled. So before beginning your research, you should clearly spell out how the authorship will be handled. If you are hiring somebody for a small part of your work, you should clearly state that you are required to do only this job, and for this, we are paying you, but you will not be the part of authorship. So that later on no such ethical conflict may arise.

If you feel you have been mistreated regarding authorship, seek the counsel of a trusted advisor. It is always right before jumping into any complaints or other things take the help of senior people in your department, in your university, in your institute, so that they can advise you appropriately that what should be the proper action because this academic world is if you talk of at a leadership position we have a limited fraternity. Therefore it is not acceptable to paint somebody in different colours thus, it is good to take proper advice before creating any kind of nuisance. So if you have this kind of grievances that I am not appropriately treated, you can advise.

Another very critical and important issue is about the ghost authorship as we already indicated that these are the professional writers. So they write n behalf of some sponsoring organization who pay for your authorship and these are unattributed contributors to data analyses may also constitute ghost authorship because they are doing some work in your research and their work is not acknowledged at any stage and they are the ghost authors. Sometimes, there are unethical practices that happen, and we know that these practices must be avoided or if you want to take the help of some kind of professional or person for your research you must give due credit and due acknowledgement in this particular contribution. So not acknowledging a writer's contribution is considered to be unethical. You need to acknowledge the contribution properly and therefore, professional writers who participated only in drafting the manuscript or did not have a role in the study's design or conduct or the interpretation of the result. This should be identified in the acknowledgement section and information about the potential conflicts of interest including whether they were compensated for the writing assistance and if so by which entity (ies). For example, these days if you submit your paper to any good journal and since we are from India; I would like to say that we get this type of comments many times that you should take the help of a professional editor for revising your manuscript. We are not from an English speaking nation, and in that situation, we are paying somebody for the professional services, for revising our manuscript for language. So he is doing the manuscript's language editing, so he is not a contributor or intellectual contributor. However, you should acknowledge that person's contribution and that it is a sponsored activity which is being carried out, so that type of thing is basically mentioned in the acknowledgement section of your manuscript. Then obviously you can consult the authorship guidelines of the journal in which you are dealing. Then there are different sources available like; Worlds Association of Medical Editors (WAME), European Medical Writers Association (EMWA), American Medical Writers Association (AMWA), etc. which can help in understanding the acknowledgement for the ghost authors.

Here we also like to conclude with the gift and guest authorship because these are the unethical practices that are increasing these days, particularly those coming from not-so-good institutions to improve their chances, either they want an excellent affiliation or a useful contributor in their paper. For that purpose, this kind of guest authorships are happening. In my organization that if I see that I am a senior person or professor of sixty-plus age then my student also faculty in the same department so many a time my student has to give my faculty collogues student they need to give my name also as a co-author in their paper so this kind of thing is gift authorship so we must avoid all these kind of gift or guest authorship. If we want to include them, we must ask them to give some intellectual contribution to the paper to remain the gift or guest author; they become an actual contributor to the paper.

Competing Interest

In previous sections, we discussed authorship that how authorship issues are fundamental mainly, we discuss some new names also ghost authors, guest authors, and gift authors. We also discussed some of the potential reasons for the conflict that you have not given the authorship to somebody, you have given the authorship which has not contributed significantly. The order of authorship, then how the benefits of authorship are shared, is trying to give some extra benefit to somebody of the authorship. So all these issues we discussed in our last two sections and the important thing is that all journals, all publication houses can only give you some broad guidelines, but it is our integrity, it is our honesty which is essential in such type of issues because how we are doing, what we are doing, who are contributing in the research in our lab. How a journal will track it, so it is to develop confidence in our research we need to have that kind of integrity, we need to have that kind of honesty. People who know us can have confidence in our research, and then only our research will be cited, and then only the research will have some meaningful impact.

If our name somehow goes into the market in the academic market, these people may follow some unethical practices. They are not so honest for their academic integrity then probably whatever good work we are doing will not have that kind of impact. So we must make proper you can say aura, we make that kind of positioning of our research that this research is an honest effort, this research results from dedicated efforts and is written with full integrity. In this section, we will discuss some more aspects of ethical behaviour, integrity in the publication, and our discussion is related to competing interest. Often, we will find that we are in a situation where we have two roles in a particular situation and how to handle those two roles, that is the matter of competing interest. On one side, I am the approver also, and on another side, I am the applicant also. So you can understand the same person is the applicant and the same person is the approver also, so there will be a kind of conflict of interest. So we need to avoid that conflict of interest in our academic activities, and that is what we are going to discuss in this particular section.

Transparency and objectivity must be essential in scientific research and the peer-review process. Many a time, it may happen with you that your own scholar writes a paper, and since you are not co-author in your paper may come to you for the review process. It becomes essential that you need to disclose that this paper should be going to some other reviewer because you have some kind of vested interest involved in that. You will not be able to do proper justice with that kind of process with that kind of review which you are following for complete blind review processes. It is very important that we need to have full transparency and objectivity in our authorship and the review processes. Going further when an investigator author, editor, or reviewer has a financial or personal interest or belief that could affect his/her objectivity or inappropriately influence his/her actions, a potential competing interest exists. Whenever with that kind of publication, we will get some kind of financial benefit, some kind of personal interest is there some kind of inappropriate influence we can create, that is the presence of competing interest. Such relationships are also known as dual commitments as I mentioned that whenever we have two roles that mean one role on one side, another role on the other side of the table, so when we have dual commitments or competing interest or competing loyalties exists. So that is the issue of the competing interest or conflict of interest also.

In this particular situation, the most obvious competing interest is financial relationships because I research how these financial issues will be implicated one is direct another is indirect. Direct will help you in terms of employment, stock ownership, grants, patents, etc. How your research may help in your employment, your organization, your particular theory may help, you are publishing something that will help a particular group of people or a particular you as an individual. The second is indirect because of your research you start getting an honorary degree, start getting the consulting assignments, and sponsor activities to start happening because you get the mutual funds ownership and some paid expertise you start getting. So these are the direct and indirect benefits of your publications which are possible.

Now, we want our research to create some kind of impact because that is an essential thing academic research should not limit only for publishing in the journal. It should create some kind of impact also but that impact should be not limited to you. If you see that impact is for me only then this whole conflict of interest, competing interest comes into the picture otherwise we won't. So there is a prominent black and white issue the research should help the larger society, the larger community it should not be directed towards me as an individual's benefit. So that is the difference between the competing interest and the real purpose of the research. So now coming to undeclared financial interest, like you might be an investigator who owns stock in a pharmaceutical company and that particular company has commissioned some kind of research. So now you will like to do, you will like to present the research so that the products of this particular company gets promoted, they get the excellent demands, and when the product of this particular company gets good into the market, they get the higher demand the stock of the company starts increasing, and that is your benefit because you are one of the stockholders in that company. So this type of interest you are using the academic platform for your interest. If you see unethical behaviour that you are not using the interest, you do not see the larger interest of society for your interest; you are using the academic platform to improve its sales.

Now for the competing interest, as I mentioned earlier, financial issues are also there and personal issues are also there. So personal relationships, academic competition, and intellectual passion are there because of which we may have some kind of competing interest like some examples I am sharing with you. A relative who works at the company whose product the researcher is evaluating is a reason for personal relationship. A self-serving stake in the research results because of your stake in the research results like you are looking for some kind of promotion, career advancement based on that outcomes so that is the academic competition and then personal beliefs that are in direct conflict with the topic he/she is researching. So like you have some kind of competition with your colleagues or people working in a similar area, that may be the intellectual passion. Some people may say that intellectual passion may create new research, but just for you can say downsizing others' research if we start working that may be we are going into the negative direction. We will not be able to create new positivity in academic research. All these issues are issues related to competing interests.

Guidelines to Declaration of Competing Interest

Now there are specific guidelines to declare the competing interest. Now the first important thing in that an undisclosed relationship that may pose a competing interest. A straightforward thing like when you are a paper setter you give an undertaking that my no blood relation appears in this particular examination. If you are not giving that declaration or giving a false declaration you understand that it is a kind of unethical behaviour, it is dishonesty and the same thing applies to the academic publications that we need to give some declaration about the competing interest. It neglects to disclose a relationship with a person or organization that could one's objectivity or inappropriately influence one's actions. Some relationships do not necessarily present a conflict. Participants in the peer review and publication process must disclose relationships that could be viewed as a potential competing interest. So that is an important thing that we have to give our own you can say declaration that can be used by the publication house so that your competing interest does not come into the picture, or does not exist when we are going for the authorship the peer-review process. What should you do in this kind of situation when submitting a paper clearly whether potential competing interests do or do not exist. That is the first important thing to give an open declaration in the submission stage itself. Indicate this in the manuscript for single-blind journals or in the title page for double-blind journals.

Investigators must disclose the potential competing interest to study participants and state whether they have done so in the manuscript. So whenever we are doing some kind of clinical trials, we must disclose this kind of statement that what are we going to do for the trials we are doing. People going through the experiment should know and that you have to declare in your manuscript also. Reviewers must also disclose any competing interest that could bias their opinions of the manuscript. So whether we are a researcher, we are an author, we are a reviewer we need to disclose our competing interest before beginning the process and undisclosed funding source that may pose a competing interest that is also a possibility of funding your research.

Many times it happens in our country also that some NGOs are funding your research, some foreign organizations are supporting your research, which is also a competing interest—neglecting to disclose the role of study sponsors in the study design, in the collection, analysis, interpretation of the data, in writing. Hence, all these are the meaning of unclosed funding sources. So undeclared financial conflicts may seriously undermine the credibility of the journal, the author, and the science itself because then you are writing that paper for a particular purpose you are trying to put the academic platform for some purpose so this is again unethical behaviour from the point of view of the competing interest.

What should you do when you are into this kind of situation when submitting a paper a declaration (with the heading' role of the funding source') should be made in a separate section of the text and placed before the reference. So when you are writing your paper, when you have completed your entire manuscript just before the reference, you should clearly, explicitly mention the role of the funding source. Describe the study sponsors' role if any in study design, wherever in the collection, analysis, interpretation of the data, in the writing of the report, and in the decision to submit the paper for publication. So that also needs to be correctly mentioned. Editors may request that authors of a study funded by an agency with a proprietary or financial interest in the outcome sign a statement such as "I had full access to all of the data in this study and I take complete responsibility for the integrity of the data and the accuracy of the data analysis". So, this type of statements can be signed by the sponsors and may take this undertaking from them. So, with this, we come to the end of this particular section on competing interest where we discussed that you need to avoid interest conflict. We may often be in dual positions, so we need to declare those things upfront so that later on we do not face any trouble, we do not have any issue from the journal side, from the editor side, or the academic fraternity side.

CHAPTER 33

Plagiarism, Simultaneous Submission, and Duplicate Publication

Prof Rajat Agarwal
Indian Institute of Technology
Roorkee-247664

Plagiarism

In this chapter, we are going to discuss the prevalent topic of similarity in the research. The similarity in the research is also known as Plagiarism. It is a widespread problem across the research regarding submission of assignments in our classrooms, submission of project report, PG dissertation, PhD thesis, paper writing, book writing, journal writing, etc. You will find that this is a big problem, and when we are talking of academic integrity and ethics in research, this is probably considered the most crucial dimension of the research's ethical behaviour. So, it needs a good understanding, and we are going to discuss this so that it gives us a different dimension and different perspectives. Nowadays, almost all Indian institutions are developing a kind of policy about Plagiarism. There are institute/university/college-level policies to minimize this kind of ethical dishonesty, unethical behaviour, and promote more and more honesty and integrity in our academic work.

Forms of Plagiarism

Plagiarism may take different forms as we all know it is not merely just copying word by word. There are different forms of it from literal copying to paraphrasing someone's works, and it can include potential forms, maybe like only data and words and phrases and ideas, and concepts. Your similarity either you have copied the data of somebody else and used that data in your research that is one type of coping. You have taken the words, phrases from somebody's else paper, somewhere some book, and then uses them in your research, or you can only take the idea and concept and based on that idea and concept you are working for your research or have developed your manuscript. So, all these are different types of Plagiarism if you are smart if you have good command on English you have good command on the language you may even do paraphrasing that the whole idea or whole concept is the same you have rewritten in your words. Sometimes, the software may not be able to check it because software checks on the word to word matching but paraphrasing is an art, but it is in fact if seen from the ethical lens it is unethical behaviour. So all these are the different types of issues related to Plagiarism.

Severity of Plagiarism

Now, as just mentioned that Plagiarism has different forms, so at the same time, it has different levels of severity, severe Plagiarism to the moderate or low level of Plagiarism. So there is a spectrum of Plagiarism where you can have a low level to high level and what it means? It means how much of someone's work was taken-a few lines, a paragraph, pages, or the full article. So, depending on that, you have a low level of Plagiarism and a high Plagiarism level. You have printed the full article of somebody by changing the title or by changing just the authorship and because that article was published at a time when the internet was not there so that article is not available on the internet and you somehow have access of that article as the printed copy of that is available with you. Now, you have typed that article and changed the author's name; you become the author and submit it for publication. So that types of things are also possible. Plagiarism's high level is a challenging situation where you have copied the full article in your name. Another issue which can give you from low level of Plagiarism to high level of Plagiarism is that what was copied results, methods or introduction section that also will help us in understanding if it is only some part of introduction you can say that it is a low level of Plagiarism but if you are copying the result that is because all researches are known for the resulting output, so that is a high level of Plagiarism. So, it is you can say that deciding the level of severity is very important.

When it comes to your work, always remember that crediting others' work is a critical part of academic writing. In our previous chapters, we already have discussed how to cite the work of others. So citing others' work is one essential tool that can help us avoid the issues related to similarities. So that is applicable even your previous work. If your work has been published previously and if you are citing your work or giving some kind of reference of your work, you need to give proper citation to your work so that it does not go into that kind of plagiarism zone. The second important thing is you should always place your work in the context of the advancement of the field and acknowledge others' findings on which you have built your research. So when we are giving the reference of others sometimes we do not use previous authors' work appropriately. It is essential that when we are writing the literature review, we need to understand the purpose of using other's work because we are trying to highlight this particular aspect that these people have done this much of work previously and now my work is an advancement to work done by all these previous researchers. If you are writing that what work has been done by the previous researchers with due credit to them, it is not a kind of Plagiarism. Thus, the intent is also very important that with which intent you are using others' work so the intent should also be clear.

Literal Copying

A critical aspect as we just discussed is literal copying when you are reproducing a work word by word or in part without permission acknowledgement of the source that is certainly a very severe kind of Plagiarism unethical behaviour not acceptable. Literal copying is only acceptable when given the source and puts quotation marks around the copied text. So, when you are putting quote-unquote marks for that copied text so like if you are writing some so can say *chaupai* from *Ramcharita Manas* and giving that *chaupai* as it is in your research paper you can use quote-unquote marks for citing that *chaupai* in your paper so that is acceptable but with proper reference and with this technique of quotation marks. You should do what in this kind of situation, keep track of sources you used while researching, and use it in your paper. Thus, you need to create a kind of database already this was discussed in the previous chapters that whenever we are using some previous material we need to create a kind of record that what material we have searched and in which particular place which material is being used so that it becomes easier to cite that previous researches. Ensure you fully acknowledge and properly cite the source in your paper and use quotation marks around the word-for-word text and reference correctly.

Substantial Copying

Besides literal copying, the second level of copying is possible, that is substantial copying. This can include research material, processes, tables, or equipment. So substantial copying can be denied as both quantity and quality of what was copied. If your work captures the essence of another's work, it should be cited so simply by citation, you can avoid the issues related to substantial copying. One important thing we are learning is that citation is a significant thing that can help us avoid Plagiarism issues. So what should you do ask yourself if your work is benefitted from the skill and judgment of the original author so you can create a kind of self-check and you ask on your own that whether the work or the publications of somebody has benefitted your research and work? If the answer comes the yes, if your inner consciousness says that yes your work is being impacted by somebody else's work, then the substantial copying has taken place, and if so you have to be sure to cite the source. So, you can create this kind of self-check mechanism if you are in doubt whether to cite or not to cite. Another important thing that we discussed during the referencing that over citation should be avoided, just to avoid Plagiarism over citation should also not happen so you can simply ask that what works are impacting your thorough and vital research.

Paraphrasing

The third critical stage is paraphrasing. Paraphrasing is "reproducing someone's else ideas while not copying word for word without permission and acknowledging the source". So, you are a smart writer and are using the idea and concept of somebody else and writing now in your language and you are not giving due recognition, citation, or due reference of the original work so that is known as paraphrasing and it should also be avoided it is also unethical. It is only acceptable if you properly reference the source and make sure that you do not change the meaning intended by this source. So you need to ensure the source meaning remains as it is and at the same time you also need to give the due recognition of the original contributor. Thus, in paraphrasing what should you do, make sure that you understand what the original author means so that you are changing the meaning of the original contributor, never copy and paste words that you do not fully understand. Sometimes, we cannot understand the proper context in which somebody has written something and we have not appropriately understood we tried to copy and paste those things in our paper so, this must be avoided. It is essential that never copy and paste those things which you have not understood properly. Think about how the essential idea of the source relates to your work, maybe this original idea is in some different context and you are trying to use that in some other context so may it is not related to your work but we see a lot of papers coming where authors, unfortunately, do not try to match the context and they use the work of somebody else. So that is another important thing and then compare your paraphrasing with the source to make sure you wrote the intended meaning even if you change the words. So, you need to give proper referencing in paraphrasing, and you need to ensure that the original meaning remains intact. You do not try to copy-paste work from the original so in that case, if you want to use some of the things of the original contributor in your language it is possible but the intent again should be clear and intent should be an ethical one. So, that is an important part for paraphrasing.

Text Recycling

Then another vital type of issue is text recycling now what is this text recycling. Text recycling is reproducing portions of an author own's work in a paper and resubmitting it for publication as an entirely new paper. So, it also happens many a time, many a time we teach our students that we should try to take benefit of multiple outputs from the same efforts. So on one side, this is also correct that how to take multiple outputs from the same efforts, but at the same time we also need to understand that it should not enter into the text recycling so if text recycling is happening that is unethical behaviour and it needed to be properly understood that what is ethical and what is not ethical. Thus, in that particular situation what should you do, put anything in quotes that is taken directly from a previously published paper even if you are reusing something in your own words a. Even it is your own paper previously published somewhere you should not use kind of paraphrasing for your own paper for resubmitting it to a new journal in the form of a new paper instead you should always use some of the work of your previous papers and highlight that this was the work which I have already published in my previous papers a. Nowt is an extension of that work it is the advancement of that work so that it does not get into the text recycling. Whenever you are using previous work you need to ensure that you have proper references cited appropriately, as we have already discussed in the referencing section.

Plagiarism 2

We were discussing ethics in research, we started with some of the common issues related to authorship. We moved to conflict of interest and in our last section we started the discussion of Plagiarism which is considered to be the most you can say widely talked issue related to academic ethics that is a widespread problem and sometimes intensely and sometimes knowingly we do it. However, many times, students do not know the ethical dimension of writing a paper and may even unintentionally without knowing may enter into the zone of Plagiarism. So it is vital to know that what is the ethics what is the general guidelines finally it is depending on you and your team that how you handle these issues but there are some broad guidelines people are continuously trying to improve the integrity in the academic research. Some of the common guidelines are also stated so that we can avoid issues related to copying. If I say simply the meaning of Plagiarism is copying the paper which is already published that copying is possible in a variety of ways, some of those critical ways we discussed in our last section and some more types of Plagiarism we will discuss in our today's session that how we can have something intentional something unintentional. So let us start our discussion from the point where we left in our last section that Plagiarism is always unethical this is without any doubt we can say it is always wrong. Now as we discussed previously that there can be a spectrum it is in principle it is wrong it is unethical but there can be a degree of Plagiarism and that is the flexibility, that is the aspect that sometimes you need to work on the idea given by somebody else as we say always that we need to work on the science of some particular philosophy and then we see that how that science is being used in being possible in our context.

In social science researches many a time we use the questionnaire for data collection which some other researcher is developing and we use that questionnaire for our research so whether it is Plagiarism or not these types of issues are there because you will not be using the exact those process of writing the entire questionnaire in your paper. However, you will give the reference that I used the questionnaire developed by XYZ. So you need to take the ideas, you need to understand the methodology, you need to take some pieces from some sources, and that's how you conduct your research but what is the intent behind all these things that is very important and therefore the extent of it can be variable. Therefore we have this flexibility it is not yes or no, its not in binary it has a lot of grey area also and as I have already saying that it can also be unintentional. These days, we find the use of various softwares for knowing the level of Plagiarism in your manuscript. So these are the text matching softwares. One of the top-rated software we use is the "Turnitin" available in almost all good higher educational institutions. So it can only alert us that Plagiarism might have taken place because of matching of the words to words but further even it gives you some kind of indication, we need domain experts to finally verify whether it is a plagiarized manuscript or not so because there may be something which word software with this text-matching software may give but that matching is only limited to a particular area. However, the actual contribution of the manuscript is much beyond. So it is quite possible that the manuscript's actual contribution, the actual outcome of the manuscript is not plagiarized, there is something you can say told by these matching software that this much matching is there. So we are basically looking for the intention of the author. We are looking more from the expert in a particular area who can tell us whether it is a copied or not a copied kind of work.

Source-based Plagiarism

Now some critical type of plagiarized activity we discussed in our last session now there are few more types of Plagiarism, one is source-based Plagiarism. what is this source-based Plagiarism, where the different types of sources are used for example, when a researcher references an incorrect source or does not exist? So you have given a source you have cited a source which is not there, or it is incorrect, so it is a kind of misleading citation that is again unethical.

Plagiarism also occurs when a researcher uses a secondary source of data or information but only cites the primary source of information, for example, we have taken some output from, let say some work of Radha Krisnan on *Bhagwad Gita*. So, rather than quoting Radha Krisnan on *Bhagwad Gita,* I am directly quoting *Bhagwad Gita*. So, this is also a kind of source-related Plagiarism that I am not giving secondary sources, I am only giving the reference of primary sources. So, what happened because of this type of issues the number of references increases in our work and this is, in turn, increases the citation number of the references so our paper has a very long list of references that is not a very appreciable thing.

Self or Auto Plagiarism

Then another widespread type of problem is that is self or auto-plagiarism. So auto-plagiarism is like you have published your manuscript earlier or is in working now you are referring your work when an author reuses a significant portion of his/her previously published work without proper citation. So this is a common problem and many a time it is again unintentional because I am doing on a particular experiment work and I have published one paper on that experiment now I am publishing the second paper, it is quite possible that 60-70 per cent of my previous papers work is again part of the second manuscript. So that is a kind of self or auto-plagiarism. Most of you can say bibliometric activities whether it is Scopus, whether it is Web of Science, whether it is Publon they all give this data that how much of you citations are self-citations. So self-citations is also not taken so good in that sense but self-plagiarism where you are not even attributing your work previously done work and you are trying that this entire work, the second paper which you are publishing, that you are trying to show that this entire work, is noble so that is self-plagiarism.

Inaccurate Authorship

In the last chapter, we have already discussed issues related to authorship. So in Plagiarism, the meaning of authorship-related aspect that an individual contributes to a manuscript but does not get credit for it. So you are part of writing the paper, but you have not been given due recognition as part of the author or the contributor. The second form is the opposite when an individual gets credit without contributing to the guest or gifted authorships. So both these things are part of you can say Plagiarism also and it is a violation of code of conduct in the research, so we have already dealt with this issue in our authorship class.

Mosaic Plagiarism

Then mosaic Plagiarism is a kind of Plagiarism that is very close to paraphrasing where mosaic Plagiarism may be more challenging to detect because it interlays someone's else phrases or text within its research. It is also known as patchwork plagiarism and it is intentional and dishonest because you are aware that the work you are copying and just to make this work looks like your work you are inserting some of your words, some of your text in the work of somebody else. So it is like a paraphrasing activity which is also known as mosaic Plagiarism. Paraphrasing we discussed already, and this is a similar type of work where partly the original text and partly you have inserted your work, therefore, this mosaic word is used here.

Accidental Plagisrism

Another type of Plagiarism which is there that is accidental Plagiarism. This accidental Plagiarism is unintended, or you have not so much intention for doing this thing, but there is no excuse as we have already begun this session with that Plagiarism of any kind is not acceptable, and the consequences, whether it is intentional or unintentional are often the same. So it is occurred because of neglect, mistake, unintentional paraphrasing. Students are likely to commit accidental Plagiarism, so universities should stress the importance of education about this form of Plagiarism. We do not know that this is unethical and therefore, we get into this paraphrasing kind of activities and we often say that ok you read the paper then you try to write this work in your language. So that is a kind of paraphrasing activity, which is often accidental Plagiarism, but it is again unethical, so that is also a type of Plagiarism. So we discuss more than ten different types of plagiarism activities, and all these are unethical. We must avoid these types of plagiarism activities at any cost, and text-matching software can help us up to an extent, but beyond that, we need expertise from the particular domain to determine whether Plagiarism is there or not there.

Simultaneous Submission and Duplicate Publication

We are discussing ethics in our scholarly research, presentations, academic writing, publications and we have done that authorship, conflict of interest. In the last two sessions, we discussed Plagiarism issues, that there are different dimensions of ethical behaviour and again and again I am emphasizing that it is our intent which is going to set the ethical standards that I want to give something new to the society. Initially, my research may not be that impactful. Initially, I may not have those number of citations for my research but integrity is more important that I have done something novel and when you do something novel people will undoubtedly recognize your research because you are giving a new dimension to the field. You are contributing in the development of the knowledge. So if your research is ethical, it is undoubtedly going to have that if your research has that standard of integrity, it will undoubtedly give you a good recognition in your field. You may not be able to write a lengthy paper but write small papers you will get more readability of your papers because if you write a paper of 30 to 40 pages, it may not be possible that many people read those papers, but if you write a paper of six to eight pages where it is very tightly written, very cohesively written chances are there that more people will read your papers. You will get more citations, and when you are writing a paper of seven to eight pages, only chances are that you are not indulging in any kind of unethical practices. So, therefore, it is essential that what type of papers we are writing. Now continuing this discussion, we are now discussing another aspect of ethical behaviour concerning publication. Many a time we are looking for more number of publications that how to increase the number of publication because my promotion is due, I am supposed to be promoted for the next level next year and for that, I want three more publication, and this is a time where I may indulge in this type of behaviour which I am going to discuss now.

So, what we may do, we may do simultaneous submission because I don't have much time, and therefore what I can do I can simultaneously submit my paper to more than one journal and I may express that usually what we think that wherever we get the acceptance first then we will withdraw our manuscript from the second journal. So we create a kind of environment or do this kind of thinking in our brain that we are doing the right thing: we are looking for multiple opportunities and whatever opportunity comes to us first we capitalize, we will use that opportunity. However, if I see the same thing from the point of view of the ethics, academic integrity, and unethical behaviour, your manuscript's simultaneous submission is a breach of publishing ethics apparent. You should wait or wait until you get a decision from the first submission and then you can proceed for the second submission of that manuscript to some other journal, but simultaneously submission is nondesirable, it is unethical, it is a breach of publishing ethics. So what is the meaning of simultaneous submission? Let us try to understand.

Simultaneous submission occurs when a person submits a paper to different publications simultaneously, which can result in more than one journal publishing that particular paper. So this is a situation because I am in a hurry I want numbers so I am submitting the same paper to two different outlets and it is quite possible that both of them accept my manuscript and then for the same paper I may have two publications in two different journals. Obviously, you can understand unethical behavior even if you withdraw the manuscript from one journal that also I say is a kind of unethical behaviour. Duplicate multiple publications occur when two or more papers without full cross reference share virtually the same hypothesis, data discussion points, and conclusions. So that is also a type of simultaneous submission. This can occur in varying degrees; literal duplication, partial but substantial duplication, or even duplication by paraphrasing. So, these are some of you can say levels of simultaneous submission relate issues.

Let us discuss them slightly more in detail. One the main reason duplicate publication of original research considered unethical is that it can result in "inadvertent double counting or inappropriate weighting of the results of a single study that distorts the available evidence". So because of duplicate publications, because of more than one publication of the same study, we are giving an inappropriate weight of the result of a single study. One study should lead to one paper that is the idea, but when one study leads to more than one paper, it gives more weight to that single study and therefore, it is considered unethical. There are certain situations in which the publishers of two journals might agree in advance to use the same work. There are some specific conditions when this same kind of work can be published in two journals. So these are those conditions when there is a combined editorial and that combined editorial will take a call that ok we will publish this paper in both these outlets so that this research can go to a broader audience because of these two journals have different types of audience and for both these audiences, this research is useful so that is one possibility. Then position, statements, guidelines, if it giving you some kind of guidelines, some kind of things which are of mass uses then again you can use multiple outlets for the same paper because it is directly in the interest of the communities so without giving you can say one benefit you need to give it multiple ways for the research so that it can quickly go. Like presently I am talking let say we are facing the problem related to coronavirus, so if good research comes in this field this should not be limited to a particular journal rather various journals may collaborate that to publish the same research at multiple places because of broader dissemination is required related to this particular thing. It is not for the academic interest it is in society's interest; sometimes, we take this type of call.

Then the translation type of issues when you are looking to give the same knowledge in local languages differs in different languages. So if I am writing a paper in English, but I also want that this paper be available in Hindi and maybe other Indian languages, it is also possible through multiple publications. However, again this needs to have a kind of prior permission before submission from the editorial board and the author so that there is no conflict of interest later on.

Avoid Simultaneous Submission

What can we do to avoid simultaneous submission? The first is to avoid submitting a paper to more than one publication, so you need to submit your paper in an ordinary condition only to one particular publication house. Even if a submitted paper is currently under review and you do not know the status wait to hear from the publisher reviewer before approaching another journal and then only if the first publisher will not publish the paper. So only in that case when you know the outcome of your review process from the first submission and the outcome is not as per your expectation then you should approach the second journal or any other journal, but without knowing the feedback from the publisher you should not approach, and that is the way we can tackle the simultaneous submission.

Duplicate Publication

Another related term with simultaneous submission is a duplicate publication. When an author submits a paper or portion of his/her paper that has been previously published to another journal without disclosing prior submissions so here you are doing the similar kind of things rather than doing it at the same time. You are getting into the plagiarism kind of activity that you have already published a significant part of your research in other journal. Now you are using the same part to the second journal so this is duplicated publication. So avoid submitting a previously published paper for consideration in another journal. Then avoid submitting papers that describe essentially the same research to more than one journal you understand. It now provides full disclosures about nay previous submissions that might be regarded as duplicate publication. This should include disclosing previous publication of an abstract during the proceedings of the meeting also. So all these things you should follow to avoid the issues related to duplicate publication.

Another possibility is translation-related issues that nowadays we are limited to English, but many journals are coming in Chinese, Japanese, which are coming in other European languages, etc. In India itself, we are very strongly looking to provide our scientific literature in our local languages. Therefore, it is again a possibility that you may go for duplicate simultaneous publications for translated manuscripts. So, it is considered unethical behaviour when we are doing it without giving proper acknowledgement to the original publication. In this particular situation, we must submit our paper to a journal published in a different country or a different language. So we need to ask the publisher permission before doing this so that no ethical issues are involved and at the time of submission disclose any detail of related paper in different languages and any existing translation of that particular paper. So this will avoid any potential issues concerning unethical behaviour. So with this, we come to the end of this section to discuss how we can avoid problems related to simultaneous submission and duplicate publication.

Further Readings

- Guidance Document "Good Academic Research Practices"; Sept. 2020, https://www.ugc.ac.in/e-book/UGC_GARP_2020_Good%20Academic%20Research%20Practices.pdf

- Self plagiarism defined by UGC: Self-plagiarism: https://www.ugc.ac.in/pdfnews/2284767_self-plagiarism001.pdf

- University Grants Commission (Promotion of Academic Integrity And Prevention Of Plagiarism In Higher Educational Institutions) Regulations, 2018 New Delhi, the 23rd July, 2018, https://www.ugc.ac.in /pdfnews/7771545_academic-integrity-Regulation2018.pdf. Semalty A, plagiarism in academic writing, https://youtu.be/X1OMEmtq57o

CHAPTER 34

Research Fraud, Slicing of Research and COPE Guidelines

Prof Rajat Agarwal
Indian Institute of Technology
Roorkee-247664

Research Fraud

We have already covered discussions on academic research and ethics, academic integrity, and we use the guidelines given by different publication houses for academic integrity. Whenever we are taking academic integrity, it is related to the publication of our work. In our earlier chapters, we discussed that academic integrity is related to submitting our assignments, project reports, postgraduation dissertation, PhD thesis, and we require an even better level of academic integrity when it is about the publication of papers publication of books. All these are the different avenues through which our academic writing goes into the market and therefore, to have a good impact, good respect for my work, I need to follow the highest order standards. Now in this particular section, we are going to discuss an exciting topic of research fraud. We have already heard about financial frauds, but research frauds are also possible so what is the meaning of research fraud and how we can avoid it, what are the different forms of research frauds all those things we are going to discuss in this particular section.

Research fraud is publishing data or conclusions not generated by experiments or observations, but by invention or data manipulation. So when we use our own some technique for generating the data, we do data manipulation. We have not conducted that experiment; we do not have that kind of capability to experiment and without doing the research, without doing the fieldwork, without doing the experiments we are creating the data that is the meaning of research fraud.

Types of Research Frauds

1. **Fabrication:** There are two types of research frauds, which are related to our interest, one is a fabrication. This is making of research data and results and recording or reporting them. So you have created some data on your

table without going to the experiment, without going to the field you have created some kind of data that is the fabrication of data.

2. **Falsification:** The second is falsification; now falsification manipulates research material, images, data equipment, or processes. Falsification includes changing or deleting data or results so that research is not accurately represented, a person might falsify data or make it fit with the desired result of a study. We have some personal interest and to fulfill that personal interest we may do some kind of you can say tinkering with the data, known as falsification. So that outcome of the result is as per my desire, that is the second type of research fraud.

Now certain instances of fraud can be easy to spot if a referee knows that a particular lab does not have the facilities to conduct the published research. The research is done in XYZ institution, and I am familiar with the capabilities of the labs available in that institution I certainly know that this data cannot be collected from the available facilities in that particular lab, so it is straightforward to spot. The data from the control experiments might be too perfect in such situations an investigation would be conducted to determine if an act of fraud was committed so when control experiments are there you may create a perfect kind of data. Digital image enhancement is acceptable; it happen so that you can have a better image. However, a positive relationship between the original data and the resulting image must be maintained to avoid creating unrepresentative data or the loss of meaningful signals. You should always refer the original image in your text so that people know that this is a digitally enhanced image. If a figure has been significantly manipulated, you must note the nature of the figure's enhancements in the material and method sections. So there are few things you have taken from the other sources and by using technology or using some of your creativity, we are improving those figures, those tables, etc. and in that case, we need to give due acknowledgement to our previous work.

General Guidelines

Now some general guidelines related to these kinds of manipulated activities. One is manipulating images, which is very commonly happening because now we have excellent digital tools to improve the images. So images may be manipulated for improved clarity only no specific feature within an image may be enhanced, obscured, moved, removed, or introduced. Adjustment of brightness, contrast, or colour balance is usually acceptable as long as they do not obscure or eliminate any information present in the original image. So when you are using technology just to create better images, clear images are acceptable when you are not changing with the content of those published images initially.

The data excess and retention are also very important, and I will mainly like to say for the Indian researcher that they must preserve the raw data in connection with a paper for editorial review. Many times because we do not submit raw data along with the manuscript, nowadays, reviewers may ask for raw data so that the issues related to fabrication and falsification can be minimized so you must keep the data readily available with you even after publication for some time. It must be available in some safe custody so that if reviewers, editors they ask the data, you should provide those necessary files to them. Then studies undertaken in human beings, for example, clinical trials have specific guidelines about the duration of data retention so in the medical field in pharmacy field we mandatorily need to keep the data all the time with us, but in the case of social sciences, we are not so habitual of keeping the data for long term. So that is need to be kept into the mind so data retention and how technology-enhanced images can be used in our research that is possible and with this we can minimize the issues related to research frauds so that you have the actual data and if you have actual data your research will also give some novel insights otherwise it will be challenging to create something new with the help of your research.

Slicing of Research

We are discussing ethics in our publications, academic writing and research also. In the last section we focused on a fascinating topic of research frauds that we create data, we may falsify the data, we may use technology for some unethical activities also and manipulating data, creating data or doing tinkering with the data for some intended result is a severe offence which you can understand. You did not carry the research, but still, you are claiming that you collected this data if that type of things happens in medical science, in biological sciences, in pharma sciences it is undoubtedly going to create a huge loss to humanity. In other fields, those directly dealing with human life in these cases, if we falsify with the data, will create huge losses to human life. Concerning product launches, concerning human behaviour for purchase, etc. if we falsify data, it will create losses for the companies. Businesses will become unsustainable. So you understand that doing the research fraud is a grave offence and if you are a smart researcher the same term, I used when I introduced the concept of paraphrasing, so that is only possible when you are a smart researcher.

Similarly, research frauds are also possible only with creative, innovative, and know-how they create the data. So my simple suggestion is that rather than going into that kind of the wrong route of creating the data, manipulating the data, why don't you develop the experiment, why don't you collect the data and that will be more grounded your studies will have more meaning, and then you will be able to present the highest order of integrity concerning data collection. One critical dimension we discussed in the last section goes further into the issues related to ethical behaviour in the research we are now introducing another exciting aspect known as "Salami Slicing".

Salami Slicing

Now, what is this salami slicing? It is an English word, or it is some other word. Salami is salute, and slicing means you are making various parts of one research. So it is that very thing that you have conducted one research, and out of that one research, you are making various slices so that you can have multiple papers from the same research. So you can think of it as part of self-plagiarism it is also part of duplicate publication. Sometimes we say that it is required so that we can take more benefit of one study for one paper one study it is too much of expectation, so we may even go that ok we will do one study and this one study we will divide into multiple parts so that we can take more benefit in the number of papers from the same study. So slicing of research that would form one meaning paper into several different papers is called salami publication or salami slicing.

So you can make one good meaningful, impactful paper but rather you are making two three four papers out of one big paper, known as slicing publication. So this is unlike duplicate publication, which involves reporting the same data in two or more publications, salami-slicing involves breaking or segmenting an extensive study into two or more publications. These segments are referred to as slices of a study. So you are making two-three slices of one significant study and that is what the breaking of a significant study into various smaller studies so all these smaller studies are connected but the beauty is that you can make one very impactful paper if you do not go into this salami slicing kind of activity. However, here is a trade-off if you are writing a big paper of thirty to forty pages and as I mentioned in the previous sections, there may be chances that people do not read that long paper. So you want to make a smaller paper, and when you think of a smaller paper, you may enter into the domain of slicing of your research. So it is a very typical trade of whether you have one big, impactful paper or you have two, three slices out of this big paper.

The point is that if those significant papers can be converted into the similar kind of impactful papers but with lesser words, it is much better than creating two three papers where you are not able to highlight the novelty, where you are not able to create the impact and each paper is dependent on some other paper. So that is the way you can identify the difference between a more impactful paper versus creating various smaller papers or various smaller slices from that big paper. As a general rule, as long as the slices of a broken-up study share the same hypothesis, population method is not acceptable. The same slice should never be published more than once otherwise, it goes into duplicate publication. There are instances where data from large clinical trials and epidemiological studies cannot be published simultaneously or are such that they address different and distinct questions with multiple and unrelated points.

In these cases, it is legitimate to describe important outcomes of the studies separately. So if this type of situation is there, you may go for slicing, or then we will not say it is slicing because each paper is independent. You have collected data from large clinical trials, but then there are multiple aspects of that clinical trial, so for each aspect, you are writing a paper that is so the data is coming from the same source, but it is being used for addressing different kind of questions, different kind of queries, objectives, etc. However, each paper should clearly define its hypothesis and be presented as one section of a much larger study. So in each paper, we must mention how data is collected from one large sample or one large clinical trial.

Guidelines for Avoiding Salami Slicing

Now the guidelines to prevent the issues related to salami slicing. Breaking up or segmenting data from a single study and creating a different manuscript for publication is the fundamental issue. How we can avoid it, avoid inappropriately breaking up data from a single study into two or more papers so don't do that. When submitting a paper be transparent. Send copies of any manuscripts closely related to the manuscript under consideration. This includes any manuscript published, recently submitted, or already accepted. So if you are writing more than one paper, so don't try to use two different outlets. So if you are using two different outlets and feel that these are unrelated, you are entering into the salami slicing. Try to use a similar kind of journal or the same journal so that the same reviewers, same editorial boards can take a call whether it comes under this slicing or it is not that kind of issue. So, this is I think the simplest way to decide the issues related to the slicing of your research that do not use channels which are very far away, use channels which are closely linked then probably you will be going to the same pool of peer reviews and then they will be able to identify whether these papers are of similar nature or different nature. So with this, we come to the end of our discussion related to the research's slicing.

COPE Guidelines

This is coming to be the last section on this discussion related to ethics in publications, academic writing, and our research activities. We discussed the various dimensions of ethical behaviour starting from authorship, conflict of interest, and how we can get into the research frauds, the simultaneous submission, and, most importantly, plagiarism. So, these are the different issues common across the board and though there may be little difference in guidelines from one university to another university, one publisher to another publisher. However, as I am again and again saying the general principle is the same, that is the researcher's intent that whether we want to do ethical work or do not want to do ethical work. So we should have that intent upfront clear that I want to research the highest order. I may not be successful in my first attempt to publish my research in very impactful journals, but if I am doing some novel research, I am hundred per cent sure that you are bound to get your place in that elite journals so academic integrity is one critical enabler to push your research to the higher level.

Now in this particular section, we will discuss some other different dimensions of academic writing and research. Most of the discussions we had that these rules are for authors, but many a time we are reviewer also, I am an editor also, I am also a guest editor, so my academic integrity and publication ethics are required in those roles also. I cannot expect all roles from the author only, but when I am on the other side of the table, I need to show and exhibit even more standards, higher standards concerning academic integrity and therefore in this particular section, we will discuss some of those aspects also. The contents of this particular section are basically taken from the processes, guidelines developed by the Committee of Publication Ethics (COPE), and this Committee of Publication Ethics they have various practices, and for each of these practices, they have a well-defined process which they write as a flow chart. So some of the flow chart we are going to discuss in this particular session, and if you go to the website of COPE and almost all leading publication houses they are signatory, they are the members of this COPE and then if you go to their website you will find that they have a lot of guidelines, they have a lot of facilities for authors also. Most of the time, as I mentioned, we expect the author to behave ethically, but if the editor does not behave ethically, the reviewer does not behave ethically than what to do, or the publisher not behaves ethically.

If the editor is rejecting a manuscript based on let say race, you are from a race that the editor does not like and the editor gives a desk rejection to your manuscript that is also unethical. If a publisher is charging money for publishing your manuscript that is also unethical, and therefore we have many discussions against predatory journals where they do not follow any review process or are charging some fee for publishing your paper. So there are unethical behaviours from other parties also who are involved in the publication of papers. So there are suspicions in the peer review process, suspicion over image manipulation in a published article, potential authorship columns also are there then manipulation of the publication processes are possible, that editors have given green signal without going through the peer-review process because that manuscript is submitted from one of his known, one of his relative. Therefore the entire publication process was bypassed to support that manuscript.

Suspected fabricated data in a submitted manuscript, reviewer or reader suspects undisclosed conflict of interest in a submitted or published article. Then it is also possible that how are you responding to a whistleblower. It may be concerns raised through social media or concerns raised directly to the publication house. These are the various aspects on which COPE has given its guidelines, processes, flow chart, how you will move the complaint, and how you will register your suspicions and then the entire activities or set of activities. I am giving you one or two examples in this particular section about what to do if you suspect an ethical problem with a submitted manuscript, so how do you help it. So in that case reviewer raises ethical concern about the manuscript and the what type of concerns maybe there lack of ethical approval, like it is related to medical science and the proper consent from the patients are not taken, and the institute ethical committees are there, you have not taken the consent of animal-related experiments. So these are the type of concerns which a reviewer may raise than what you do.

Then you thank to the reviewer and say you plan to investigate. Then author supplies the relevant detail and the authors will provide you the consent obtained from respective committees about which reviewer has raised the concern. Now there are two possibilities either it is a satisfactory answer or an unsatisfactory answer or no response. So when it is a satisfactory answer, the reviewer may thank for or give a kind of apology and continue the review process. Another possibility is that the author informs the author that the review process is suspended until the case is resolved means in case the answer is not satisfactory or the publisher does not receive a response, the review process will be suspended. Then the forward concern to the author's employer or person responsible for research governance at the institution. If the reviewer and the publication house are not satisfied with the response, it will escalate to a higher level. The author is working somewhere, and then this complaint will be forwarded to those people who are governing research in that particular institute then issue resolved maybe because the complaint is raised at a higher level, it may result into the solution of the problem, and then again there will be a kind of apology, thanks and the review process will continue. If it is not satisfactory then again contact the institute at three or six months period and seeking the concluding remark of the investigation and again if there is no satisfactory response. In case of any satisfactory response, the review process will continue. If a satisfactory response is not received, refer to other authorities like the entire body of researchers, the professional committees, etc.

so that it is informed in a larger body and that larger body takes the issue into its cognizance. People know that what other people are doing concerning ethical problems and the institute is also not supporting or not providing proper, satisfactory answers so this is one issue.

Another issue is related to peer review manipulation as an author. I feel that the peer-review process is not up to the mark. Now here what happens that a third party comes to you that says the author that I will give you the assured acceptance, which is possible against some fee. So there is a manuscript with authors A and B that suggests reviewers with false email addresses and submits to a journal. Now peer review invitations go to the third party via false email addresses. The third-party generates false favourable peer review reports. The manuscript is accepted, and authors A and B pay this third party, and the manuscript with authors A and B is published in some journals. So this type of issue is also happening where you are creating a kind of totally parallel review process than the actual review process because you need to show to the world that we are following the process, but we are following a process that the genuine reviewers not do. Some paid reviewers do it and are creating false reports, so this is an example of unethical behaviour in the review process. Then authorship for sale is also there you have the third party, and his authorship is saled like I want to give a guest or gift authorship, but now I am charging I am expecting some benefit to give you the authorship. So this is also a kind of unethical behaviour in the field of research publication.

Systematic Manipulation of the Publication Process

In this final section, we will see what the systematic manipulation of the publication process are. Now a large number of activities are there, plagiarism is just one but it is related to authorship, it is related to the review process, it is related to editorial offices, it is related to publication processes. All these stakeholders involved in this process of publication at any of these levels' unethical behaviour are possible. So this committee on publication ethics deals with all the possible dimensions. You can see that all those dimensions are available in this particular diagram that you may have a suspicious activity that may raise a red flag. This can include that manuscript on unrelated topics which is reviewed by somebody who is not an expert in that particular area and from that to various issues related to authors and authorship. So with this, we can understand that ethical behaviour in research is fundamental, and it is a kind of tool. If you are doing good research, and if you follow the ethical standards, it will increase the bar and increase the acceptance of research to a higher level.

Further Readings

- Guidance Document "Good Academic Research Practices"; Sept. 2020, https://www.ugc.ac.in/e-book/UGC_GARP_2020_Good%20Academic%20Research%20Practices.pdf

- Self plagiarism defined by UGC: Self-plagiarism: https://www.ugc.ac.in/pdfnews/2284767_self-plagiarism001.pdf

- University Grants Commission (Promotion of Academic Integrity and Prevention Of Plagiarism In Higher Educational Institutions) Regulations, 2018 New Delhi, the 23rd July, 2018, https://www.ugc.ac.in/pdfnews/7771545_academic-integrity-Regulation2018.pdf

CHAPTER 35

Research Proposal Writing

Dr Ajay Semalty
H.N.B Garhwal University (A CentralUniversity)
SrinagarGarhwal-246174

Welcome dear learners, in this chapter of Research Proposal Writing, we will learn about the basic concepts of proposal writing.

Learning Outcome

After completing this chapter, you will be able

- to know the basics of writing an effective project proposal. And
- to write an effective project proposal and communicate to the funding agency.

Lesson Plan

This chapter will be covered under the heads

- What are research projects?
- Why are they Needed?
- When to prepare the research project? Preplanning
- Types of projects
- Targeting funding agencies. And
- Basic principles and basic steps of proposal writing
- Proposal writing: preliminary requirements
- Choosing and defining problem area/ objectives
- Reviewing literature and identifying the gap
- Planning and methodology
- Expected outcome
- Timeline

So, let us begin writing research project proposals. Very firstly, I would like to define the research before moving to the research proposal project writing. What is research?

Research

"Any creative systematic activity undertaken in order to increase the stock of knowledge, including knowledge of man, culture and society, and the use of this knowledge to devise new applications" is called research.

Let us take another definition

"Research is a process of steps used to collect and analyze information to increase our understanding of a topic or an issue". It consists of three steps: pose a question, collect data to answer the question, and present an answer to the question.

So, these are some major definitions of research. However, the perception of research may vary little bit, methodology may vary but the concept is same.

Research Projects

Now, let us move to the research projects. What are research projects?

You won't find any specific definition of the research project- What is the project? So, what are research projects?

"Research projects are the specific, time bound, rational, and result oriented group of activities funded by any grant agency which focus on studying, exploring, and developing a new theory, product or service which has the practical application or utility."

It may also include developing or establishing understanding of facts and observation.

So, the last word was utility and remember the quote from the Edison-

"Anything that won't sell, I don't want to invent it because sale is the proof of utility, and utility is success."

So, to do need based research is the focus point and the demand of current era. We cannot waste our money in things which don't have any applicability, which don't have any utility. Now, let us answer the question- why? Why we should do the research projects?

Need of the research projects

Let us discuss Why we need to do the research projects.

- To be updated with new challenges/ research
- To develop culture of innovation/ IPR
- To develop the national and international collaborations
- To get recognized in the academics and research

- To develop an attitude of continuous learning (which reflects in your teaching).

- To cater the need of research work

- To establish your own lab

- To develop your own team (PF, co-investigators, collaborators, service providers)

- The first answer is to be updated with the new challenges and research. Until and unless we are updated with the advancement of research, we cannot do the utility-based research. That's why research projects need to be done.

- The second answer to our question why we should do the research project is to develop a culture of innovation/IPR (Intellectual Property Right). Developing the culture of innovation means IPR based research in which component of novelty is there. We do the research project so that we can have funds to do the research which can have the patentability. And we can protect our intellectual property with that thing that is called a patent. And it fulfills the demand of research that is utility based.

- The next thing is to develop the national and international collaborations. We do the research projects, so that we can have good interaction globally with the experts, with the researchers and have collaborations with them. I remember, I would like to just give this example. I was in need of phosphatidylcholine for my project and for my research. I was going through the opportunity but was not having any funds with me. I discussed this thing with some of the collaborators some of the experts of Europe. They suggested me and then I managed to have a sample of phosphatidylcholine worth more than 1 lakh rupees that too twice from a source. So, this was the beauty of collaborations. You can have much more advantage with developing the collaborations.

- The next thing is to get recognized in academy and research. Research projects give the recognition, the number of projects you have done give rise to number of publications and number of other things. It might lead to patents. So, they are the..you can say the foundation of your research career or academic career. So, they have very vital role in giving recognition in academics.

- The next reason is to develop an attitude of continuous learning. In the teaching learning process, a teacher who is actively involved in research apart from the teaching is well upgraded with the body of knowledge in comparison of teacher who is not doing the research. And it gives a value addition to academics to teaching learning process. And this learning is reflected in your teaching.

- The next reason is to establish your own lab. You can have your own lab if you are doing the research projects. You need not to depend upon the institutional laboratories or departmental laboratories. If you are doing the projects, you can develop your own labs. You can do your small projects without any funding maybe.

- And the last reason but not the least, to develop your own team. You can develop your own research team with the project fellow with the collaborators with the co-PI to the research project. To do the further research to plan the future research.

So, this was all about the concept of research, research projects. We can have the idea that why we should do the research project in this section. Now, we will move to next part of research project- planning in the next section.

After going through the introduction of research projects and answering why to do the research projects. We will move to time and planning now.

Timing

There is no specific timing in doing research projects. For every section, for every part of your academic and research career you can plan research projects.

- Early (Fresh entrant, <35 yrs),
- Seed grant (UGC)
- Project proposal sent along with job application
- Mid (Mid career, around 40-50 yrs of age)
- Senior Level (Professor level)
- Expert/Post retirement level
- The first stage is early career. For fresh entrant in academic career. For the researchers below 35 yrs, young researchers. Seed grant is available by the UGC and in these particular instances after recruitment the seed grant is provided by the UGC or by the university itself. In some institutions, project proposals are sent along with the job applications. In many foreign universities, it is the culture.
- The next is mid-career. In mid- career around to 40-50 yrs of age there are some other different set of research projects which are available for you.
- For the senior level (professor level) different set of projects are available.
- And also for the expert and post retirement level the projects are available by the various funding agencies in that scenario.

Now, let us move to the very first step pre-planning. For pre-planning: -

- The first point is brain storming. Brainstorming an idea. I will give you an example that how interdisciplinary, multidisciplinary, or out of the box it may be. John snow, he was a physician (doctor). In London outbreak of cholera, he was the person who identified that cholera is a water borne disease. Before that, it was believed that it has got some supernatural source. What he did? He used DOT mapping, a population dynamics statistical tool for indicating the cause of cholera (Fig. 35.1).

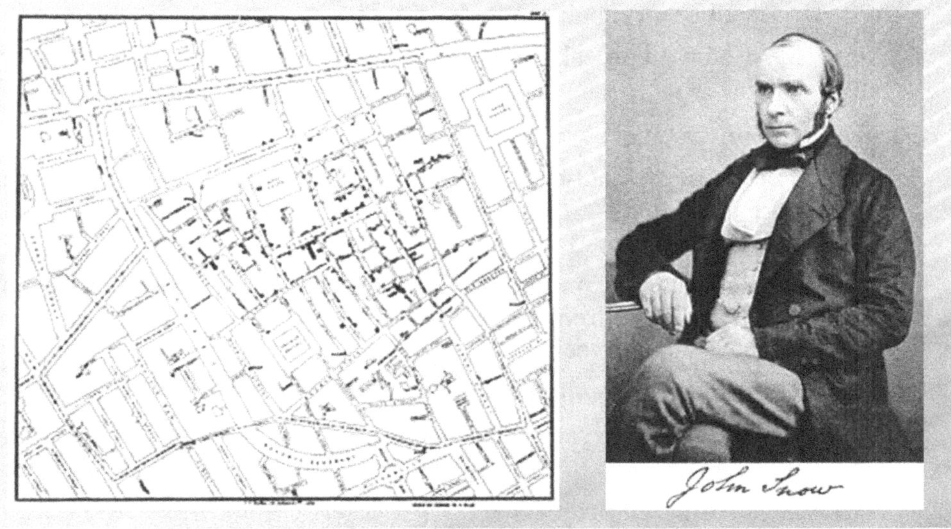

Fig. 35.1 Dr John Snow: Used DOT mapping (A population dynamics statistical tool) for indicating the cause of Cholera

The mapping was done as you can see here and he observed that casualties in and out in Thames area and the area where the water was dirty were higher as compared two the area in which water was of good quality. So, it was the statistical tool which yielded into the identification of cause of a disease. So, anything can help. What is the complementary remark? Anything can help. You should brainstorm the idea and develop the idea.

- For pre-planning, after the brainstorming refer the available resources. What are the resources you are having, refer them? Books, journals, go to the online resources. Refer the previous work done your self-work and the work by others. Critically analyze them just like the literature review these pre-planning parts are there. Chalk out every aspect. Write them down and synthesize your own idea, what you want to do? On the basis of summaries of previous work. Then visualize the methodology. The main methodology and alternate course of action. If plan A doesn't work than what will be my plan B, it should be effectively done and that too within the realm of available resources.

- Now, you will enlist the requirements. What is the requirement with respect to the manpower with respect to the consumers, contingency, and other resources whatever you need? You will chalk them down.

When you have pre-planned the things see what kind of research you want to apply. What kind of research project you want to do?

Types of research and projects

- Whether it will be an Empirical study (field study) or,
- It may be a Non-empirical study or laboratory-based study or exploratory study

The research projects may be

- **Extra mural:** Extra mural projects are those projects which a funding agency provides funds to any person any faculty member who is not associated with that agency.
- **Intramural** means that one person will be mandatorily from that agency and one person will be from any other agency who is applying here. These are called intramural projects.
- **Inter institutional and collaborative projects**
- **Industry-academia interactive projects** and lastly the major projects that are,
- **National coordinated projects**

So, you may apply depending upon your scope your level and identify a particular type of project. Then focus to most important thing- funding agency.

Funding Agencies

To which funding agency you are going to apply?

In the national level, in the Indian scenario the major funding agencies are: -

- UGC
- DST
- ICMR
- ICSSR
- DBT
- AICTE and so on

Depending upon the subjects there are so many agencies. So, go through the list as followed.

http://icmr.nic.in/guide.htm

Indian Council for Medical Research

www.csir.nic.in

Council for Scientific and Industrial Research

www.dbtindia.nic.in

Department of Biotechnology, Govt. of India

www.dod.nic.in

Department of Ocean Development

Identify any funding agency and chalk your plan according to that (Table 35.1).

Table 35.1 Important Links of Funding Agencies

S. No.	Funding Agency and links
1	UGC's Faculty Research Promotion Schemes, https://ugcfrps.ac.in/uohyd/
2	UGC: https://www.ugc.ac.in/pdfnews/2089255_STRIDE_FINAL_BOOK.pdf
3	DST/SERB: https://www.serbonline.in/SERB/emr?HomePage=New
4	ICSSR:http://impress-icssr.res.in/wp-content/uploads/2019/09/Guidelines-Project.pdf
5	ICSSR: https://icssr.org/impactful-policy-research-social-science-impress
6	AYUSH: https://main.ayush.gov.in/schemes/extra-mural-research
7	DBT: https://www.dbtepromis.nic.in/Login.aspx
8	Sample format AICTE Project (COLLABORATIVE RESEARCH SCHEME (CRS) UNDER TEQIP): https://www.aicte-india.org/sites/default/files/Compiled%20Sample%20application%20form_0.pdf
9	AICTE: https://www.aicte-india.org/opportunities/faculty/overview
10	ICMR: Link for the online submission http://icmrextramural.in

Targeting funding agencies

On which basis we should target the funding agencies.

- On the basis of subject area of research, you should identify the funding agencies
- Level of research
- Funding ceiling (allowed or required). What I meant to say- what is the maximum limit allowed and what is your expectation. If you are expecting a bigger project, you should apply for the bigger only. If you go for the minor one, you won't get that much fund.
- On call or round the year proposal submission

- What is your expectation and self-assessment of your own project? You should define, what do you expect? Then target the funding agency

- See the eligibility conditions. The eligibility condition is a major thing for targeting whether you are eligible or not for that scheme by that agency.

- Time taken in the processing is the last factor which is very crucial. Sometimes agency takes lot of time to process the things which may be beyond your expectations.

No, we will start writing the proposal

Proposal writing: begins with

Preliminary requirements. What are the preliminary requirements?

- Eligibility criteria. Have you check the eligibility criteria?

- Have the biodata of principle investigator PI and co-PI in the prescribed format with the length required, highlighting the expertise available, achievements, publications, patents, projects completed etc.

- Keep ready your photo and signature scanned for online proposals.

- Keep ready for forwarding through head of the institution. It is very important aspect. You should be having that forwarding in your hand before applying the project.

- Download guidelines and read it carefully.

- Download the proposal form. Check it carefully.

- Collect the things needed and make a rough draft.

- Check the word limit of each section/head/subhead.

- Check the price of equipment needed or get the quote for the same. If you need any equipment you must have the tentative idea about the budget.

- Now plan the body of the proposal

So, in this section we focused on timing, when you should plan, you can plan at any point of your career. The brainstorming is very important. Pre-planning your idea systematically, covering all the aspects, covering the methodology, and laying down your aim and objectives.

Cover all the aspects. Check the eligibility criteria. Download the form. Collect all the things and make a rough draft.

We will move ahead with each section in next part of the chapter.

Now let us focus on standard heads in the project proposal

Standards heads in a proposal

- PI's Information Title of the project Summary
- Origin of the idea/ Background/ Introduction/ rational
- Aim and objectives Literature review Methodology
- Plan of action
- Time-line
- Budget with justification
- Expected Outcome (benefit, IPR/ industrial potential) of the project Expertise/ facilities available in the institute
- Suggested reviewers, sometimes asked

Taking the very first one

PIs information

Provide all the information required like

- Name
- Academic qualifications
- Address for

  communication

  DOB
- Institutional address
- Upload your current CV with publications, projects, patents etc.
- Mention your H index (for sciences and life sciences) Total Citations details, If you are having, it helps

Then, we will move to the

Title of the project

- Should be representative
- Related to the need-based research/ hot topic/ within the scope of the funding agency. You can go through the titles which have been granted funds by the agency earlier and see what kinds of projects are prioritized or preferred for funding. Plan them accordingly
- Specific
- Title should be unambiguous
- Clear, error free. No punctuation error should be there
- Plan two/ three titles and finalize after completing the first draft

Tips:

The tips I would like to give this time for this title:

That never write your title in upper case because in uppercase the spell check doesn't work, and then before submitting or finalizing check it alphabet by alphabet correctness because if the title is wrong nobody is going to evaluate or even if it is being evaluated, the rejection is sure. Because it shows that the proposal is made carelessly.

Summary

Should be of one page/ 300-500 words. It should be concise and provide complete roadmap of research. It should contain need, Aim/objective, methodology, plan and expected outcome in a concise way. It should give the power packed complete story. It must be flawless, and should be cohesive and coherent.

Tip:

What is the tip? Can you tell the whole story in one page or half page without any flaw? This is the tip for writing the summary.

Introduction

- Give background or origin of the idea in a cohesive and coherent manner. Present the problem logically, sequentially, systematically introduce it and give your hypothesis.
- Focus on justifying the rationality of your work.
- Give proper citation to the latest and important previous studies. Don't use very old references

Tips:

What are the tips?

- Be concise
- Develop the cohesion and coherence
- Follow the Logical transition
- The reader is an expert you know, unlike the research article or review articles here the first and the last reader is the expert. You are writing for the expert so there is no chance of any mistake.

Next is

Origin/Idea/Background/Introduction/Rational

Give a brief background (within word limit allowed) How the idea originated?

How much back do you need to go?

As much as tiger take back steps while on prey

As much as run up taken by a bowler. You will have to decide the optimum length of the background.

Generate the rationale and need of the study.

Try to answer: What is the problem and what is your hypothesis

Aim & Objectives

Give one aim 4-5 objectives

Objectives should be SMART. SMART stands for
- SPECIFIC
- MEASURABLE
- ATTAINABLE
- RELEVANT
- TIME BOUND, we have discussed it earlier also

Tips
- Do not use the vague, casual or unpractical language
- Give precise objectives and the method should be directed to achieve each objective
- Identify precise subject or the tentative list of subjects/domain etc.

So, in summarizing the section we can say that after identifying a funding agency, you prepare the documents and finalize your title. Write a good summary, effective summary. Introduce your topic and give aims and objectives which are strong, convincing, specific with respect to the scope of the funding agency.

Literature review

The next part is going through the Literature review. We do the literature review just like the literature review we have done but here the scope is limited. You have to concise the literature review within the word limits of the research project proposal.
- Our last 5-10 years are the most important and relevant studies.
- Give the international-national status
- Give proper citations
- Try to present the gap left behind for bridging with the project.

Tips:

Develop the need of the study through literature review and establish or strengthen the idea/hypothesis.

Remember! it's not a formality as in Ph D literature. It might be the base of accepting or rejection of your project proposal. So, be cautious.

Now, after the literature review methodology comes

Methodology

- How you are searching the way out? What is the blueprint of methodology? What methods are planned to be used. What is the sequence?
- Methods: classical or modified?

Tips:

Now, what are the tips?

- Methods should be capable to address the aim and objectives logically.
- Maintain connection with the main idea.
- Avoid unnecessary methods

Plan of action

It Should be in good tune with objectives and methodology. It Should be idea centric.

It should be result oriented.

Tips:

What is the tip here?

The plan should be neither too defined nor too gross. Don't constrict it or make it too gross.

Now comes the

Time-line

The whole project must have a clock associated with it.

Prepare the PERT chart (performance evaluation review technique)/activity-time graph (Fig. 35.2 & Table 35.1).

Time Schedule of activities giving milestones through BAR diagram.

Tips:

What are the tips?

Plan practically

Neither be too optimistic nor be too pessimistic regarding time-line.

Plan wisely, the progress of the project will be benchmarked with this time-line.

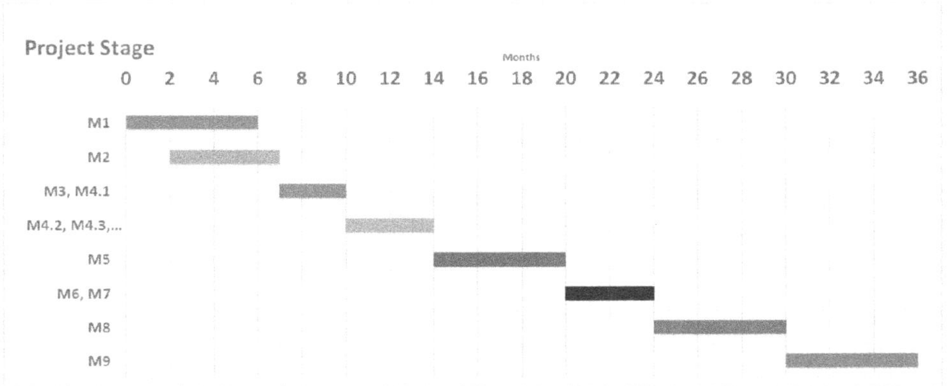

Fig. 35.2 Example Activity Time Graph

Table 35.1 Activity Time Table

Sl. No.	Activity Block	Time required (in months)
1	M1	06 (0-6)
2	M2	05 (2-7)
3	M3, M4.1	03 (07th-10th)
4	(M4.2, M4.3, M4.4)	04 (11th-14th)
5	(M5)	06 (15th-20th)
6	(M6, M7)	03 (21th-24th)
7	(M8)	06 (25th-30th)
8.	Compilation of results and discussion (M9)	06 (30-36)
	Total Time 36 Months	

As we can see in this example that we have seen with respect to the methodology. We have given the timeline that what methodology or what part of the methodology will be covered in which time frame in the tabular and graphical way (Table 35.1 & Fig. 35.2).

Budget with justification

Fund demand must be practical. See the guidelines and previous years' granted projects.

HR requirements and qualifications must be clarified for budgetary requirements. Equipment demand must be logical and supported with justification. Consumables and contingency budget should be as per the need and well justified.

Tips:

What are the tips

- Ask as per the trend.
- Do not make over ambitious budget

And lastly what are the

Expected Outcome

- Think to yourself, why should money be given?
- Point out the expected outcome: new product, service, technology, and its utility How can society be benefitted?
- Possibility of patentability an industrial application of the research must be seen.

Tips:

What are the tips?

One or more important benefits to society or industry must be there. Patentability is always a plus.

Expertise/ facilities available

You can give the Expertise/ facilities available in your institute

Show what you already have for executing the project.

Enlist the equipment, infrastructure and other facilities available at your institute or department.

Your previous experience with the field of research and your familiarity with relevant know how supported by the previous publication records.

Tips:

What are the tips?

Give your relevant and related prior publications.

Give the detail of completed or ongoing related projects by PI or Co PIs.

Suggested reviewers

You can give the list of suggested reviewers

- 2 to 5 reviewers for your project (if asked to do so).
- Each suggested reviewer must be well known in that field and expert of that field.

- Each suggested reviewer must not be your supervisor or the faculty member of your own institute.

Tips:

What is the tip?

Do not suggest your guide or your any coauthor. Only suggest a good academician/ researcher

References

The last is references

- Follow a uniform style of referencing. Give brief and only relevant references Cite only relevant and quality papers
- Cite from good journals and good indexing agencies only of good impact factor / prestige. If not from good impact, must be from a journal with good indexing like SCI, ESCI, SCOPUS, Web of Science.

  Have a control on self-references.

  Have a checklist on your hand

Checklist

PI's detail/ CV uploaded? Endorsement from Head included Double check the Title

Double check the summary Check the objective

Check the consistency between objectives and methodology/plan of action. Check the ceiling and justifications for budget.

So dear learners,

This was all about the research project proposal writing. You should write in the rational way for the rational problem. Ask for the budget in the practical way. Plan a good, sound utility-based project and double check whether your aim and objective, methodology, and expected outcome are worthy to be focused or not. Hope you enjoyed this chapter and you will be conditioned to apply for research grants after going through this chapter. Happy learning.

Defending a project proposal

- Online/offline
- If online: check the number of slide limit
- Focus on objectives, rational and methodology.

- If offline: Prepare well, study the complete literature, if asked what you can cut from budget?; define priority (equipment v/s project fellow v/s consumables v/s TA/DA).

- Do not be overconfident.

To begin with Project proposal writing

- Do not jump in high value projects

- First do some preliminary work and get it published

- Then try to apply and do the projects with lower budget. Always focus on one major theme.

- Anticipate the future need and plan the project on time Prepare a good team.

- Each project is evaluated by outcome at least in the form of publications. So, publish in reputed journals with acknowledgement to the grant.

Major Challenges

- Major cut in funding by MHRD

- Teaching workload

- Lack of academic writing skills

- Not much support from employers/institutions except the forwarding/ endorsement Lack of efficient research support cells

- Reluctant seniors who are not willing to pass on the legacy of writing the project or may be do not have the time for this

But recently, NEP 2020 (https://www.education.gov.in/sites/ upload_files/mhrd/files/NEP_Final_English_0.pdf) has come up with a National Research Foundation (NRF) with a huge budget (Rs 50,000 Crores) in year 2021. NRF will provide research grants apart from major funding agencies like DST, DBT, ICMR etc.

Activity TIME.....

- See the website of funding agencies of your subject/expertise/interest.

- Check out and enlist the timing and name of schemes.

- Plan a project proposal on your topic of interest.

- Get it checked and approved by your seniors/friends.

Further Reading

- Guidance Document "Good Academic Research Practices"; Sept. 2020, https://www.ugc.ac.in/e-book/UGC_GARP_2020_Good% 20Academic%20Research%20Practices.pdf

- NEP 2020https://www.education.gov.in/sites/upload_files/mhrd /files/NEP_ Final_English_0.pdf. Semalty M, Grant Writing, https://youtu.be/ pGte75DauRg (English); https://youtu.be/IiYnNoBu3Uc (Hindi)

References

- How to Get Started With a Research Project https://www.wikihow.com/Get-Started-With-a-Research-Project
- Creswell, J. W. (2008). Educational Research: Planning, conducting, and evaluating quantitative and qualitative research (3rd ed.). Upper Saddle River: Pearson.
- http://www.mnit.ac.in/research/admin/Final_LIST.pdf

CHAPTER 36

Abstract/Conference Paper/ Chapter/Book Writing

Dr Ajay Semalty
H.N.B Garhwal University (A Central University)
Srinagar Garhwal-246174

Welcome dear learners! we are here, with the chapter on some versatile type of academic writing like abstract/conference paper/ chapter/ book writing.

We learned about the basic concept of review paper, research paper, thesis and research proposal writing. Now let us discuss abstract writing and book writing in this chapter.

Learning Outcome

After learning this chapter, you will be able to write and plan

- Abstract for Conferences
- Chapter/Book writing

Lesson Plan

- Why to attend conferences?
- Writing abstracts for conferences
- Presentation of conference paper
- Chapter/ book writing
- proposal submission
- Salient features of book /book chapter writing

Let us open up the chapter.

Attending Conferences

It is one of the very important academic exercise and event. You need to attend conferences

- Sharing your research on national and international platform
- Exchanging ideas and updating knowledge and ideas
- Basis for collaboration
- Career opportunities

Steps for conference abstracts

1. **Gather information and be updated:** Enrol yourself in the web platforms which provide time to time information on your subject specific or research field specific national and international conferences.
2. **Read the call for abstract for details: theme of conference, subject /section, format, deadline, etc.**
3. **Prepare abstract**
4. **Submit**
5. **Prepare for presentation: PowerPoint slides are required both for oral and poster presentation**

Planning abstract:

1. Targeting the section to which paper is to be submitted is the first step

   Plan two or three tentative titles: representative, in good tune with theme of conference, catchy, brief

Title 1: With aniasomnifera herbal formulation on hair growth promotion: Preparation & Characterization

Title 2: Indian ginseng Withania herbal hair growth formulation for alopecia

1. Plan authors and coauthors/ research team members, presenting author must be marked and the main author and/or presenting author's email id is given after affiliation.
2. Body of abstract: Write the abstract structured or unstructured as directed in the guidelines.

Structured abstract contains

Background:

Aim:

Materials and Method:

Result & Discussion:

Conclusion

References:

The abstract must give concise information with the highlight on your new work. Give some quantitative result also. You can add one figure one table some time. The defined word limit must be followed and should be able to give the complete message.

References maximum 3-5 may be added if allowed by the guidelines.

Finalize the title:

Title 1: With aniasomnifera herbal formulation on hair growth promotion: Preparation & Characterization

Title 2: Indian ginseng With ania herbal hair growth formulation for alopecia

[BRIEF, CATCHY, COMPLETE]

1. Finalize the type of presentation : Oral/ Poster

2. Plan the slides for presentation

For poster no. of slides: 9 to 12 (refer the size of poster and take the assistance of poster designer)

Time

Oral presentation time : 5-8 minutes generally with or without time for interaction

Poster Presentation: should be able to describe your study in about 3-minutes, if asked.

Key points for PPT slides

1. Not more than 10- 15 slides for 10 minutes' presentation

2. Font size: minimum 18 or more, arial/times new roman: basically it must be visible in the presentation hall.

3. Color: Avoid green, Donot over use color it is an academic presentation

4. Animation: Use animation if required for better understanding, but do not over use it.

5. Content: Do not use more than 3 lines in a slide, donot overload slide with figure, convert text into figure or table for easy presentation.

6. Give attribution in foot note if required.

7. Conclude effectively in tune with the objectives laid down.

8. Donot exaggerate the work done.

9. Be prepared for queries and leave due time for allowing the audience to question.

So major elements are

Complete, concise, time bound, systematic presentation.

Summary

Let me summarize the section for you. We studied why we need to attend the conferences.

It is needed for exchanging ideas / research collaborations/ career opportunities etc.

Abstract writing is important aspect. The basics were covered in review and research paper writing in previous chapters. Here the key feature were discussed only. Plan the presentation effectively. The presentation must be Effective, systemic and time bound. The abstract must be concise effective, complete.

In the next section we will cover chapter and book writing.

Chapter/ Book Writing

For creative writing there is no age, no qualification required. You can write at any time at any age. Here we will discuss only the academic chapter writing/ book writing.

Books may be

- Text
- Reference
- Review book of a standard text
- Edited text/ reference book
- Practical books

Research groups can plan good chapters or Book. A good team needed. You can contribute a book chapters in an edited book. If your supervisor/ mentor/ collaborator is the editor of a book under development, you can plan a chapter. Sometimes major publishers invite potential authors for contributing in a book.

Preplanning (Target book subject/level/type and publisher) → *proposal writing* → *submission and peer review* → *including and addressing the comments of publisher after review* → *Copyright agreement* → *Time bound writing* → *submission* → *formatting by publisher* → *proof checking* → *revision* → *Final print*

Preplanning (Target book subject/level/type and publisher)

Considering that you have decided topic after literature review preplanning is to be done for following factors.

Type of Book/chapter: Preplanning starts with planning type of book i.e. text/ reference, authored or edited. For book chapter it also the type of chapter i.e. thematic review/ critical research paper must be planned firstly.

Targeting Publisher

- Select the publisher on the basis of geographical coverage or presence of publisher on national and international level. Check the reputation, experience, previous books published on your subject and decide to target.

- Policy/ royalty: Check the publisher's policy about copyright, conditions of reprint, e-book conversion and rate of royalty.

Book Proposal writing

Sample proposal form

Working title Titles and subtitles should be focused to include key terms that readers would use if searching for information on this topic

Keywords Include key terms (not already included in the title/subtitle) that readers would use if searching for information on this topic

Author/Editor information

- Specify if this book is authored or edited

- Include address and contact details, qualifications and experience, and a short biography for all book authors and/or editors

Primary audience: Indicate the most relevant target audience

Secondary audience: Please describe your entire audience in as much detail as possible, e.g. industry sector, job role, level, subject specialism. If the book could be used for a course please provide details, including program and level.

Background and purpose: • Provide a brief description of this book, similar to what you would find on the back cover of the published book • What is your purpose in writing this book? Why is there a need for a new resource in this area? • What problem does this book solve for readers?

Benefits to audience: With reference to the target audience(s) listed above, please give details of: • The information needs and daily challenges of the audience relating to the subject of the book and how your book will address these needs, challenges or main points • List three key features and content in your book that will be most valuable to the reader

Competition: List the books, websites or other information sources that would compete most closely with your book and briefly describe how they compare and why readers would choose your book over the competition

Table of contents: Please provide here or attach separately the planned contents of your book, including chapter titles and a sentence or two related to the scope/initial content plan for each chapter If your book is edited, please include a tentative list of the chapter authors (they do not need to be finalized at this time)

Publishing timeline:

- When would you be able to submit a complete manuscript?

- Are there any events or conferences where we can market this book?

- Would be appropriate to publish in view of factors such as regulatory updates, medical advances, etc.?

Specifications: Address each of the following with your best estimates:

- Number of double-spaced Microsoft Word pages you anticipate producing OR the total number of words

- The approximate number of illustrations and figures you anticipate including, and whether any of these would require color

- What other multimedia content (audio/video files), maps, etc. can you include to increase the market value of the book?

Reviewers: Suggest experts in the field who can review your work with details like: Name: Email address:

Sample chapter – Be prepared to produce a sample chapter (or part of a chapter), if asked, to show the level, approach and style of writing of the book.

Submission and peer review: Submit the proposal for peer review to the publisher.

Addressing comments and feedbacks: After receiving comments from publisher forwarded by reviewer try including and addressing the comments.

Copyright agreement: If agreed copyright agreement is to be signed by publisher and author. Read it properly and ensure your rights, other conditions and royalty conditions.

Summary

An Effective proposal needs good rational and must have complete information. Then submit the proposal to publisher. Address reviewer's comments. Fill Copyright form in discussion with publisher. We will cover further detail in the next section.

Writing Book

Basic concept is similar to review and research paper wiring.

Just remember specially for text books

- Stick to outline/chapter planning

- Student/ reader friendly

- Easy language: Write to inform, not to impress

- Easy to follow presentation by infographics, quick notes, summary, FAQs or glossary at end of each paper.

- Comply to syllabi in text book

For reference books

- Readers: PG students, research scholars, scientist, faculty members.
- Style: Free
- Level: Next
- Reader friendly
- Good bibliography

For book chapters

- Readers: UG/PG students, research scholars, scientist, faculty members.
- Style: determined by editor
- Level: determined by editor (text/ reference)
- Reader friendly
- Good bibliography

Submission and post submission of manuscript of book/ chapters: After final checking submit the manuscript of book/ chapter within time frame to publisher. Publisher again sends it for review. Modify as per the review comments. Have a final check on grammatical error/ typo error and then submit. Proof read the final print given by producer carefully. Check location and accuracy of text and figure. Also check cover page, back page, preface etc. If all OK. You can ask publisher to go ahead with book production.

Universal traits/ USPs of a good book/chapter

- Reader friendly
- Simple and effective presentation
- Cohesion and Coherence
- Rationality
- Focussing
- Info graphics to condense text and to understand better

Further Reading

- https://researcheracademy.elsevier.com/writing-research/book-writing/get-book-published
- WECP M L3A Organising and Planning a Technical Conference Paper Part 1, https://youtu.be/4j9_FnraPHM
- WECP M L3B Organising and Planning a Technical Conference Paper Part 1, https://youtu.be/ybQPqxpdY2w

- WECP M L8B Organizing and Planning a Technical Conference Paper Part 2https://youtu.be/SMABo5IBZQk

- https://writingcenter.unc.edu/tips-and-tools/conference-papers/

References

- https://journa linsights.elsevier.com/journals/0888-613X/authors

- https://www.elsevier.com/en-in/authors/book-authors

- Adler, Abby. "Talking the Talk: Tips on Giving a Successful Conference Presentation." Psychological Science Agenda 24.4 (April 2010). American Psychological
Association. http://www.apa.org/science/about/psa/2010/04/presentation.asp x.

- Kerber, Linda K. "Conference Rules, Part 2." The Chronicle of Higher Education. March 21, 2008. http://chronicle.com/article/Conference-Rules-Part-2/45734.

CHAPTER 37

Open Educational Resources-I

Dr Ajay Semalty
H.N.B Garhwal University (A Central University)
Srinagar Garhwal-246174

Learning Outcome

After completing this chapter you will be familiar with the basic concept of Open Education Resources and using the basic open licences for your academic and research endeavours

Lesson Plan

- What is OER?
- Concept
- Why Needed?
- The open licenses

"OER" The Term

The term open educational resources first came into use at a conference hosted **by UNESCO in 2002**, defined as "the open provision of educational resources, enabled by information and communication technologies, for consultation, use and adaptation by a community of users for noncommercial purposes"

In general we just know that it includes e-learning....e-content....

What?

Though there is no exact definition of OER some of the popular definitions are......

OER are freely accessible, openly licensed text, media, and other digital assets that are useful for teaching, learning, and assessing as well as for research purposes. There is no universal usage of open file formats in OER. (Wikipedia)

"OER are digitised materials offered freely and openly for educators, students and self-learners to use and reuse for teaching, learning and research." OER includes learning content, software tools to develop, use and distribute content, and implementation resources such as open licenses." - (OECD)*

Organisation for Economic Co-operation And Development

Concept of OER

Emergence of OER

It is about LEAVING IMPACT

"A student can recall a number of examples about his/her teachers of school but he never talks about his college teachers."

In Higher education, we have been failure to leave impact on students (as compared to school teachers).

Text books v/s OER

Push –Pull **Theory** of Learning

Text Books are about **pushing** Knowledge and questions to Learners

OERs are about pulling Knowledge and Questions by **Learners**

Open: How much

The openness of any OER may be in terms of freedom to use, in the technical domain or as a characteristics of the resource (Fig. 1). Every OER cannot be completely free.

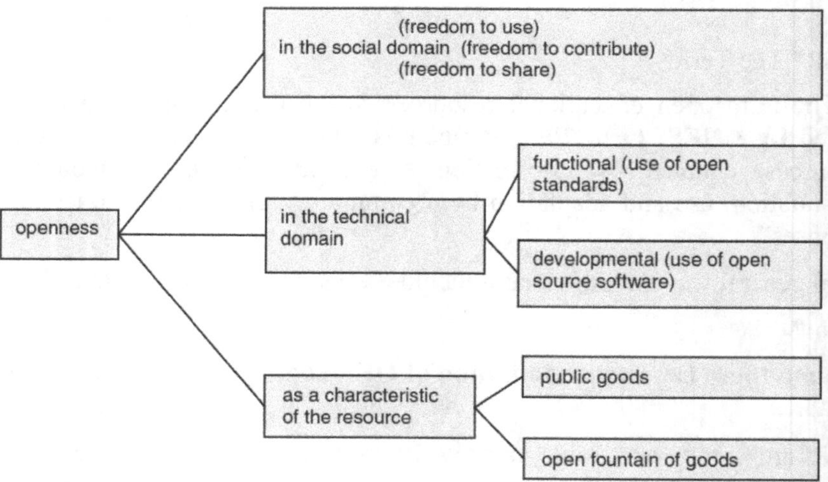

Fig. 37.1 Openness in OERs (Adapted from Organisation for Economic Co-operation and Development (OECD), Giving knowledge for free the emergence of open educational resources, 2017)

5 R's of OER

The 5 R's are associated with every OER. These are

Retain: Make and Own copies

Reuse: Use in wide range of ways

Revise: Adapt, modify and improve

Remix: Combine two or more

Redistribute: Share with others

OER include

- Full courses,
- course materials,
- chapters,
- textbooks,
- streaming videos,
- tests,
- software, and
- any other tools, materials, or techniques used to support access to knowledge.

Section Summary

We discussed

- What are OERs?
- Open Education Resources available for free for
- 5 R's (Retain, Reuse, Revise, Remix and Redistribute)
- Any material which is used for educational purpose can be developed as OER

In the next section we will discuss: Timeline & other aspects of OERs.

In this section we will discuss

- **Why OERs are Needed?**
- **Timeline of OERS**

    There are no slow learners,

    There are only slow teachers...   - Bezamin Franklin

    Therefore, OERs are needed to cater to variety and diversity of learners.

Why OER?

Why do we need OER? Because,

- **It is about the presentation**
- **The way we communicate**
- **Which make the learning successful**
- **This is the prime aim that we want to communicate in effective manner.**

Why?

Major factors which indicate why OERs are needed.

- **Globalization**
- **Demography**
- **New approaches to governance**
- **Technology**

Globalization

Growing collaboration and growing competition among countries and among institutional providers, researchers mobility etc.

Globalization has triggered OER. Online conferences, Chatting, Getting lecture from experts sitting abroad are the examples. OER provides global access for everyone for using for learning.

Demography

Huge impact of demographic factors on higher education:

Reductions in the traditional 18-to-25-year-old student age group will affect institutions in a number of OECD countries.

This decline may be offset by increased participation rates, the flow of foreign students (numbers of young people are rising in many non-OECD countries where demand for education is not fully satisfied) and by the increasing tendency of older adults to enter or return to education and the provision of programmes for them.

A person who is already entered the profession wants to learn some more basics. Will he able to enter into the regular courses? No! What does he need?Open courses/ distance courses. That's why OERs are needed

New Approaches to Governance

Authorities need quality governance; OER initiatives can be said to cater for improved quality control through enhanced transparency and comparability between institutions, departments or individual faculty members as well as direct feedback from both enrolled and informal learners.

Technology

Continuous development of ICT in knowledge economy is vital for improved quality education and easy/ widespread access by learners.

OER origins

The chronological development of OERs is shown in Table 37.1.

Table 37.1 The chronological development of OERs

Year	Milestone
1998	Open content term coined by Wiley, (2006); first open content license (OpenContent Principles and License) released.
1999 - June	Second open content license (Open Publication License) released , includes license options concerning attribution, derivative works, and commercial use of materials
1999 - October	Connexions project is born at Rice University
2000 - March	Nupedia, predecessor of Wikipedia launched by Jimmy Wales and Larry Sanger[4], using original "OpenContent Principles and License"
2001 - Jan	Wikipedia, launched by Jimmy Wales and Larry Sanger
2002 - July	Open Educational Resources (OER) term coined by UNESCO
2002 - December	Established in 2001, Creative Commons release the first set of licenses.
2006 - October	Open University of the United Kingdom launches OpenLearn
2006 - November	FLOSS4Edu launches for African OER
2006 - Feb	WikiEducator domain name registered and prototype server installed
2006 - August	First Multi-Country Education Curriculum using OER: Virtual University of the Small States of the Commonwealth
2007 - February	OER Commons launches
2007 - March	First commercial training provider launches OER courses: Novell
2007 - July	First U.S. OER College Consortium: Community College Consortium for Open Educational Resources
2007 - September	First Public Declaration: Cape Town Declaration
2007 September???	First university to adopot CC-BY-SA default Intellectual Property Policy: Otago Polytechnic, New Zealand
2007 - September	First US statewide OCW consortium launches, Utah OpenCourseWare Alliance
2007 - November	First university to publish over 90% of its courses as OER: MIT OpenCourseware, established in 2002
2008 - February	OER Africa launched

Contd...

2008 - April	First global OER Consortium legally established: OpenCourseware Consortium
2008 - July	First OER Handbook v1.0 published
2008 - July	First For-Profit Publishing Company based on OER: FlatWorld Knowledge
2009 - April	First Masters Thesis published as OER
2009 - April	First OER Non-Profit Foundation: OER Foundation
2009 - May	First-in-U.S. Initiative to Develop Free Digital Textbooks for High School Students: California Governor Schwarzenegger
2009 - May	First For-Profit University Publishes OER: Kaplan University
2009 - July	US President Obama Announces Federal OER Initiative for Community Colleges
2009 - August	First High School based on OER launches: Open High School of Utah
2009 - September	First OER University: University of the People

First major project: MIT OpenCourseWare (OCW) Project.

In 2000 MIT faculty and administrators asked:

"How is the Internet going to be used in education and what is our university going to do about it?"

MIT faculty answered:

"Use it to provide free access to the primary materials for virtually all our courses. We are going to make our educational material available to students, faculty, and other learners, anywhere in the world, at any time, for free."

2002: Proof of concept with 50 courses

2014 : Materials from 2150 courses and 125 million visitors

www.ocw.mit.edu

Materu (2004) is probably the first comprehensive report on what is later called OER.

MIT is still the world leader in OER development.

Over 3000 open access courses (opencourseware) are currently available from over 300 universities.

Section Summary

We discussed in this section

- Why we need the OERs?
- Factors: demographic, globalization, new approaches to governance & technology

- How OER developed with time.
- Journey started from MIT with OCW
- OER term was coined
- Wikipedia came
- Now in India

Next section: Open Licences

How much open?

How much open OERs are? What does the free mean?

The "Free" license means

- ***The freedom to study and apply the information.***
- ***The freedom to redistribute copies.***
- ***The freedom to distribute modified versions.***

    But not all OER are completely free.

    Conditions to license may differ even in Open contents.

Licensing

Step 1 Ensure that you have copyright for the resource

Step 2 Choose a license

Step 3 Include the license details in the resource

Step 1 Ensure that you have copyright for the resource

- Establish the copyright owner of the text, graphics, video etc.
- If there is copyrighted material within the resource that belongs to someone else (3rd party copyright), then this person or agency needs to be contacted before the resource can be released.

Step 2 Choose a license (1)

Understand the **4 conditions**:

- **Attribution** - You let others copy, distribute, display, and perform your copyrighted work — and derivative works based upon it — but only if they give credit the way you request.
- **Share-alike** - You allow others to distribute derivative works only under a license identical to the license that governs your work.
- **Non-commercial** - You let others copy, distribute, display, and perform your work — and derivative works based upon it — but for noncommercial purposes only.

- **No Derivative Works** - You let others copy, distribute, display, and perform only verbatim copies of your work, not derivative works based upon it.

http://creativecommons.org/about/licenses

Legal openness

Step 2 Choose a license (2)

Choose one of the 6 licences (Fig. 37.2)

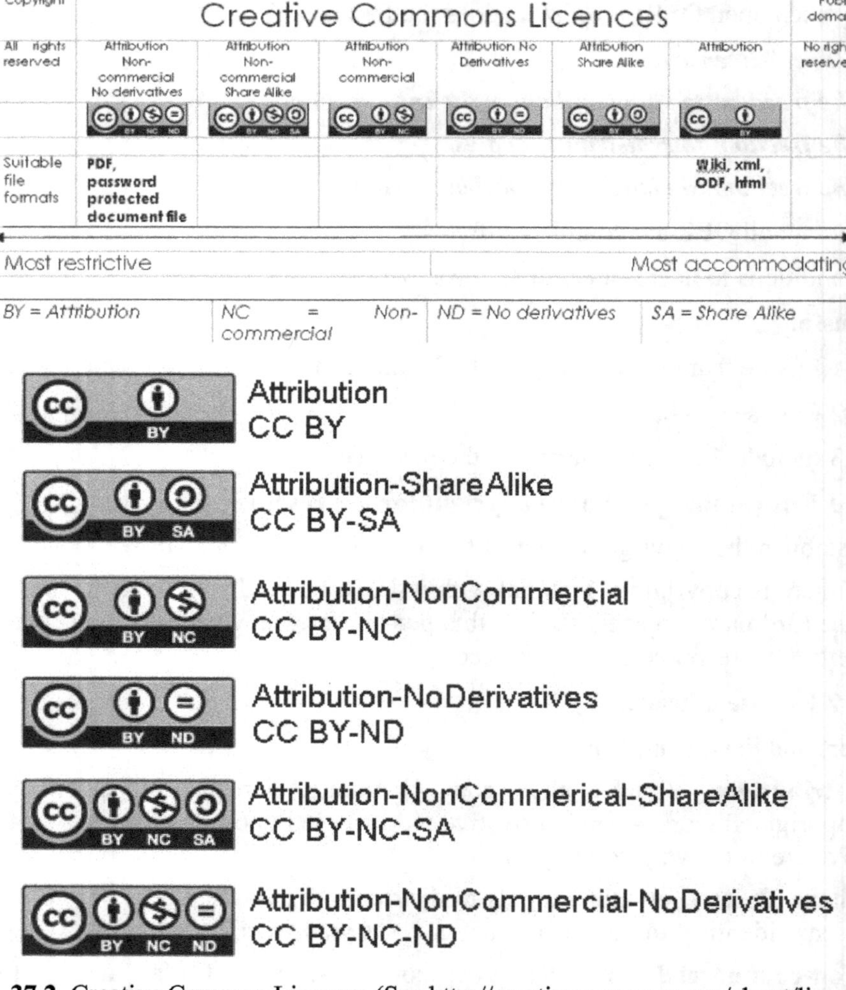

Fig. 37.2 Creative Common Licences (See http://creativecommons.org/about/licenses)

Step 2 Choose a license (3)

- Visit the Creative Commons Licence (http://creativecommons.org/ license/) page and use their simple licence chooser to select a Creative Commons licence that indicates how others may use your creative content. *(These responses to the questions will be used to automatically generate HTML text which includes all these details for an electronic version of the Creative Commons licence that you have chosen. The HTML code will display an icon as well as a link to the full license deed hosted at the Creative Commons site. Note that you also need to select a legal jurisdiction (country). South Africa is listed at the end of the drop down list).*

Step 3 Include the licence details in the resource

- For electronic works: Cut and paste this HTML text on your website (Fig. 37.3).

- For non-electronic works: Select the option "Mark a document not on the web, add this text to your work." (this is only available once you have chosen a licence) In addition you might like to note the icon that they suggest and download the appropriate CC icon (http://creativecommons.org /about/downloads/) and paste it onto your word processed document for a paper-based cc licence.

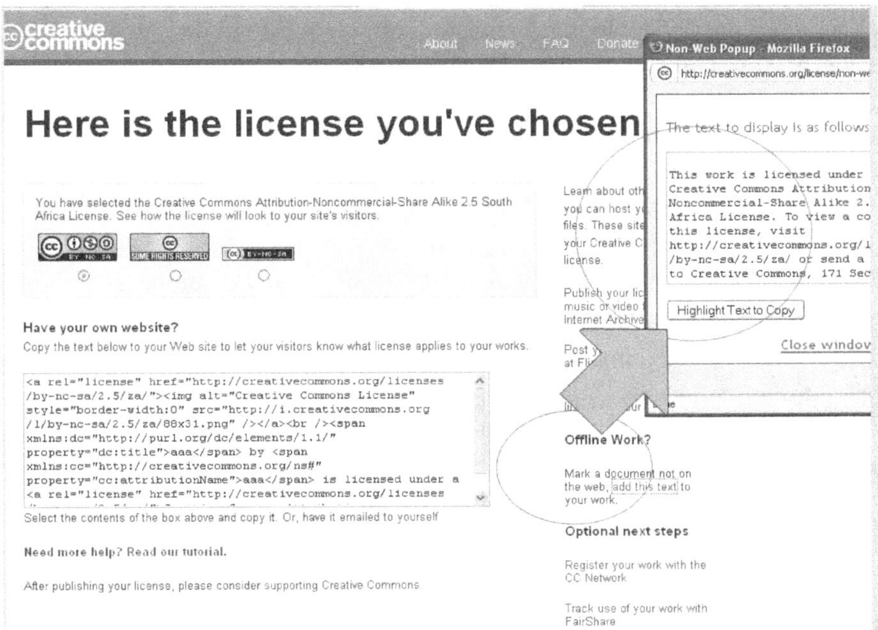

Fig. 37.3 Including the licence details in the resource

Summary

In this chapter we discussed

- What are OERs?
- Open educational material with 5 R's
- Retain, Reuse, Revise, Remix, Redistribute
- Journey started with MIT's OCW
- UNESCO termed OER
- Indian OERs
- How to choose suitable Open license for OERs

Further Reading

- Open Educational Resources (OER): Dr Ajay Semalty at FDC, HNB Garhwal University Srinagar garhwal https://youtu.be/K7xg4gRXMN8
- Open Educational Resources (OER) - II : Sharing- What, where, who, and why?: Dr Ajay Semalty, https://youtu.be/3Br_3I4oCgk
- Understanding Open Educational Resources, Commonwealth of learning, https://oerknowledgecloud.org/sites/oerknowledgecloud.org/files/2015_Butcher_Moore_Understanding-OER.pdf
- Guidelines for Quality Assurance and Accreditation of MOOCs, Commonwealth of learning, http://oasis.col.org/bitstream/handle/11599/2362/2016__Guidelines-QAA-MOOCs.pdf?sequence=6&isAllowed=y
- Semalty A, Increasing the reach of Digital Learning in India, (Guest Column) The daily Pioneer, 25 April 2020 (http://tiny.cc/pioneerswayam).
- Semalty A, Digital learning: The Uttarakhand Plan, https://pharmastate.blog/digital-learning-the-uttarakhand-plan/
- Semalty A., Digital learning through SWAYAM MOOCs: SWOT Analysis, CEC News, Sept 2019, 5-6 (http://tiny.cc/digitallearningCEC).
- Semalty M, Adhikari L, Semalty A., Benchmarking SWAYAM MOOCs, Taking learners from enrollment to exams, CEC News, March 2020: 3-6 (http://tiny.cc/benchmarkingSWAYAM).
- Semalty M., Ten commandments of Digital learning through MOOCs: SWOT Analysis, CEC News, Nov. 2019, 5-6 (http://tiny.cc/TENCOMMAND).

References

- Wikipedia, Open educational resources, https://en.wikipedia.org/wiki/Open_educational_resources
- Cheryl Hodgkinson-Williams, Michael Paskevicius, Roger Brown, A culture of sharing: Open education resources an introduction, Teaching with Technology Seminar, UCT, 18 May 2009; cheryl.hodgkinson-williams@uct.ac.za
- Organisation for Economic Co-operation and Development (OECD), Giving knowledge for free the emergence of open educational resources, 2017, e book, https://www.oecd.org/edu/ceri/38654317.pdf
- Wolfenden F, Open Educational Resources (OER), www.TESS-India.edu.in
- Kanwar A and Uvalic´-Trumbic´ S (Eds), Butcher N, A Basic Guide to Open Educational Resources (OER), http://unesdoc.unesco.org/images/0021/002158/215804e.pdf
- Open Educational Resources (OER)". CoL.org. Commonwealth of Learning. https://www.col.org
- http://creativecommons.org/
- https://www.oercommons.org/
- https://swayam.gov.in/
- Open Educational Resources: Global Report 2017, http://oasis.col.org/handle/11599/2788; COL, 2017.
- UNESCO. (2016). Open Educational Resources (OER) Road Map Meeting: Outcome Report. Paris: UNESCO.

CHAPTER 38

OERs for Learning and Research

Dr Ajay Semalty
H.N.B Garhwal University (A Central University)
SrinagarGarhwal-246174

Welcome dear learners, welcome in the last section of our book "Academic Writing". In the last three chapters we will be discussing OER. And in this chapter we will be discussing uses of OERs.

Learning Outcomes

After completing this chapter, you will be familiar with

- Reasons of using/developing Open Educational Resources(OER)
- Indian OER basket
- Using OER in learning &Research

Lesson Plan

- Who is involved?
- Who uses?
- What to develop as OER?
- Why to share?
- Why institutes/ teachers be involved?
- OER in India
- Using and searching OERs in various OER platforms

Who is involved:

In the OER development; the major players are

- International agencies: OECD, COL, UNESCO, WHO etc.
- National agencies and Universities and OER platforms: MIT, National OER repository India, CEMCA, MHRD (India)etc.
- Online platforms: **YouTube, Wikipedia, OER commons, Creative commons, Moodle etc.**

Who uses OER?

- Students with in institutions
- Students external to institutions
- Self Learners
- Teachers/Professors/Academics/Researchers

Wide variety of learners are there traditional and non-traditional, school going, college going and who are working or who are just doing their household work, so it has got wide spread approach, coverage.

Now moving to the types of OER's

What to develop as OER?

Let us discuss what to develop as OER in brief. It may be-

- Course/Instructor Resources (MIT, OCW, Coursera, Edx, Moodle)
- Full Distance Course Chapters (Open Learn UK)
- Course Chapters/seminars
- Learning Objects
    - o   Images(www.flickr.com)
    - o   Videos (www.academicearth.com,YouTube)
    - o   Audio(http://itunes.stanford.edu)
    - o   Open Textbooks (www.wikibooks.org)
    - o   Journals(www.doaj.org)
- OER can be as small as an individual picture or as large as an entire course.
- Whatever you know and which is factually correct can be developed in to OER.

Reasons to adapt an OER include:

- To address a teaching style or learning style
- To adapt for a different grade level
- To adapt for a different discipline
- To adjust for a different learning environment
- To address diversity needs
- To address a cultural preference
- To support a specific pedagogical need
- And to address either a school or district's standardized curriculum.

So, for these reasons we need to adapt the OERs. We want to communicate effectively. We want to individualize the demand. We want to personalize the demand of the educational material. Various educational portals are very popular nowadays "Topper learning", "Byjus" these are the huge success stories. These platforms of personalizing the education with respect to the learner.

Major Drivers of OERs- Technological and Economic drivers

- Increased broadband availability.
- Increased hard drive capacity and processing speeds coupled with lower costs.
- Rise of technologies to create, distribute and share content.

- Provision of simpler software tools for creating, editing and remixing.
- Decreased cost and increased quality of consumer technology devices for audio, photo and video.

  So, the technology has made it easier and cost effective to develop OER

  Now the next driver is

Social driver

- OERs are needed to expand access to learning for everyone- for traditional and non-traditional students.
- To promote lifelong learning for both the individual and the government, we need the OERs. The OERs are needed for these reasons.
- And to bridge the gap between the non-formal, informal, and formal learning.

  In OER development the teachers, researchers, and institutions should be involved.

  Why? Let's see

Why teachers/researchers should be involved?

- Number one factor is **to get the altruistic motivation of sharing** (as for institutions), which is again supported by traditional academic values.
- **For personal non-monetary gain** like publicity, reputation within the open community.
- **For free sharing because free sharing can be good for economic or commercial reasons**, as a way of getting publicity, reaching the market more quickly, gaining the first-mover advantage, etc.
- **Sometimes it is not worth the effort to keep the resource closed.** That's why the teachers should keep the resources open. If it can be of value to other people one might just as well share it for free.
- **And with the OERs there is the possibility of increased opportunities for collaboration and academic alliances.**

Why institutes are involved?

- Sharing knowledge is in line with academic traditions and a good thing todo.
- Educational institutions (particularly those publicly financed) should leverage taxpayers' money by allowing free sharing and reuse of resources.
- Quality can be improved, and the cost of content development is reduced by sharing and reusing.

- It is good for the institution's public relations to have an OER project as a showcase for attracting new students.

- There is a need to look for new cost recovery models as institutions experience competition.

- Open sharing will speed up the development of new learning resources, stimulate internal improvement, innovation and reuse and help the institution to keep good records of materials and their internal and external use.

We discussed why there is a need of adapting OERs. We will focus on the Indian OER basket in the next section.

Dear learners, before moving to the other OER repositories, before introducing the other OER repositories I will focus on our national OER basket. What the government of India is doing for developing the OERs.

Indian OER Basket and National OER repositories

The major repository of India with respect to the OER is the NROER- National Repository of Open Educational Resources. Before moving to the other platforms, very first thing which we little bit introduced in earlier module that is inflibnetcentre facilities. The cluster of platforms available by the Inflibnetcentre. In the inflibnetcentre websites if you go there you will find a bunch of OER repository. First of all we must focus on these. Whenever you are moving, we have studied rather little bit about that the Shodhgangotri, Shodhganga, we have studied about the INDCAT, ICSSR, these are major repository for your literature review. Apart from that we will move to various OER platforms. Please go through these websites and search the suitable resource for you.

Now moving to

SWAYAM

"Study Webs of Active-Learning for Young Aspiring Minds" (SWAYAM) is the indigenous ICT based platform for hosting Massive Open Online Courses (MOOCS) developed under the aegis of NME-ICT. An initiative of Ministry of Human Resource Development (MHRD), Government of India, MOOCs supplement the formal education system in the country from high school to higher education offering courses based on curriculum, continuing education and skill.

See https://swayam.gov.in/

SAWAYAM is an indigenous developed IT platform that facilitates hosting of all the courses, taught in classrooms from 9th class till post-graduation to be accessed by anyone, anywhere at anytime.

Objectives of SWAYAM

- Initiated and designed to achieve the three cardinal principles of Education Policy viz., access, equity and quality.

- The objective of this effort is to take the best teaching learning resources to all, including the mostdis advantaged.

- To bridge the digital divide for students who have hitherto remained untouched by the digital revolution and have not been able to join the mainstream of the knowledge economy.

Contents of SWAYAM

- Interactive quality MOOCs for the students of 9th toPG.

- Each MOOC may be credit based or noncredit based with free access to content (for certificate an exam is to be passed after payment of a nominal fee)

- Each MOOC on SWAYAM is in 4 quadrants–
    - video lectures,
    - specially prepared reading material that can be downloaded/printed
    - self-assessment tests through tests and quizzes, and
    - An online discussion forum for clearing the doubts.

National Coordinators for MOOC

- AICTE (All India Council for Technical Education) for self-paced and international courses

- NPTEL (National Programme on Technology Enhanced Learning) for Engineering

- UGC (University Grants Commission) for nontechnical post-graduation education (Now, PG Courses are being coordinated by CEC)

- CEC (Consortium for Educational Communication) for under-graduate and post graduate education

- NCERT (National Council of Educational Research and Training) for school education

- NIOS (National Institute of Open Schooling) for school education

- IGNOU (Indira Gandhi National Open University) for out-of-school students

- IIMB (Indian Institute of Management, Bangalore) for management studies

- NITTTR (National Institute of Technical Teachers Training and Research) for Teacher Training programme

So, the courses in the SWAYAM with the quality contents prepared by the experts and need based courses are here. Faculty member can apply for developing the MOOCs, for developing these contents. And get the grant for developing that thing. Now let us have a quick look on the major national coordinators.

UGC MOOC

First of all, this is the website of UGC MOOCs: https://ugcmoocs. inflibnet.ac.in/; which deals with the PG MOOCs. In this platform we can get all the information. Even the faculty members can apply for the new projects from this website particularly. (But please note the National Coordinator CEC is taking care of all PG MOOCs also along with the UG MOOCs. So, the website of CEC should be consulted for current proposals and MOOCs). However, the courses are available on the SWAYAM platform. You need not to go to the individual coordinator's websites.

So the next coordinator is

CEC MOOC

Consortium for Educational Communication (CEC) deals with MOOCs of various subjects nontechnical UG subjects. (But please note the National Coordinator CEC is taking care of all PG MOOCs also along with the UG MOOCs. So, the website of CEC should be consulted for current proposals and MOOCs). Refer annexure of the book. For UG and PG nontechnical MOOC the proposal can be submitted to the online portal: https://swayam. inflibnet.ac.in/

NPTEL MOOC

After that, for engineering. NPTEL is website for courses developed by NITs and ICs experts. These courses are available through swayam platform itself. NPTEL is way ahead in development of MOOCs.

IIM Bangalore: It provides quality MOOCs and it got the collaboration with Edx also.

So, IIMB is at international level and is showing the potential.

So, these are the major national coordinators which I have covered and which are relevant to your tertiary education. Rest of the OER repositories can be accessed through the Inflibnetcentre website.

Other Indian OER repositories

- SHODHGANGOTRIhttp://shodhgangotri.inflibnet.ac.in
- SHODHGANGAhttp://shodhganga.inflibnet.ac.in/
- E-shodhsindhuhttp://ess.inflibnet.ac.in
- ICSSR DATA SERVICEhttp://icssrdataservice.in
- INDCAThttp://indcat.inflibnet.ac.in
- VIDWAN: EXPERTDATABASE
- http://vidwan.inflibnet.ac.in
- SWAYAMPRABHAhttps://www.swayamprabha.gov.in/ (is a 24-hour channel providing the educational contents which is freely available in the DTH.)

A huge quality content is available by these government initiatives. So, you enjoy and use them with utmost efficiency and disseminate the information to all the stakeholders. We will be studying using and searching the other OER repositories in the next section.

Dear learners, after covering the national repository of India let us discuss the international repositories.

Major OER repositories (International)

- Google scholar- source for scholarly article, free to use.
- Google classroom
- Creative commons
- OER Commons
- Wikipedia
- Moodle
- SlideShare
- Edx
- Others: Coursera, Blackboard

Google Scholar

- Google is the most widely used search engine.
- Google scholar is a ready source of scholarly articles and it is free to use.

- You must use the advanced search as we have earlier also mentioned and filter your result as per the time or the licenses.

For just a quick recap. You just have to go to the scholar not google, please remember refer scholar (Fig. 38.1). Type whatever you want. Let's say for example- hair loss. We got a list of articles (Fig. 38.2). In the left navigation bar, you can filter them with respect to the time, type of articles. You can also get the related article links, citation information. You can also access the author profile and access all the publication of a particular author to whom you want to refer in a single platform. You can come to the know the potential of author by using this profile as we have discussed in detail earlier also.

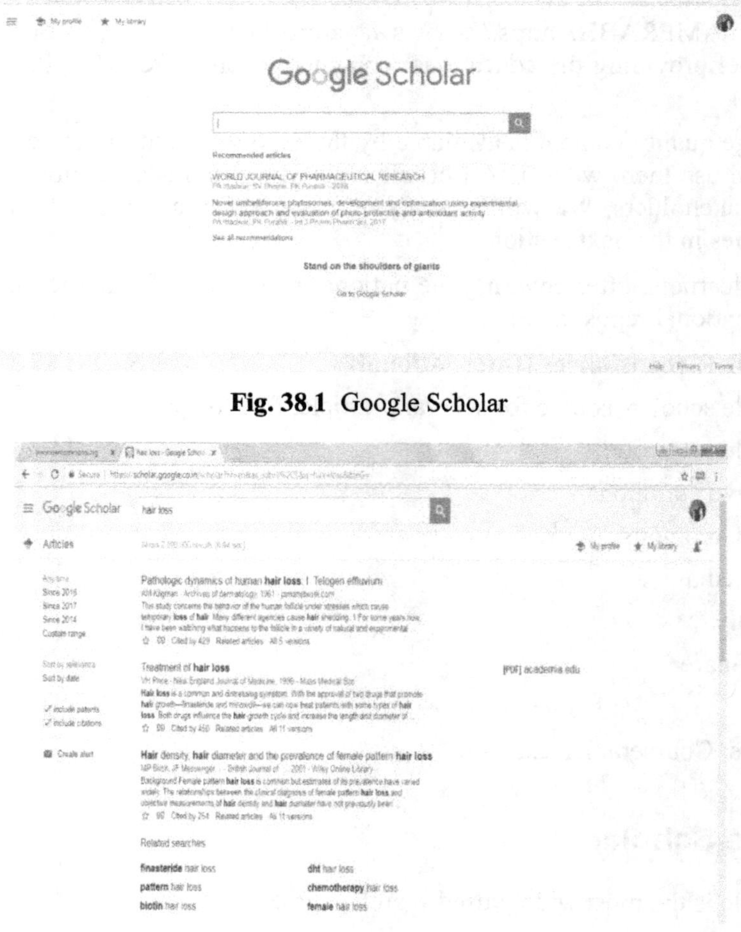

Fig. 38.1 Google Scholar

Fig. 38.2 Searching article in google scholar

The next feature is that you can search the images also like if you want to search for Himalayan deer image, it will show a number of images. If you want to go for the free to reuse image you have to go to advance search and then click that particular filter. Then you will get the image which you can reuse without violating any copyright. The creative commons figures will be there.

You can search related author, images and filter it according to your choice.

Google Classroom

As the scholar, a classroom is another feature by the google. How to access? If you are having a Gmail id you will get option in right top corner on bullet, Click there and you will find google classroom (Fig. 3). Open it. This will show a page. You can identify yourself as a teacher or student. In the google classroom, the beauty is that you can create very easily any kind of assignment, question, reuse reports and projects. You can sort them topic wise. It is a very effective and user-friendly interface in which you can develop the assignments, reports, questions and offer to your students. You can prepare the things in your research. You can prepare the documents for your research collaborators.

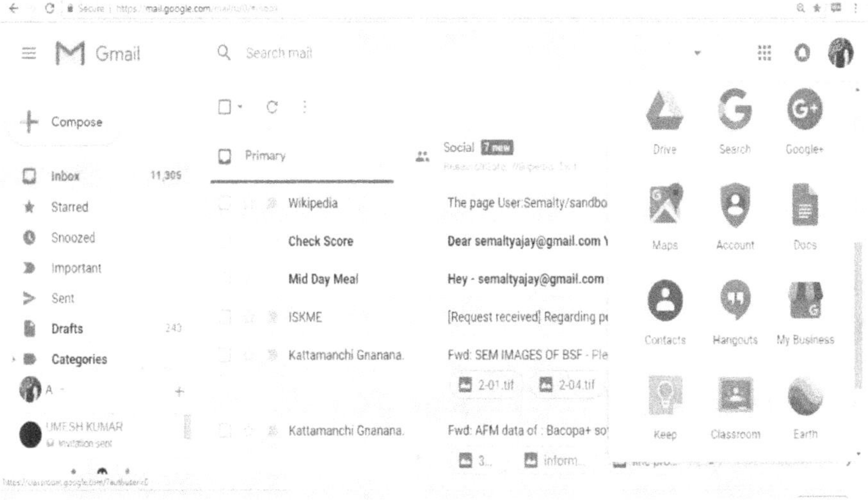

Fig. 38.3 Locating Google Classroom

Creative commons

As we have discussed from copyright to copyleft, the creative commons is that NGO that foundation which is working for copyleft movement.

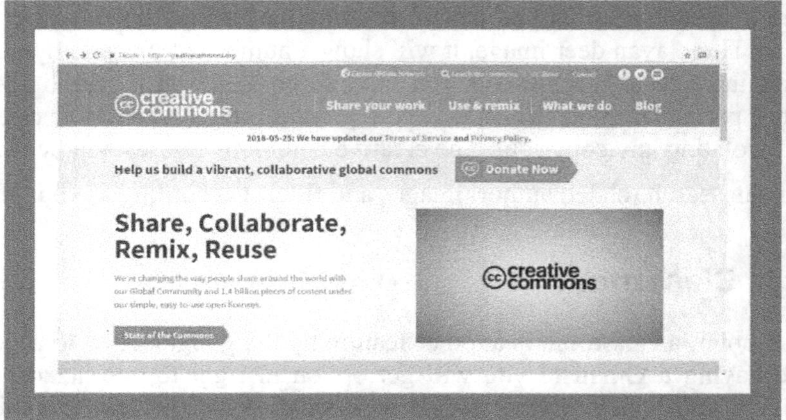

Fig. 38.4 Creative Commons

This is the web page of creative commons (Fig. 38.4). In the creative commons you can share your work. Use and remix of the available work and have your own account there. In the creative commons as you can see on the website, so many available features or available platforms links are there which you can use freely like flickr, bandcamp, wikipedia, Youtube, vimeo, PLOS, etc.

So, what is the advantage?

Advantages of CC

- You can get all the free platforms under one single umbrella.

- It would make it easy to search Creative Commons items or the contents which are not covered in a particular copyright.

In the similar way another platform is there-

OER Commons

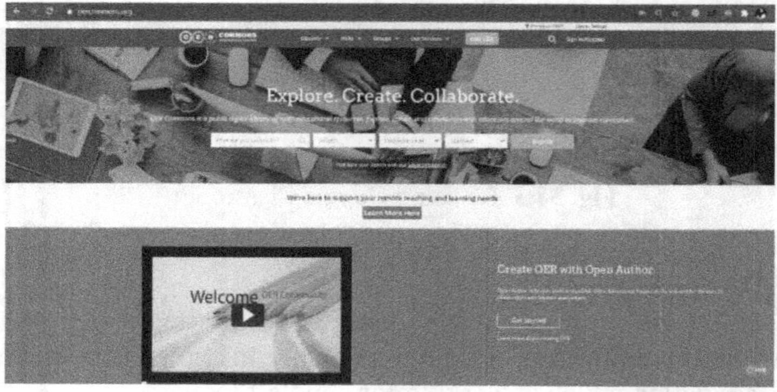

Fig. 38.5 OER Commons

This is the website of OER Commons: https://www.oercommons.org/ (Fig. 38.5). It is meant for exploring, creating, and collaborating. You can create your own account in the OER Commons. For that you will have to register yourself by putting your entry of name, email, and password (Fig. 38.6).

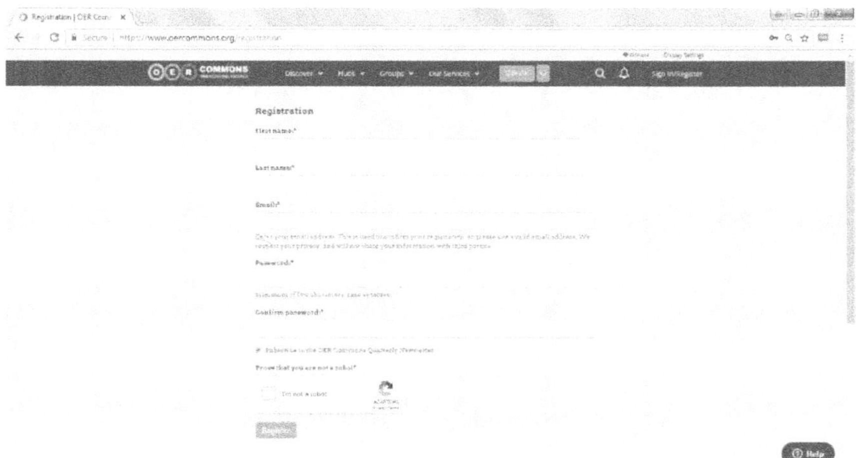

Fig. 38.6 Registration ion OER Commons

After that you will have your own page own account in which you can refine your search on the basis of educational standards, subject area, material type, conditions of use, content, source, primary use, media format ,educational use, language etc (Fig. 38.7). It covers lot of subjects like arts and humanities, business and communication, career and technical education, English language, history, law, mathematics, physical science etc.

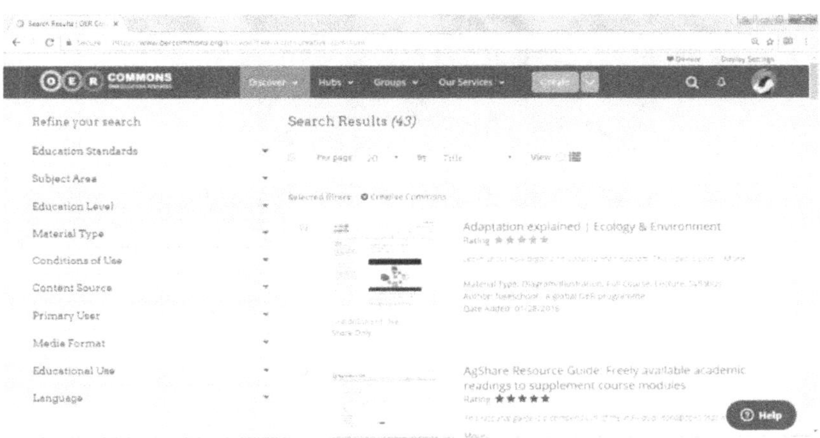

Fig. 38.7 Refining search in OER Commons

It also provides various network hubs by the various universities (Fig. 38.8). Links to free repositories by the major universities and institutions are given in the OER Commons.

Fig. 38.8 Network Hubs in OERCommons

What you can do?

You can develop your OER. You can submit it in to the OER Commons in the form of resource builder or in the form of resource, lesson or the module (Fig. 38.9).

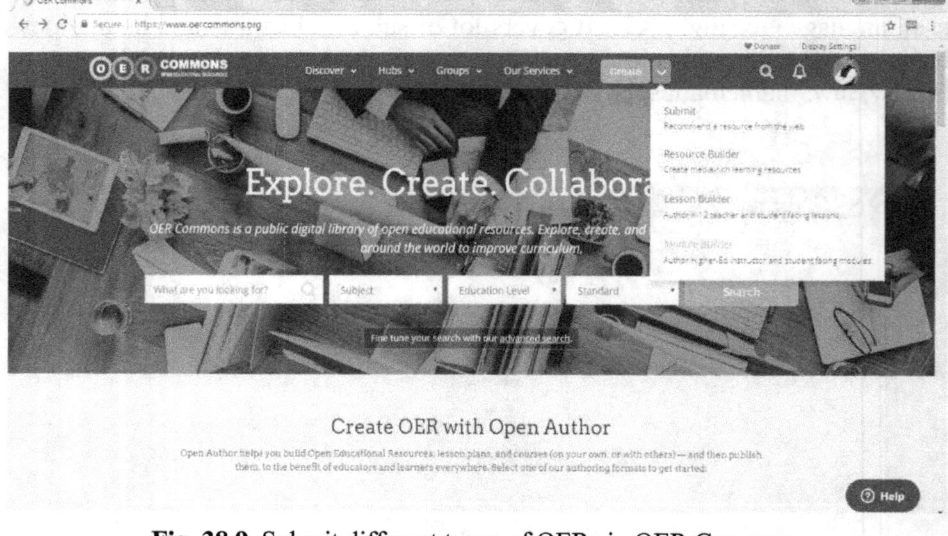

Fig. 38.9 Submit different types of OERs in OER Common

We have given the link- how to search in the OER Commons (https://youtu.be/JXFUOVxv0gY).After you develop modules, your page looks like this. This is my page of OER Commons (Fig. 38.10).

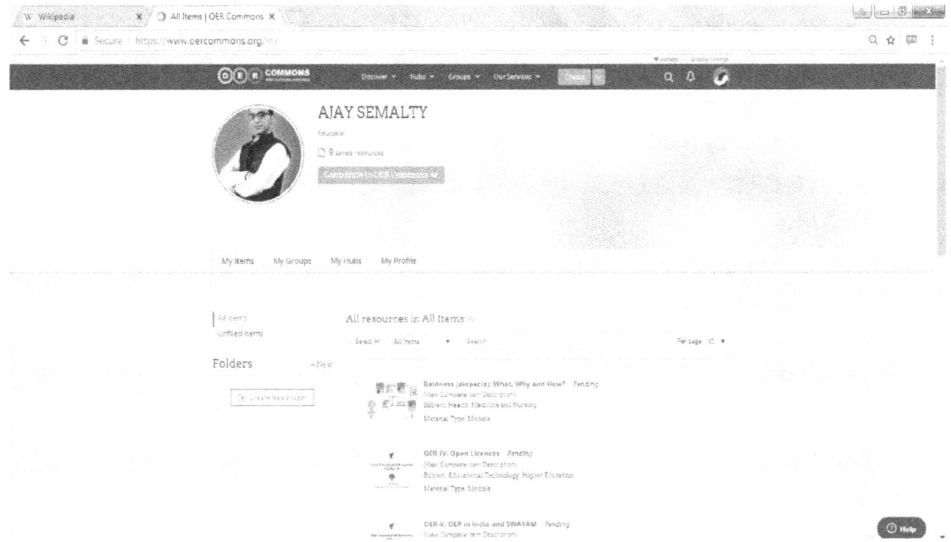

Fig. 38.10 Standard OERCommon Page with OERs

The modules are available with free access.

Summary

So, let me summarize this section for you that we covered the

- Major repositories like Google scholar, Google classroom, Creative Commons, and OER Commons in this section.

Further Reading

- Open Educational Resources (OER): Dr Ajay Semalty at FDC, HNB Garhwal University Srinagar garhwal https://youtu.be/K7xg4gRXMN8

- Open Educational Resources (OER) - II : Sharing- What, where, who, and why?: Dr Ajay Semalty, https://youtu.be/3Br_3I4oCgk

- Understanding Open Educational Resources, Commonwealth of learning, https://oerknowledgecloud.org/sites/oerknowledgecloud.org/files/2015_But cher_Moore_Understanding-OER.pdf

- https://wikieducator.org/Educators_care/Defining_OER
- Bell, Steven. "Research Guides: Discovering Open Educational Resources (OER): Home". guides.temple.edu.2017-12-05.
- http://www.oecd.org/education/ceri/38654317.pdf
- Sanchez, Claudia. "The use of technological resources for education: a new professional competency for teachers". Intel® Learning Series blog. Intel Corporation. Archived from the original on 29 March 2013. 23 April2013.
- "What is OER?". wiki.creativecommons.org. Creative Commons. 18 April2013.
- Guidelines for Quality Assurance and Accreditation of MOOCs, Commonwealth of learning, http://oasis.col.org/bitstream/handle/11599/2362/2016__Guidelines-QAA-MOOCs.pdf?sequence=6&isAllowed=y
- How to search in the OER Commons, https://youtu.be/JXFUOVxv0gY
- How to Use Open Author on OER Commons, https://www.youtube.com/watch?v=kaFbQcvF9r4
- Semalty A, Increasing the reach of Digital Learning in India, (Guest Column) The daily Pioneer, 25 April 2020 (http://tiny.cc/pioneerswayam).
- Semalty A, Digital learning: The Uttarakhand Plan, https://pharmastate.blog/digital-learning-the-uttarakhand-plan/
- Semalty A., Digital learning through SWAYAM MOOCs: SWOT Analysis, CEC News, Sept 2019, 5-6 (http://tiny.cc/digitallearningCEC).
- Semalty M, Adhikari L, Semalty A., Benchmarking SWAYAM MOOCs, Taking learners from enrollment to exams, CEC News, March 2020: 3-6 (http://tiny.cc/benchmarkingSWAYAM).
- Semalty M., Ten commandments of Digital learning through MOOCs: SWOT Analysis, CEC News, Nov. 2019, 5-6 (http://tiny.cc/TENCOMMAND).
- SWAYAM Guidelines: https://swayam.inflibnet.ac.in/home_assets/download/Guidelines.pdf
- Financial norms of SWAYAM MOOCs: https://swayam.inflibnet.ac.in/home_assets/download/Financial_norms_for_swayam_MOOCs.pdf
- MOOC Proposal submission portal: https://swayam.inflibnet.ac.in/
- SWAYAM Online Courses: www.swayam.gov.in

References

- Wikipedia, Open educational resources, https://en.wikipedia.org/wiki/Open_educational_resources
- Cheryl Hodgkinson-Williams, Michael Paskevicius, Roger Brown, A culture of sharing: Open education resources an introduction, Teaching with Technology Seminar, UCT, 18 May 2009; cheryl.hodgkinson-williams@uct.ac.za
- Organisation for Economic Co-operation and Development (OECD), Giving knowledge for free the emergence of open educational resources, 2017, e book, https://www.oecd.org/edu/ceri/38654317.pdf
- Wolfenden F, Open Educational Resources (OER), www.TESS-India.edu.in
- Kanwar A and Uvalic´-Trumbic´ S (Eds), Butcher N, A Basic Guide to Open Educational Resources (OER), http://unesdoc.unesco.org/images/0021/002158/215804e.pdf
- Open Educational Resources (OER)". CoL.org. Commonwealth of Learning. https://www.col.org
- http://creativecommons.org/
- https://www.oercommons.org/
- https://swayam.gov.in/
- Open Educational Resources: Global Report 2017, http://oasis.col.org/handle/11599/2788; COL, 2017.
- UNESCO. (2016). Open Educational Resources (OER) Road Map Meeting: Outcome Report. Paris: UNESCO.
- https://classroom.google.com/h

CHAPTER 39

OER Development-I

Dr Ajay Semalty
H.N.B Garhwal University (A Central University)
Srinagar Garhwal-246174

Welcome dear learners, welcome in this chapter on Open Educational Resources. In previous chapter we studied:

- What are Open Educational Resources?
- What is the need of Open Educational Resources?
- What are the drivers for Open Educational Resources?

We studied the importance and then started going through various platforms of Open Educational Resources.

In continuation, now we will be dealing with some of the more Open Educational Resource platforms and creating the content in those platforms. -

Learning Outcome

After completing this chapter, you will be able to -

- Use/search Open Educational Resources(OER)
- Creating contents for OER platforms

Lesson Plan

As far as the lesson plan is concerned, we will start with the

- Wikipedia
- Moodle
- Slide share
- Edx
- COL/CEMCA
- Creating OERs

Wikipedia

As we know that Wikipedia is a platform for free web resources and free content which is available in public domain. Wikipedia is a non-profit organization which provides free knowledge in the public domain. It consists of bundle of various platforms which provides different kinds of resources.

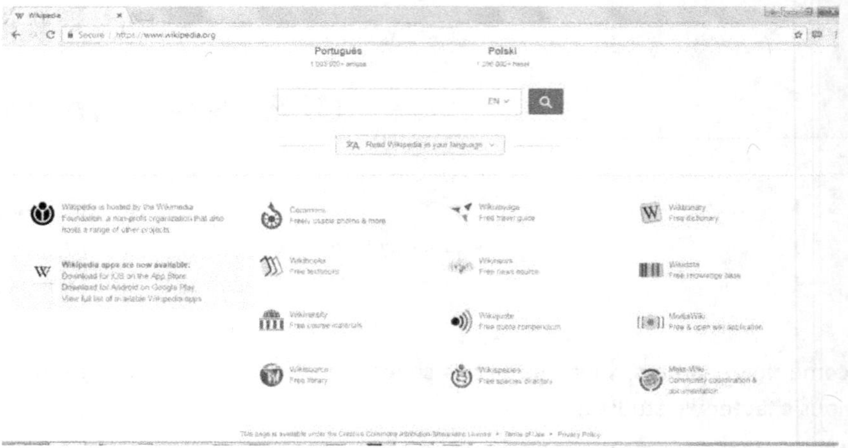

Fig. 39.1 Various sub-platforms of Wikipedia by Wikimedia Foundation

For example, as you can see in the wikipedia page itself the site consists of Wiki commons, Wikibooks, Wikivarsity, Wikiresource, Wikinews, Wikimedia, Wikidata like that (Fig. 39.1). So many platforms are available which are ready source of information about various segments of the knowledge domain.

In this free encyclopedia you just type whatever you need in search bar. You will get the things that too with the list of references at the end of the article. Here, a particular article is written by someone and edited by lot of persons. So, having a wikipedia page means that for a particular article you will have only a single page. If you want to contribute, you can edit in that particular page only.

How to Contribute to Wikipedia

- Create a Wikipedia account. Account creation might not require however, if you register for an account, you will be given more privileges than a non-registereduser.

- You can expand the subjects there.

- You can add a photo.

- Write a new article.

- Removes spam.

- Do some maintenance.

- Revertvandalism(Theactofeditingtheprojectinamaliciousmannerthatisintenti onally disruptive) Now let's discuss steps to creating a Wikipediapage.

Steps to creating a Wikipedia page

- Do your research first. Before creating any content on Wikipedia, learn about the Wikipedia community and how it works....

- Create an account.
- Start small.
- Gather your sources.
- Write the copy.
- Submit the page for review.

Please do remember that creating Wikipedia page is not free now. it was free few years ago. But now if you want to have a Wikipedia page you have to spend ranging from $250 to $3,000; and average cost is $400 to $800. What is the beauty of Wikipedia? Wikipedia's beauty is that whatever is there you can take them because it is in public domain provided you give attribution. It is a free resource. It is not bound to any copyright.

MOODLE

MOODLE is a learning management system LMS. As you can see this is the website of Moodle (https://moodle.org). If you want to go into detail- what is moodle actually? What moodle is providing? We have given this link (http://youtube.com/watch?v=3ORsUGVNxGs) for the YouTube video- What is moodle?

Moodle provide various MOOC courses on their platform. There are two categories courses

- Which you can download
- Which you canenroll.

You can search courses by adding keywords on website. You can also enroll using moodle cloud and have your own website to develop a course. In Moodle cloud you can create your own learning environment and it is very easy and user friendly.

It is simple and maintenance free, auto updated it is flexible and affordable. Affordable means up to 50 users it is free, if more than 50 user you have to pay. Still it is good to learn in small learning group specially for PG students. Again, we are giving this link (https://www.youtube.com/watch?v=DXSZEON_WpU&t=104s) for "creating course in moodle".

So, for a quick overview you can generate your own website in Moodle cloud as we are having one in the Moodle. You just have to enter your Moodle cloud site name, your password. Log in and you can develop your own course in that platform (Fig. 39.2).

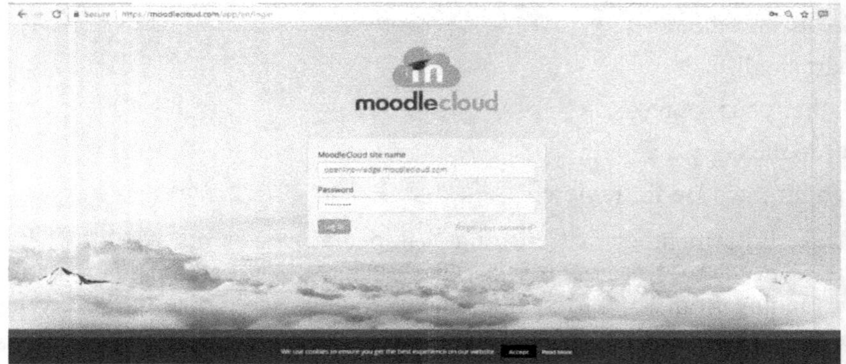

Fig. 39.2 Moodle Cloud login

The platform is very user friendly not only to teachers but also for learners. As you can see in the dashboard of the website, the list of courses available as we are having one course to the introduction to Open Education Resources (Fig. 39.3).

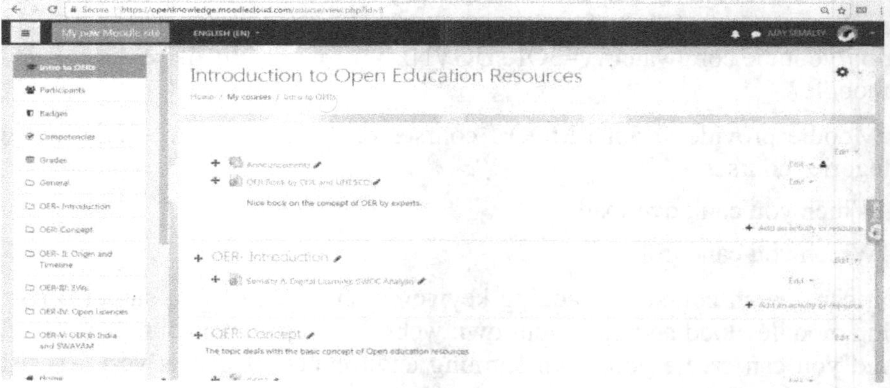

Fig. 39.3 Online Course on moodle

We are having in my website a course introduction to Open Education Resources which is available though not for learning but for download only. In the course you can see variety of contents can be given. It may be a link to a book, ppt, word file, pdf file, quiz, and various kind of evaluation assessment techniques, grading techniques are available there.

Section Summary

So dear learners, let me summarize this section. We learned

- Wikipedia is a free public domain in which you can get the OER and you can create your own OER and contribute there but it is paid.

- Next one is moodle, which is a learning management system. A complete package in which entire course is available with the variety of resources by which you can develop the contents for your online course. It is free up to 50 users.

Dear learner after covering the Wikipedia and the Moodle as the learning management system. We will be moving to the

SlideShare

Slide Share is linked with Linkedin. Many of the students do know about the slide Share. Here you can upload and publish your PowerPoint presentation and can download required file from slide share. Slide share platform is meant for discussion, sharing and learning (Fig. 39.4).

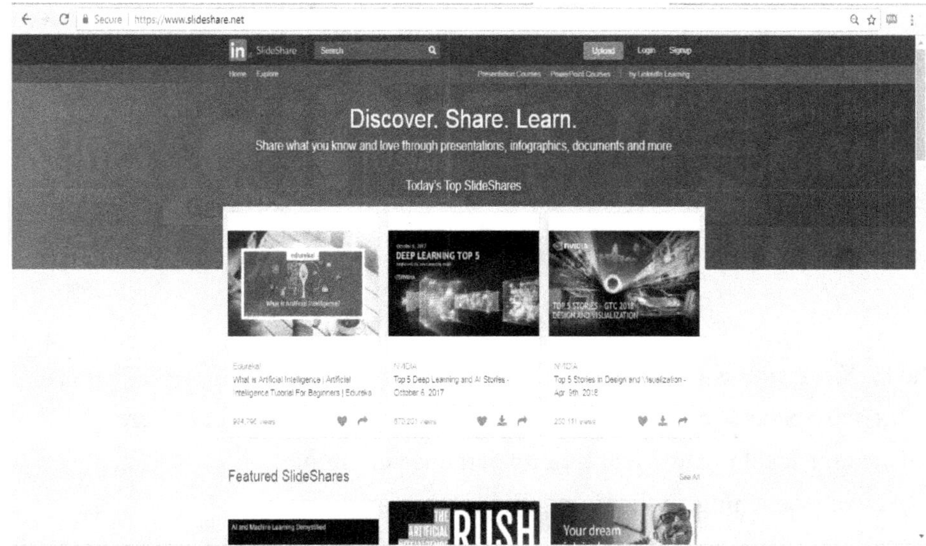

Fig. 39.4 Homepage of Slide Share

Like other platforms you can register here, you can plan or develop your own Power Point presentation. Just you have to upload it and it will be shown in your domain or in account and it will be available in the slide share. Slide share also provides the link and you can also join your YouTube videos to the presentation. So, it will be marked specially that embedded with the YouTube video. Whatever you are uploading in the ppt form if you are having audio-video presentation you can also give the link. For that presentation, the ppt will run first and it will be followed by the YouTube video (But this function has recently been removed by Slideshare). So, Slide share is ready source of information and the students as well as the faculty members can upload their presentation on this platform at free of cost. It is a completely free platform.

EdX (Learning management system)

EdX is the online learning destination founded by Harvard and MIT in 2012.EdX offers courses from the world's best universities and institutions to learners everywhere (Fig. 39.5). It is the only leading MOOC provider that is both nonprofit oriented and available as an open source.

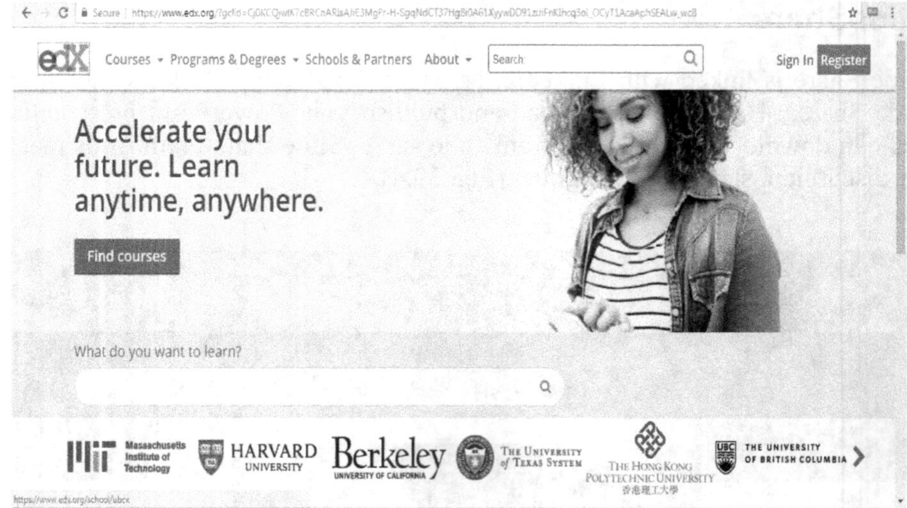

Fig. 39.5 Homepage of edX

The EdX mission is to-

- Increase access to high-quality education for everyone, everywhere
- Enhance teaching and learning on campus and online.
- Advance teaching and learning through research.

You can go to website can search courses available in different areas. You can enroll for any courses. It gives advantage of that once you enroll/ register you can enroll to any of courses learning is free of cost but for certificate you have to pay and cost of certification ranges 200to300dollar.So, if you want to learn about the paper writing there is a course. Just type and you will get a course from the internationally reputed professors and universities with quality content in the edX platform.

Activity time-

I am giving a little exercise to you that whatever we are discussing in any platform. What can you do?

Choose any platform. Prepare your own presentation. For students it may be on particular topic. For faculty members it may be on a particular lecture which they want to share with students and research community. Upload it. It may be on the slide share or any other platform. So, try it and don't limit yourself with just uploading the contents. Share the contents and link of particular resource to all your students, faculty members, research team members. Then it will solve the purpose of OER.

Next, we will be dealing with creating OER, conceptual creation points we will be discussing in brief. Apart from edX there are several similar platforms for open learning UDACITY, coursera, black board, khan academy, open learning. You just go through them. Huge number of these learning management systems are available. Many of these are free to learn and need charges only for certifications. So, for learning, please do enjoy these platforms.

The next is

CEMCA

Commonwealth Educational Media Centre for Asia (CEMCA) is a part of Commonwealth of Learning. This is the most powerful rising source of OER. In the website of CEMCA you can find the various course material which are hosted in their main website Commonwealth of Learning (COL). Various free texts are also available like this one which I am giving you is a very nice book on the Open Educational Resources is made available by the commonwealth of learning. A free course on understanding Open Educational Resources. This course is free to learn as well as you will get the free certificate after completing the course. The course contains audio-video, ungraded quiz and lastly the graded quiz and you will get the certification after that. So, do please attempt completing this course.

Section Summary

So, in this section we covered

- SlideShare

- edX and

- The OER Initiative by Commonwealth of Learning through CEMCA

What you have to do is just plan a small presentation or small OER. Enroll or register in to one of these platforms. Upload your contents and share with students, faculty members or research team and enjoy the basic concept of Open Educational Resources by sharing.

Dear learners, after going through the various OER platforms and learning management systems. let us move to how to create an OER.

Creating OER

Before moving to how to create an OER the first question comes in our mind-

What is required for creating an OER? What are the requirements?

- The first and the most important factor is the desire to share.
- The second is tools like directories, platform and learning management system.
- Then content and digital content creators are needed.
- And lastly the implementation resources, that licensing models which enables us to share. We can see the major parameters or requirements for OER creation in this conceptual map- tools, content, and implementation resources.

Now analyze yourself -

1. Do you want to be updated?
2. Do you want to be global?
3. Do you want to attract collaborator?
4. Do you want to be recognized in your field?
5. Do you want to share your knowledge and work? Tools: directories/platforms/LMS

- Open platforms and LMS (we have discussed earlier in previous chapter and in this chapter also. So choose the most suitable one. You can choose any platforms like Creative Commons, OER Commons, SlideShare, Moodle, etc.)
- LinkedIn, RG
- YouTube
- FBpage
- Own institutional repository
- Governmental repository
- SWAYAMMOOCs

However, in SWAYAM MOOCs the students cannot contribute. Students can go for other platforms like YouTube, SlideShare, OERs commons to explore. But faculty member should plan ahead and go for MOOCs on SWAYAM. We will be discussing this in short in later section.

SWAYAM MOOC

Experienced faculty members and researchers can contribute. Certain guidelines and minimum eligibility criteria are there (see further reading links) and you can go through the UGG MOOCs website or CEC website (www.cec.nic.in) for tertiary education and plan your MOOCs. In a concise way we will be addressing this section also.

Contents

- Content should be Factually correct the utmost important criteria.
- It should have good quality content. If the person is referring to an OER, it must be of good quality.
- It must be plagiarism free (Take care of copy right issue). You cannot go and just copy paste.
- The content should be easy to follow and understand. The reader friendliness must be there. What is the need? Because the learner wants easy to follow and easy to understand things.
- Simple and effective organization
- The next point is Cohesion and coherence. The content must be flowing like a fountain like a river. That systematic flow must be there.
- After that, the content must be concise for each topic and subtopic.
- And last but not the least the next point is it must be engaging. It should have the feature of engaging the learner. Then and only then it will solve its purpose of its creation.

Summary

So, dear learners, let me summarize this chapter for you. We learned about

- OER platforms like Wikipedia, Moodle, Slide share, edX, CEMCA
- Then we moved to the creating OERs.
- Plan content for the any of the platform, register there, upload it, and share with your friends.
- The contents must be factually correct.
- And your presentation must be good.

We will be discussing other aspects in the next chapter with respect to the developing OERs.

Further Readings

- https://en.wikipedia.org/wiki/Main_Page
- https://docs.moodle.org/dev/Releases
- https://dougiamas.com/archives/edmedia2003/
- Open Educational Resources (OER): Dr Ajay Semalty at FDC, HNB Garhwal University Srinagar Garhwal https://youtu.be/K7xg4gRXMN8
- Open Educational Resources (OER) - II : Sharing- What, where, who, and why?: Dr Ajay Semalty, https://youtu.be/3Br_3I4oCgk
- Understanding Open Educational Resources, Commonwealth of learning, https://oerknowledgecloud.org/sites/oerknowledgecloud.org/files/2015_Butcher_Moore_Understanding-OER.pdf
- Guidelines for Quality Assurance and Accreditation of MOOCs, Commonwealth of learning, http://oasis.col.org/bitstream/handle/11599/2362/2016__Guidelines-QAA-MOOCs.pdf?sequence=6&isAllowed=y
- Semalty A, Increasing the reach of Digital Learning in India, (Guest Column) The daily Pioneer, 25 April 2020 (http://tiny.cc/pioneerswayam).
- Semalty A, Digital learning: The Uttarakhand Plan, https://pharmastate.blog/digital-learning-the-uttarakhand-plan/
- Semalty A., Digital learning through SWAYAM MOOCs: SWOT Analysis, CEC News, Sept 2019, 5-6 (http://tiny.cc/digitallearningCEC).
- Semalty M, Adhikari L, Semalty A., Benchmarking SWAYAM MOOCs, Taking learners from enrollment to exams, CEC News, March 2020: 3-6 (http://tiny.cc/benchmarkingSWAYAM).
- Semalty M., Ten commandments of Digital learning through MOOCs: SWOT Analysis, CEC News, Nov. 2019, 5-6 (http://tiny.cc/TENCOMMAND).
- SWAYAM Guidelines: https://swayam.inflibnet.ac.in/home_assets/download/Guidelines.pdf
- Financial norms of SWAYAM MOOCs: https://swayam.inflibnet.ac.in/home_assets/download/Financial_norms_for_swayam_MOOCs.pdf
- MOOC Proposal submission portal: https://swayam.inflibnet.ac.in/
- Dr. Mona Semalty, Concept of MOOCs & Proposal Submission Dec 14, 2020, Vishvabharti University, https://youtu.be/y4EYNCfyTDU

References

- https://moodle.org
- https://en.wikipedia.org/wiki/Main_Page
- https://www.slideshare.net
- Wikipedia, Open educational resources, https://en.wikipedia.org/wiki/Open_educational_resources
- Cheryl Hodgkinson-Williams, Michael Paskevicius, Roger Brown, A culture of sharing: Open education resources an introduction, Teaching with Technology Seminar, UCT, 18 May 2009; cheryl.hodgkinson-williams@uct.ac.za
- Organisation for Economic Co-operation and Development (OECD), Giving knowledge for free the emergence of open educational resources, 2017, e book, https://www.oecd.org/edu/ceri/38654317.pdf
- Wolfenden F, Open Educational Resources (OER), www.TESS-India.edu.in
- Kanwar A and Uvalic´-Trumbic´ S (Eds), Butcher N, A Basic Guide to Open Educational Resources (OER), http://unesdoc.unesco.org/images/0021/002158/215804e.pdf
- Open Educational Resources (OER)". CoL.org.Commonwealth of Learning.https://www.col.org
- http://creativecommons.org/
- https://www.oercommons.org/
- https://swayam.gov.in/
- Open Educational Resources: Global Report 2017, http://oasis.col.org/handle/11599/2788; COL, 2017.
- UNESCO. (2016). Open Educational Resources (OER) Road Map Meeting: Outcome Report. Paris: UNESCO.

CHAPTER 40

OER Development-II

Dr Ajay Semalty
H.N.B Garhwal University (A CentralUniversity)
SrinagarGarhwal-246174

Learning Outcome

After completing this chapter, you will be able to

- Develop Open Educational Resources(OER)
- Plan to develop OER/MOOC

Lesson Plan

- What to develop as an OER?
- Student/Faculty perspective of OER development
- Types of OER
- Designing OER
- Submitting proposals for MOOC

What to develop as OER?

OER can be as small and simple and it can be in the form of as an individual picture or in the form of an entire course. Whatever you know and which is factually correct can be developed in the form of OER.

Types of OER

- Course/Instructor Resources (MIT, OCW, Coursera, Edx, Moodle)
- Full Distance Course Chapters (Open Learn UK)
- Course Chapters/seminars
- Learning Objects
- Images (www.flickr.com)
- Video (www.academicearth.com,YouTube)
- Audio (http://itunes.stanford.edu)
- Open Textbooks (www.wikibooks.org)
- Journals (www.doaj.org)

Students' perspective: OER

- A student can develop an OER on the advancement in the field of knowledge/research.
- A research scholar may present the quick summary or key points: the know how & basics of research methods in presentable and or easy to learn form.

- Own viewpoints on atopic may be converted into a form of an OER. It may be your own creation. It may be critical analysis of the previous body of knowledge.

- Any small lab based new workable model of research/method/process can be developed into the form of an OER. For example- chemistry and pharmacy students can show new techniques or methods by recording it and develop it into an OER and post. It may be simple protocols of chromatographic techniques, practical aspects of TLC development. These may be recorded and developed into the form of an OER.

- It may also be highlighting the new use of old content/equipment /product/service.

- It may be helpful in improving the Altimetric of students and researchers.

- The way you express yourself in social media is made the basis of selection by big companies. Dear learners, most of the time nowadays it has become a trend that big companies and MNCs are analyzing the way you express yourself on the social media. The way you present the things, the OERs. It is not just about the educational part, it is all about your overall personality. So, why not to add the knowledge, expertise, efficiency that what you can do in the form of an OER and share in social media. It will help you in your placements.

Faculty perspective

- Making quick/smart class notes/ outline/ reminders for students
- Makingroutine/semesterinstructions/SOPsashandyreferencee.g.Labmanual/r eferring literature
- Making specific topic-based content
- Making the complex topics easier to understand
- Personalizing learning for student or group of students

I remember an example, when I was doing my post-doctoral research in Japan in Meijo University. I was given a handout as a ready reference book of around 20-30 pages which was written in Japanese as well as English. I went through it and found ready protocols for doing the basic things like preparing the column for column chromatography, seeing the meniscus. Pictorial presentations were there about how to hold the conical flask below the burette. Basic techniques were given and the standard protocols like switching off the lights, ACs, using instruments present in the room. So, these kind of things can be developed into an OER.

Section Summary

Dear learners, let me summarize this section for you

- What to develop as an OER?

It can be a simple individual picture or as large as an entire course.

- Students can plan an OER to show their potential for quick learning notes.
- The teachers can develop the OER for sharing their knowledge, quick notes, quick summaries and for personalized learning.

Let us discuss other aspects of OER development in the next section.

Major component: content writer

Experienced academicians like in case of MOOC development as per as UGC guideline for UGC MOOC faculty member of approved institutions with at least 5 year PG teaching experience can be subject teacher expert or course coordinator where in case of other platforms there is no such rule like if you want to create an OER. You can plan OER if you are scholar, faculty or student. You can prepare good quality of content and upload it in OER platforms.

For project like MOOCs a single content writer is not enough, a good and efficient content writing team is needed depending on subject, topic and subtopics. Depending on experience in the field where they are working. A content writer must be comfortable with use of computer (use of MS office), expert in planning the quality infographics.

Content writer should be familiar with the concept plagiarism because he/she is the person who knows from where the text has to be taken. Whether they are copied or written by a himself and content writer must familiar with technique to avoid plagiarism.

Major type of OER -

- Simple image and / or text
- Multimedia based
    - o Video
    - o Audio
    - o Info graphics
    - ✓ Simple image and / or text
- Write a page or pages / draw a figure or blend of both
- Scan the same, convert into the desired file format and Post to any OER platform (it may be FB/ WhatsApp/ Google classroom etc.)
    - ✓ Multimedia based

- Use screen recording with or without voiceover

  E.g. presentation tube, DU recorder etc.

- Prepare ppts, use infographics, run ppts along with your voiceover.

- Select the platform, upload the same to the platform. (e.g. Try YouTube: very simple to upload in your ownchannel)

Note-Remember to use creative commons license while uploading in YouTube

✓ Audio/ Video

- Simple audio podcast is considered very effective due to the element of focusing and require not much efforts. You just need a good voice. (limited to the topics which are theoretical completely or do not require infographics).

- Record audio in your smart phone, upload in suitable platform and share with CClicense.

- Video presentations are the most challenging. But it simplifies the topic most effectively.

Types of video presentation

- Talking head- As I am addressing all people this is talking heads.
- Presentation slides with voice over.
- Picture in picture
- Text Overlay
- Tablet Capture (Khan Style/ UdacityStyle)
- Actual paper/ Whiteboard/glassboard
- Screencast
- Animation
- Classroom lecture
- Recorded seminar
- Conversation- may be one to one and it may be as group discussion.

However, depending upon the requirement of the OER for the MOOC you can plan a good mix of video lecture presentation. Video e-learning increases the ability of online course to effectively transfer knowledge. When video is leveraged for explaining topic it helps by

- visualizing complex concepts,
- reduces the reading load on learners
- breaks the monotony of plain text by presenting it as an engaging visual

- It presents information as a narrative helping the learner in retaining information more effectively
- Lastly, it helps in breaking complex topic into digestible pieces of information

Infographics

- Visual representation of information
- Data, presentations, concepts
- Information, timeline, process
- Use word art, SmartArt, vector images, self-drawn pictures, photos, animation etc.
- Info graphics reduce the text content load, increase the engagement and help in the easy understanding.

Section Summary

So, we discussed in this section

- What is expected from a content writer of an OER?
- What may be types of OER depending upon the size?
- Use of audio-video presentations for the OER
- Effective use of infographics for making them engaging

Key features of OER

Some key features for OER development are as-

- Freely available educational content
- Quality
- Simple and effective
- Engaging
- Short/condensed with tight outline
- Factually correct
- Plagiarism free

Steps of development of OER are-

How should we develop an OER?

1. Planning (outline, tool, content, presentation style)
2. Content writing

3. Design presentation
4. Presentation (multimediaetc.)
5. Uploading and sharing
6. Feedback/assessment/certification along with a parallel discussion forum

Massive Open Online Courses (**MOOCs**)

- In the MOOCs, the students can enroll in any course and get certificate and credits. As per the latest UGC guidelines, SWAYAM MOOCs are available for credit transfer for upto 40 % of the papers of a semester.
- All the courses are free to learn from the peers.
- Faculty members can either apply for a course development or be a part of developing MOOC as presenter or content writer.

Now we will be addressing an important aspect for the faculty members.

Applying for MOOCs

- Got to the website of national coordinator for example in case of UG go for CECMOOCs (Refer annexure)
- You can go to CEC, UGC and NPTEL WEBSITE for referring the process. NPTEL takes the course proposal from IITs only.
- In CEC website a form is available for expression of Interest for developing MOOC for UG nontechnological MOOCs. Fill it and submit with intro video to CEC throughemail. Now the online portal is available for submitting UG and PG proposals (MOOC Proposal submission portal: https://swayam.inflibnet.ac.in/)

Basic features of proposal-

A proposal of MOOC contains: -
- Course Coordinators'(CC's) detail, institutes detail
- Course name, level, type
- Rational of developing the course
- Aim and objective
- Outline and time schedule
- Course team: SMEs
- Evaluation/assessment plan
- Letter of consent from host institute

Introvideo-You must be specific for the limited time and you should upload the introductive video to drive and share link with national coordinator with proposal for submission.

- Time- 3-5minute
- Introduce CC,
- Course name, level, type
- Rational of developing the course
- Aim and objective
- Target learner
- Outline or Coverage

So, be very specific and cover all the material in 3-5 minutes.

For proposal submission

- Simply prepare with any mobile camera or digital camera
- Insertslides
- Or record in presentation tube
- Voice clarity
- Overall presentation/ body language is vital

To begin with OERs

- Do not jump in creating OERs
- First use OERs for flipped and blended learning
- Use WhatsApp, and Facebook like social media for teaching learning
- Screen and select suitable OERs from YouTube/ Wikipedia/ SlideShare/ UGC e content and share the link with students; and students should use these OERs for exam preparation and quality learning.
- Enrich your knowledge in enrolling yourself in SWAYAM MOOCs (ARPIT Courses) and get extra advantage in CAS as per UGC regulation 2018 (equivalent to orientation/ refresher course).
- Make small OERs and share with CC license
- Make the proposal for developing MOOCs with a good team

Major Challenges-

- Lack of users' capacity to access, reuse and share OER
- Language and Cultural barrier
- Insufficient inclusive and equitable access to quality content
- Lack of appropriate policy solutions
- People are still working in isolation although collaboration is promoted.

- Awareness and use of available OER repositories are significantly low.

- Not much support from employers/institutions

- Lack of skills

There are so many challenges and always limitations are going to be there, but I feel, as a researcher, student, and teacher and I agree with the famous quote by the John Cotton Dana that-

"Who dares to teach, must never cease to learn."

Over 100 hundred years ago, Indian poet and Laurette Gurudev Rabindranath Tagore articulated his vision of the future. When he wrote of a world-

Where the mind is without fear and the head is held high..

Where knowledge is free

Can OER and MOOCs help us to get to a closure world. Yes of course. So, try OER and MOOCs. It is an advanced form of digital writing. Now, SWAYAM MOOCs are available for credit transfer for upto 40 % of the papers of a semester.

Further Readings

- https://wikieducator.org/Educators_care/Defining_OER

- Bell,Steven."ResearchGuides:DiscoveringOpenEducationalResources (OER):Home". guides.temple.edu.2017-12-05.

- http://www.oecd.org/education/ceri/38654317.pdf

- Sanchez, Claudia. "The use of technological resources for education: a new professional competency for teachers". Intel® Learning Series blog. Intel Corporation. Archived from the original on 29 March 2013. 23 April2013.

- "What is OER?". wiki.creativecommons.org. Creative Commons. 18 April2013.

- Semalty A, Increasing the reach of Digital Learning in India, (Guest Column) The daily Pioneer, 25 April 2020 (http://tiny.cc/pioneerswayam).

- Dr. Mona Semalty, Concept of MOOCs & Proposal Submission Dec 14, 2020, Vishvabharti University, https://youtu.be/y4EYNCfyTDU

- Semalty A., Digital learning through SWAYAM MOOCs: SWOT Analysis, CEC News, Sept 2019, 5-6 (http://tiny.cc/digitallearningCEC).

- Semalty M, Adhikari L, Semalty A., Benchmarking SWAYAM MOOCs, Taking learners from enrollment to exams, CEC News, March 2020: 3-6 (http://tiny.cc/benchmarkingSWAYAM).

- Semalty M., Ten commandments of Digital learning through MOOCs: SWOT Analysis, CEC News, Nov. 2019, 5-6 (http://tiny.cc/TENCOMMAND).
- SWAYAM Guidelines: https://swayam.inflibnet.ac.in/home_assets/download/Guidelines.pdf
- Financial norms of SWAYAM MOOCs: https://swayam.inflibnet.ac.in/home_assets/download/Financial_norms_for_swayam_MOOCs.pdf
- MOOC Proposal submission portal: https://swayam.inflibnet.ac.in/
- SWAYAM Online Courses: www.swayam.gov.in

References

- https://swayam.gov.in
- http://cec.nic.in/Pages/Home.aspx
- https://wikieduator.org/Educators_care/Defining_OER

Important Guidelines and Links

1. University Grants Commission (Promotion of Academic Integrity and Prevention of Plagiarism In Higher Educational Institutions) Regulations, 2018 New Delhi, the 23rd July, 2018, https://www.ugc.ac.in/pdfnews/7771545_ academic-integrity-Regulation2018.pdf

[भाग III—खण्ड 4]   भारत का राजपत्र : असाधारण   9

3. Objectives

3.1 To create awareness about responsible conduct of research, thesis, dissertation, promotion of academic integrity and prevention of misconduct including plagiarism in academic writing among student, faculty, researcher and staff.

3.2 To establish institutional mechanism through education and training to facilitate responsible conduct of research, thesis, dissertation, promotion of academic integrity and deterrence from plagiarism.

3.3. To develop systems to detect plagiarism and to set up mechanisms to prevent plagiarism and punish a student, faculty, researcher or staff of HEI committing the act of plagiarism.

4. Duties of HEI:

Every HEI should establish the mechanism as prescribed in these regulations, to enhance awareness about responsible conduct of research and academic activities, to promote academic integrity and to prevent plagiarism.

5. Awareness Programs and Trainings:

(a) HEI shall instruct students, faculty, researcher and staff about proper attribution, seeking permission of the author wherever necessary, acknowledgement of source compatible with the needs and specificities of disciplines and in accordance with rules, international conventions and regulations governing the source.

(b) HEI shall conduct sensitization seminars/ awareness programs every semester on responsible conduct of research, thesis, dissertation, promotion of academic integrity and ethics in education for students, faculty,

7. Similarity checks for exclusion from Plagiarism

The similarity checks for plagiarism shall exclude the following:

    i.   All quoted work reproduced with all necessary permission and/or attribution.

    ii.   All references, bibliography, table of content, preface and acknowledgements.

    iii.   All generic terms, laws, standard symbols and standards equations.

Note:

The research work carried out by the student, faculty, researcher and staff shall be based on original ideas, which shall include abstract, summary, hypothesis, observations, results, conclusions and recommendations only and shall not have any similarities. It shall exclude a common knowledge or coincidental terms, up to fourteen (14) consecutive words.

8. Levels of Plagiarism

Plagiarism would be quantified into following levels in ascending order of severity for the purpose of its definition:

    i.   Level 0: Similarities upto 10% - Minor similarities, no penalty

    ii.   Level 1: Similarities above 10% to 40%

    iii.   Level 2: Similarities above 40% to 60%

    iv.   Level 3: Similarities above 60%

9. Detection/Reporting/Handling of Plagiarism

12. Penalties

Penalties in the cases of plagiarism shall be imposed on students pursuing studies at the level of Masters and Research programs and on researcher, faculty & staff of the HEI only after academic misconduct on the part of the individual has been established without doubt, when all avenues of appeal have been exhausted and individual in question has been provided enough opportunity to defend himself or herself in a fair or transparent manner.

12.1 Penalties in case of plagiarism in submission of thesis and dissertations

Institutional Academic Integrity Panel (IAIP) shall impose penalty considering the severity of the Plagiarism.

    i.   **Level 0: Similarities upto 10% -** Minor Similarities, no penalty.

    ii.   **Level 1: Similarities above 10% to 40% -** Such student shall be asked to submit a revised script within a stipulated time period not exceeding 6 months.

    iii. **Level 2: Similarities above 40% to 60% -** Such student shall be debarred from submitting a revised script for a period of one year.

    iv. **Level 3: Similarities above 60% -**Such student registration for that programme shall be cancelled.

Note 1: Penalty on repeated plagiarism- Such student shall be punished for the plagiarism of one level higher than the previous level committed by him/her. In case where plagiarism of highest level is committed then the punishment for the same shall be operative.

Note 2: Penalty in case where the degree/credit has already been obtained - If plagiarism is proved on a date later than the date of award of degree or credit as the case may be then his/her degree or credit shall be put in abeyance for a period recommended by the IAIP and approved by the Head of the Institution.

12.2 Penalties in case of plagiarism in academic and research publications

    I.   **Level 0: Similarities up to 10% -** Minor similarities, no penalty.

    II.   **Level 1: Similarities above 10% to 40%**

        i)   Shall be asked to withdraw manuscript.

    III. **Level 2: Similarities above 40% to 60%**

        i)   Shall be asked to withdraw manuscript.

        ii)   Shall be denied a right to one annual increment.

        iii) Shall not be allowed to be a supervisor to any new Master's, M.Phil., Ph.D. Student/scholar for a period of two years.

    IV. **Level 3: Similarities above 60%**

        i)   Shall be asked to withdraw manuscript.

        ii)   Shall be denied a right to two successive annual increments.

        iii) Shall not be allowed to be a supervisor to any new Master's, M.Phil., Ph.D. Student/scholar for a period of three years.

Note 1: Penalty on repeated plagiarism - Shall be asked to withdraw manuscript and shall be punished for the plagiarism of one level higher than the lower level committed by him/her. In case where plagiarism of highest level is committed then the punishment for the same shall be operative. In case level 3 offence is repeated then the disciplinary action including suspension/termination as per service rules shall be taken by the HEI.

Note 2: Penalty in case where the benefit or credit has already been obtained - If plagiarism is proved on a date later than the date of benefit or credit obtained as the case may be then his/her benefit or credit shall be put in abeyance for a

2. Self plagiarism defined by UGC: Self-plagiarism: https://www.ugc.ac.in/pdfnews/2284767_self-plagiarism001.pdf

UNIVERSITY GRANTS COMMISSION
BHADURSHAH ZAFAR MARG
NEW DELHI-110 002

N.F.1-1/2020(SECY)   20th April, 2020

PUBLIC NOTICE

SELF-PLAGIARISM

In the interests of Indian academia, to promote Indian research among the nations, and to ensure credibility and quality, from time to time the UGC has instituted various measures. In its efforts to curb plagiarism the UGC issued the University Grants Commission (Promotion of Academic Integrity and Prevention of Plagiarism in Higher Educational Institutions) Regulations, 2018, so that plagiarised work does not acquire any credibility or value in evaluation. In continuation of its initiative, and in line with global standards of ethical publishing established by leading institutions and Committee on Publication Ethics (COPE), the UGC draws the attention of the academic community to the following:

(i)   Reproduction, in part or whole, of *one's own previously published* work *without adequate citation* and proper acknowledgment and claiming the most recent work as new and original for any academic advantage amounts to 'text-recycling' (also known as 'self-plagiarism') and is **not** acceptable.

(ii) Text-recycling/self-plagiarism includes:
  - republishing the same paper already published elsewhere *without due and full citation*;
  - publishing smaller/excerpted work from a longer and previous *without due and full citations* in order to show a larger number of publications;
  - reusing data already used in a published work, or communicated for publication, in another work *without due and full citation*;
  - breaking up a longer/larger study into smaller sections and publishing them as altogether new work *without due and full citation*;
  - paraphrasing one's own previously published work *without due and full citation* of the original.

(iii) Self-citations do not add any number/s to the individual's citation index or h-index in global academia.

(iv) Vice Chancellors, Selection Committees, Screening Committees, IQACs and all/any experts involved in academic performance/evaluation and assessment are hereby strongly advised that their decisions in the case of promotions, selections, credit allotment, award of research degrees must be based on an evaluation of the applicant's published work to ensure that the work being submitted for promotion/selection is not self-plagiarized.

The UGC will be issuing a set of parameters to evaluate instances of text recycling/self-plagiarism soon.

(Prof. Rajnish Jain)
Secretary

3. A new subject Research and Publication Ethics introduced (https://www.ugc.ac.in/pdfnews/9836633_Research-and-Publication-Ethics.pdf)

Course structure

- The course comprises of six modules listed in table below. Each module has 4-5 units.

Modules	Unit title	Teaching hours
Theory		
RPE 01	Philosophy and Ethics	4
RPE 02	Scientific Conduct	4
RPE 03	Publication Ethics	7
Practice		
RPE 04	Open Access Publishing	4
RPE 05	Publication Misconduct	4
RPE 06	Databases and Research Metrics	7
	Total	30

<u>**Syllabus in detail**</u>

THEORY

- **RPE 01: PHILOSOPHY AND ETHICS (3 hrs.)**

    1. Introduction to philosophy: definition, nature and scope, concept, branches
    2. Ethics: definition, moral philosophy, nature of moral judgements and reactions

- **RPE 02: SCIENTIFICCONDUCT (5hrs.)**

    1. Ethics with respect to science and research
    2. Intellectual honesty and research integrity
    3. Scientific misconducts: Falsification, Fabrication, and Plagiarism (FFP)
    4. Redundant publications: duplicate and overlapping publications, salami slicing
    5. Selective reporting and misrepresentation of data

- **RPE 03: PUBLICATION ETHICS (7 hrs.)**

    1. Publication ethics: definition, introduction and importance
    2. Best practices / standards setting initiatives and guidelines: COPE, WAME, etc.
    3. Conflicts of interest
    4. Publication misconduct: definition, concept, problems that lead to unethical behavior and vice versa, types
    5. Violation of publication ethics, authorship and contributorship
    6. Identification of publication misconduct, complaints and appeals
    7. Predatory publishers and journals

PRACTICE

- **RPE 04: OPEN ACCESS PUBLISHING(4 hrs.)**

1. Open access publications and initiatives
2. SHERPA/RoMEO online resource to check publisher copyright & self-archiving policies
3. Software tool to identify predatory publications developed by SPPU
4. Journal finder / journal suggestion tools viz. JANE, Elsevier Journal Finder, Springer Journal Suggester, etc.

- **RPE 05: PUBLICATION MISCONDUCT (4hrs.)**

  ### A. Group Discussions (2 hrs.)

    1. Subject specific ethical issues, FFP, authorship
    2. Conflicts of interest
    3. Complaints and appeals: examples and fraud from India and abroad

  ### B. Software tools (2 hrs.)

  Use of plagiarism software like Turnitin, Urkund and other open source software tools

- **RPE 06: DATABASES AND RESEARCH METRICS (7hrs.)**

  ### A. Databases (4 hrs.)
    1. Indexing databases
    2. Citation databases: Web of Science, Scopus, etc.

  ### B. Research Metrics (3 hrs.)
    1. Impact Factor of journal as per Journal Citation Report, SNIP, SJR, IPP, Cite Score
    2. Metrics: h-index, g index, i10 index, altmetrics

4. Guidance Document "Good Academic Research Practices"; Sept. 2020,https://www.ugc.ac.in/e-book/UGC_GARP_2020_Good% 20Academic%20Research%20Practices.pdf

Contents

5. API Scoring: UGC Regulations on Minimum Qualification for appointment of Teachers and Other Academic Staff in Universities and Colleges and Measures for the Maintenance of Standards in Higher Education 2018https://www.ugc.ac.in/pdfnews/4033931_UGC-Regulation_min_Qualification_Jul2018.pdf

<u>Table 2</u>

Methodology for University and College Teachers for calculating Academic/Research Score

(Assessment must be based on evidence produced by the teacher such as: copy of publications, project sanction letter, utilization and completion certificates issued by the University and acknowledgements for patent filing and approval letters, students' Ph.D. award letter, etc,.)

S.N.	Academic/Research Activity	Faculty of Sciences /Engineering / Agriculture / Medical /Veterinary Sciences	Faculty of Languages / Humanities / Arts / Social Sciences / Library /Education / Physical Education / Commerce / Management & other related disciplines
1.	**Research Papers in Peer-Reviewed or UGC listed Journals**	08 per paper	10 per paper
2.	**Publications (other than Research papers)**		
	(a) Books authored which are published by ;		
	International publishers	12	12
	National Publishers	10	10
	Chapter in Edited Book	05	05
	Editor of Book by International Publisher	10	10
	Editor of Book by National Publisher	08	08
	(b) Translation works in Indian and Foreign Languages by qualified faculties		
	Chapter or Research paper	03	03
	Book	08	08
3.	**Creation of ICT mediated Teaching Learning pedagogy and content and development of new and innovative courses and curricula**		
	(a) Development of Innovative pedagogy	05	05
	(b) Design of new curricula and courses	02 per curricula/course	02 per curricula/course
	(c) MOOCs		
	Development of complete MOOCs in 4 quadrants (4 credit course)(In case of MOOCs of lesser credits 05 marks/credit)	20	20
	MOOCs (developed in 4 quadrant) per module/lecture	05	05
	Content writer/subject matter expert for each module of MOOCs (at least one quadrant)	02	02
	Course Coordinator for MOOCs (4 credit course)(In case of MOOCs of lesser credits 02 marks/credit)	08	08
	(d) E-Content		
	Development of e-Content in 4 quadrants for a complete course/e-book	12	12
	e-Content (developed in 4 quadrants) per module	05	05
	Contribution to development of e-content module in complete course/paper/e-book (at least one quadrant)	02	02
	Editor of e-content for complete course/ paper /e-book	10	10
4	**(a) Research guidance**		

	Ph.D.	10 per degree awarded 05 per thesis submitted	10 per degree awarded 05 per thesis submitted
	M.Phil./P.G dissertation	02 per degree awarded	02 per degree awarded
	(b) Research Projects Completed		
	More than 10 lakhs	10	10
	Less than 10 lakhs	05	05
	(c) Research Projects Ongoing :		
	More than 10 lakhs	05	05
	Less than 10 lakhs	02	02
	(d) Consultancy	03	03
5	**(a) Patents**		
	International	10	10
	National	07	07
	(b) *Policy Document (Submitted to an International body/organisation like UNO/UNESCO/World Bank/International Monetary Fund etc. or Central Government or State Government)		
	International	10	10
	National	07	07
	State	04	04
	(c) Awards/Fellowship		
	International	07	07
	National	05	05
6.	***Invited lectures / Resource Person/ paper presentation in Seminars/ Conferences/full paper in Conference Proceedings (Paper presented in Seminars/Conferences and also published as full paper in Conference Proceedings will be counted only once)**		
	International (Abroad)	07	07
	International (within country)	05	05
	National	03	03
	State/University	02	02

The Research score for research papers would be augmented as follows :

Peer-Reviewed or UGC-listed Journals (Impact factor to be determined as per Thomson Reuters list) :

i)   Paper in refereed journals without impact factor   -   5 Points

ii) Paper with impact factor less than 1   -   10 Points

iii) Paper with impact factor between 1 and 2   -   15 Points

iv) Paper with impact factor between 2 and 5   -   20 Points

v)   Paper with impact factor between 5 and 10   -   25 Points

vi) Paper with impact factor >10   -   30 Points

(a)   Two authors: 70% of total value of publication for each author.

(b)   More than two authors: 70% of total value of publication for the First/Principal/Corresponding author and 30% of total value of publication for each of the joint authors.

Joint Projects: Principal Investigator and Co-investigator would get 50% each.

Note:

- Paper presented if part of edited book or proceeding then it can be claimed only once.
- For joint supervision of research students, the formula shall be 70% of the total score for Supervisor and Co-supervisor. Supervisor and Co-supervisor, both shall get 7 marks each.
- *For the purpose of calculating research score of the teacher, the combined research score from the categories of 5(b), Policy Document and 6. Invited lectures/Resource Person/Paper presentation shall have an upper capping of thirty percent of the total research score of the teacher concerned.
- The research score shall be from the minimum of three categories out of six categories.

6. National Education Policy (NEP) 2020, https://www.education.gov.in/sites/upload_files/mhrd/files/NEP_Final_English_0.pdf

7. UGC-CARE Reference List of Quality Journals: https://ugccare.unipune.ac.in/ apps1/home/index

8. HNBGU OER repository: www.swayamhnbgu.in/repo

9. SWAYAM Guidelines: https://swayam.inflibnet.ac.in/home_assets/download/Guidelines.pdf

10. Financial norms of SWAYAM MOOCs: https://swayam.inflibnet.ac.in/home_assets/download/Financial_norms_for_swayam_MOOCs.pdf

11. MOOC Proposal submission portal: https://swayam.inflibnet.ac.in/

12. SWAYAM Online Courses: www.swayam.gov.in

13. SWAYAM Online Courses: https://storage.googleapis.com/uniquecourses/online.html

14. UG/PG MOOCs: http://ugcmoocs.inflibnet.ac.in/ugcmoocs/courses.php

15. e-PG Pathshala: www.epgp.inflibnet.ac.in

16. e-Content courseware in UG subjects: http://cec.nic.in/

17. SWAYAMPRABHA: https://www.swayamprabha.gov.in/

18. CEC-UGC YouTube channel: https://www.youtube.com/user/cecedusat

19. National Digital Library: https://ndl.iitkgp.ac.in/

20. Shodhganga: https://shodhganga.inflibnet.ac.in/

21. e-Shodh Sindhu: https://ess.inflibnet.ac.in/

22. ICT initiative by MHRD: https://mhrd.gov.in/ict-initiatives

23. Vidyamitra YouTube Channel: https://www.youtube.com/channel/UCCUr096WDp86n62CXBeHlQw

24. Free lectures on Academic writing on YouTube Channel:
    https://cutt.ly/OKlive

25. Dr A Semalty, Academic Writing, Live Lectures, recordings:
    http://tiny.cc/AWSemalty

26. About SWAYAM, http://tiny.cc/SWAYAM

27. About OERs: http://tiny.cc/OERlectures

28. Lectures on NDDS: http://tiny.cc/SEMALTYNDDS

29. Lectures on Cosmetics: http://tiny.cc/SEMALTYCOSMETICS

30. Lectures on Pharmaceutics I: http://tiny.cc/SEMALTYCEUTICS1

31. Academic Writing's Open Podcasting Service:
    http://bit.do/awpodcastsemalty

32. Dr. Mona Semalty, Concept of MOOCs & Proposal Submission Dec 14,
    2020, Vishvabharti University, https://youtu.be/y4EYNCfyTDU

33. CH15 SWAYAM Prabha IIT Madras, (Physics, Chemistry, Engineering,
    Biomedical, Management and pharmacy);
    https://www.youtube.com/channel/UC3LYjJBtmmUsC2t-VlcCkJw

34. Ch-16 SWAYAMPRABHA, Humanities, Social Sciences and
    Management,
    https://www.youtube.com/channel/UC5Mwfc56bllPWmDAoLAPtjQ

35. SWAYAMPRABHA, CH-08:ARYABHATT [Mathematics, Physics,
    Chemistry],
    https://www.youtube.com/channel/UCBMvdXXJ7BcZcTKGPj9WxKg

36. Lectures on ARPIT/ Refresher Course: http://tiny.cc/ARPIT

37. Author's Online Courses:
    https://pharmastate.academy/profile/ajaysemalty/

Academic Writing MOOC

Myself, Dr Ajay Semalty, from HNB Garhwal University (A Central university) Srinagar Garhwal, Uttarakhand, welcome you on behalf of our entire team in the course. Four successful runs of the course are already over. We are proud to present India's most popular SWAYAM MOOC which got maximum number of exam registration (among all the SWAYAM MOOCs) and has been listed in top 30 MOOCs worldwide (http://tiny.cc/pib27april 2020). So far, more than 38000 learners across more than 95 countries have taken the advantage of the course.

In academic and research, who does not want the publications? In spite of being the vital requirement in academic & research career there is no comprehensive set up of learning academic writing in the knowledge domain. This course aims to fill this gap by providing the fundamental knowledge required for effective and result oriented academic writing. It is a foundation course and the application of this knowledge completely depends on an individual learner and his or her area of research.

Objectives of Course

Upon completing the course, one would be able

1. To differentiate between various kind of academic writings.
2. To identify and avoid the plagiarism.
3. To practice the basic skills of performing quality literature review.
4. To practice the basic skills of research paper, review paper and thesis writing.
5. To target the research work to suitable journal and communicate for publication
6. To practice the Time and team management.
7. To practice digital writing or develop Open Educational Resources (OER).
8. To write research proposals, conference abstract and book chapters/ book proposals.

Target Group

Any student or learner who wants to learn the academic writing. The course shall be helpful for PG students, research scholars, young scientists and faculty members (of any discipline or subject) for their career growth.

Team AW

Dr. Ajay Semalty

(Course Coordinator & Subject Matter Expert)

Dr. Mona Semalty

(Co-Course Coordinator & Subject Matter Expert)

Prof Rajat Agrawal

(Subject Matter Expert)

Dr. Lokesh Adhikari

(Instructional Designer)

Pandavaas Creations Pvt. Ltd.

Production

Course Layout

Course Duration 15 week   Credits : 04

Week 1: Academic & research writing: Introduction; Importance of academic writing; Basic rules of academic writing

Week 2: English in academic writing I & II; Styles of research writing

Week 3*: Plagiarism: Introduction; Tools for the detection of plagiarism; Avoiding plagiarism

Week 4*: Journal Metrics

Week 5*: Author Metrics

Week 6*: Literature review: Introduction, Source of literature; Process of literature review

Week 7*: Online literature databases; Literature management tools

Week 8: Review Paper Writing, I & II

Week 9: Research paper writing I, II, III

Week 10: Referencing and citation; Submission and; Post submission

Week 11: Thesis Writing I, II & III

Week 12: Empirical Study I, II & III

Week 13*: Challenges in Indian research & writing; Team management (mentor and collaborators); Time Management

Week 14: Research proposal writing; Abstract/ Conference Paper/ Book/ Book Chapter writing; OERs: basic concept and licenses

Week 15: Open Educational Resources (OERs) for learning & Research; OERs development I & II

Additional Week*: Ethics in Research: Research fraud, competing interest, authorship, slicing research, FFP, COPE guidelines.

** The contents of these weeks in total cover the UGC's newly introduced subject "Research & Publication Ethics".*

Introductory Video: https://youtu.be/uA3OcRy7eMY

Major Features:

- Free to learn at SWAYAM (www.swayam.gov.in).

- More than 38000 learners across 95 countries

- Ranked in top 30 MOOCs worldwide in 2019 appreciated by Press Information Bureau and Ministry of Education, GoI(http://tiny.cc/ pib27april2020)

- Listed in all time ranking of MOOCs in 2020

- Ranked #1 in exam registration in last two cycles back to back.

- Fifth cycle is going to be started on July 12, 2021.

- More than 40 hrs of Total Teaching Learning with more than 20 hrs of video lectures

- More than 20 hrs of live interactions/ webinar

- Vibrant discussion forum

- Certification from Ministry of Education, Govt. of India

Course in News: http://tiny.cc/AWnews